New ICMI Study Series
VOLUME

Published under the auspices of the International Commission on
Mathematical Instruction under the general editorship of

Ferdinando Arzarello, President   Abraham Arcavi, Secretary-General
Cheryl E. Praeger, Vice-President

For further volumes:
http://www.springer.com/series/6351

Information on the ICMI Study programme and on the resulting publications can be obtained at the ICMI website http://www.mathunion.org/ICMI/ or by contacting the ICMI Secretary-General, whose email address is available on that website.

Anne Watson • Minoru Ohtani
Editors

Task Design In Mathematics Education

an ICMI Study 22

Editors
Anne Watson
Department of Education
University of Oxford
Oxford, United Kingdom

Minoru Ohtani
School of Education
Kanazawa University
Kanazawa, Ishikawa, Japan

ISSN 1387-6872 ISSN 2215-1745 (electronic)
New ICMI Study Series
ISBN 978-3-319-09628-5 ISBN 978-3-319-09629-2 (eBook)
DOI 10.1007/978-3-319-09629-2

Library of Congress Control Number: 2015945100

© The Editor(s) (if applicable) and the Author(s) 2015, corrected publication 2021
Open Access This book was originally published with exclusive rights reserved by the Publisher in 2015 and was licensed as an open access publication in March 2021 under the terms of the Creative Commons Attribution-NonCommercial-NoDerivatives 4.0 International License (http://creativecommons.org/licenses/by-nc-nd/4.0/), which permits any noncommercial use, sharing, distribution and reproduction in any medium or format, as long as you give appropriate credit to the original author(s) and the source, provide a link to the Creative Commons licence and indicate if you modified the licensed material. You do not have permission under this licence to share adapted material derived from this book or parts of it.

The images or other third party material in this book may be included in the book's Creative Commons license, unless indicated otherwise in a credit line to the material or in the Correction Note appended to the book. For details on rights and licenses please read the Correction https://doi.org/10.1007/978-3-319-09629-2_13. If material is not included in the book's Creative Commons license and your intended use is not permitted by statutory regulation or exceeds the permitted use, you will need to obtain permission directly from the copyright holder.

This work is subject to copyright. All commercial rights are reserved by the author(s), whether the whole or part of the material is concerned, specifically the rights of translation, reprinting, reuse of illustrations, recitation, broadcasting, reproduction on microfilms or in any other physical way, and transmission or information storage and retrieval, electronic adaptation, computer software, or by similar or dissimilar methodology now known or hereafter developed. Regarding these commercial rights a non-exclusive license has been granted to the publisher.

The use of general descriptive names, registered names, trademarks, service marks, etc. in this publication does not imply, even in the absence of a specific statement, that such names are exempt from the relevant protective laws and regulations and therefore free for general use.

The publisher, the authors and the editors are safe to assume that the advice and information in this book are believed to be true and accurate at the date of publication. Neither the publisher nor the authors or the editors give a warranty, expressed or implied, with respect to the material contained herein or for any errors or omissions that may have been made. The publisher remains neutral with regard to jurisdictional claims in published maps and institutional affiliations.

Preface

This book is the outcome of ICMI Study 22: *Task design in mathematics education*. The proposal was presented to the International Commission on Mathematics Instruction (ICMI) in 2011 and accepted. Co-convenors and an International Programme Committee (IPC) were appointed from across the world and met in Oxford early in 2012. Following this, a discussion document and call for papers were prepared for the study conference to be held in July 2013. It was decided to organize the study under five themes, each theme having shared responsibility among two or three of the IPC members. Introductory text and orienting questions for each theme were prepared by the IPC members.

The conference was announced for July 2013 in Oxford and a call for papers issued in May 2012. There was a robust response, and we are grateful to Ellie Darlington for the academic administration. The size of the study conference was limited strictly to 80 attenders in order to ensure effective working groups, so papers were reviewed not only on the basis of quality but also to ensure an optimal coverage of relevant issues. The acceptance rate for papers was about 45 %. It was explained to authors that the final book would not consist of a collection of their papers, but there would be online publication of the conference proceedings. IPC members would develop a book which synthesised the papers and discussions that took place within each theme. Authors of high-quality papers which could not be accepted because of overlap of contents were encouraged to seek publication elsewhere. One author from each paper was invited to attend the conference. In addition, four plenary speakers were invited, each representing a particular approach to the principles, practice, and implementation of task design from across the world. Each of them has contributed a chapter to this book. Presentations were invited from the Shell Centre, the Freudenthal Institute, and the TDS/ATD approach to didactic engineering, and a panel presentation was given by attenders who use variation theory as a design tool. The majority of the conference time was spent in working groups organised and led by IPC members. Two cross-theme meetings provided an opportunity to explore contrasts and overlaps in content and approach.

The IPC decided to structure the book as five main chapters, one for each theme, to be co-authored by IPC members drawing on the papers accepted for the conference, high-quality papers submitted or published elsewhere, relevant literature from the research field, and discussions at the study conference itself. The contribution of co-authors of accepted papers and authors who were prevented from attending is also recognised in each chapter. Four more chapters were provided by the plenary speakers; there is no separate chapter about variation theory as it informs ideas in several chapters. Michèle Artigue and Ken Ruthven were invited to provide commentaries.

The five theme chapters were reviewed internally by the IPC and also by the study co-convenors. Through this process, cross-referencing was developed as much as possible, and there was careful examination of any overlaps. Where different chapters have treated similar ideas in different ways, we have tried as far as possible to clarify the differences and indicate cross-references.

We want to acknowledge the particular roles played by three members of the IPC: Claire Margolinas edited the online proceedings which can be found at https://hal.archives-ouvertes.fr/hal-00834054; Peter Sullivan reviewed all the theme chapters in order to comment on overall coherence; and Denisse Thompson, also a member of the IPC, undertook a considerable amount of technical editing. We are indebted to these colleagues for their support and practical wisdom. We are also grateful to Bill Barton who, as president of ICMI when the proposal was submitted, was encouraging and supportive, and to Lena Koch who manages the ICMI website. Finally, we wish to thank the task designers and researchers who have been bringing their work to publication more and more during the last two decades and those who supported and advised the original proposal. We are committed to the idea that task design is at the heart of effective teaching and learning of mathematics and are delighted to be part of the growth of research attention to this focus.

Oxford, United Kingdom Anne Watson
Kanazawa, Ishikawa, Japan Minoru Ohtani

ICMI Study 22: Task Design in Mathematics Education

Convenors

Anne Watson, University of Oxford, UK
Minoru Ohtani, Kanazawa University, Japan

International Programme Committee

Janet Ainley, School of Education, University of Leicester, UK
Janete Bolite Frant, LOVEME Lab, UNIBAN, Brazil
Michiel Doorman, Utrecht University, Netherlands
Carolyn Kieran, Université du Québec à Montréal, Canada
Allen Leung, Hong Kong Baptist University, Hong Kong
Claire Margolinas, Laboratoire ACTé, Université Blaise Pascal, Clermont Université, France
Peter Sullivan, Monash University, Australia
Denisse Thompson, University of South Florida, USA
Yudong Yang, Shanghai Academy of Educational Sciences, China

Attenders at the Study Conference Held in Oxford, July 2013

Janet Ainley	Toshiakira Fujii	Anthony Or
Einav Aizikovitsh-Udi	Javier García	Irit Peled
Abraham Arcavi	Kimberley Gardner	Birgit Pepin
Ferdinando Arzarello	Vince Geiger	João Pedro da Ponte
Mike Askew	Joaquim Gimenez	Peter Radonich
Marita Barabash	Maria J. Gonzales	Elisabetta Robotti
Berta Barquero	Merrilyn Goos	Gila Ron
Bärbel Barzel	Lulu Healy	Annie Savard
Angelika Bikner-Ahsbahs	Patricia Hunsader	Martin Simon
Christian Bokhove	Heather L. Johnson	Sophie Soury-Lavergne
Janete Bolite Frant	Keith Jones	Susan Staats
Peter Boon	Marie Joubert	Michelle Stephan
Laurinda Brown	Carolyn Kieran	Heidi Strømskag Måsøval
Orly Buchbinder	Libby Knott	Peter Sullivan
Hugh Burkhardt	Boris Koichu	Malcolm Swan
Maria G. Bartolini Bussi	Kotaru Komatsu	Mike Thomas
James Calleja	Angelika Kullberg	Denisse Thompson
Yu-Ping Chan	Kyeong-Hwa Lee	Wim Van Dooren
Yip-cheung Chan	Allen Leung	Yuly Marsela Vanegas
Haiwen Chu	Pi-Jen Lin	Geoff Wake
David Clarke	Johan Lithner	Anne Watson
Alison Clark-Wilson	Anna Lundberg	Floriane Wozniak
Alf Coles	Katja Maass	Sun Xuhua
Jan de Lange	Kate Mackrell	Yudong Yang
Jaguthsing Dindyal	Ami Mamolo	Michal Yerushalmy
Liping Ding	Claire Margolinas	Orit Zaslavsky
Michiel Doorman	Francesca Morselli	
Paul Drijvers	Minoru Ohtani	

Contents

Part I Introduction

1 Themes and Issues in Mathematics Education Concerning Task Design: Editorial Introduction........ 3
Anne Watson and Minoru Ohtani
   1.1 Rationale 3
   1.2 Structure of the Book 4
   1.3 Editorial Overview of Chapters from the Thematic
        Groups of the Study 5
        1.3.1 Frameworks and Principles for Task Design.......... 5
        1.3.2 The Relationship Between Task Design,
                Anticipated Pedagogies, and Student Learning 6
        1.3.3 Accounting for Student Perspectives in Task Design 7
        1.3.4 Design Issues Related to Text-Based Tasks 8
        1.3.5 Designing Mathematics Tasks: The Role of Tools.......... 9
   1.4 Overview of the Plenary Chapters 10
   1.5 Final Comments and Recent Developments 11
   References 14

Part II Theme Group Chapters

2 Frameworks and Principles for Task Design....................... 19
Carolyn Kieran, Michiel Doorman, and Minoru Ohtani
   2.1 Introduction 19
   2.2 Emergence and Development of Frameworks
        and Principles for Task Design 20

	2.2.1	Brief History of the Emergence of Design-Related Work from the 1960s to the 1990s	20
		2.2.1.1 Influences from Psychology at Large	21
		2.2.1.2 Early Design Initiatives of the Mathematics Education Research Community	22
		2.2.1.3 The 1990s and Early 2000s: Development of Research Specifically Referred to as *Design Experiments*	25
		2.2.1.4 An Example of Design Work by Educational Technologists...	26
		2.2.1.5 From Early 2000 Onward...	27
		2.2.1.6 Two Key Issues	28
	2.2.2	A Conceptualization of Current Theoretical Frameworks and Principles for Task Design in Mathematics Education Research	29
		2.2.2.1 Introduction	29
		2.2.2.2 Grand Theoretical Frames and Their Affordances	30
		2.2.2.3 Intermediate-Level Frames....................................	31
		An Example of a Theory-Based Intermediate-Level Frame: Theory of Didactical Situations	32
		An Example of a Craft-Based Intermediate-Level Frame: Lesson Study...................................	33
		2.2.2.4 Domain-Specific Frames..	36
		A Domain-Specific Frame for Fostering Mathematical Argumentation Within Geometric Problem-Solving....................................	37
		A Domain-Specific Frame for Proof Problems with Diagrams	38
		A Domain-Specific Frame for the Learning of Integer Concepts and Operations	40
		2.2.2.5 In Drawing This Section to a Close.......................	41
2.3	Case Studies Illustrating the Relation Between Frameworks and Task Design		42
	2.3.1	Introduction	42
	2.3.2	Cases	43
		2.3.2.1 Case 1: Anthropological Theory of Didactics	43
		2.3.2.2 Case 2: Variation Theory..	45
		2.3.2.3 Case 3: Conceptual Change Theory	47
		2.3.2.4 Case 4: Conceptual Learning Through Reflective Abstraction	49
		2.3.2.5 Case 5: Realistic Mathematics Education	51
		2.3.2.6 Case 6: Formative Assessment for Developing Problem-Solving Strategies...................................	54
		2.3.2.7 Case 7: Japanese Lesson Study	56
	2.3.3	Discussion	59

	2.4	Frameworks and Principles Do Not Tell the Whole Story in Task Design	62
		2.4.1 Introduction	62
		2.4.2 Additional Factors Related to Task Design	62
		2.4.2.1 The Tension Between *Design as Science* and *Design as Art*	62
		2.4.2.2 Values in Task Design	64
		2.4.3 Diversity of Design Approaches Through the Lens of Basic Versus Applied Research	65
		2.4.3.1 A Two-Dimensional Scheme for Classifying Research	65
		2.4.3.2 Design Frames in the Light of Distinctions Between Basic and Applied Research	66
		2.4.4 Design Activity Across Professional Communities	68
		2.4.4.1 The Stakeholder Approach as a Foundation for Thinking About Collaborative Design Activity	68
		2.4.4.2 Task Design Involving Practitioners and Researchers	68
		2.4.5 Toward the Resolving of Perceived Tensions	70
	2.5	Concluding Discussion: Progress Thus Far and Progress Still Needed	71
	References		74
3	**The Relationships Between Task Design, Anticipated Pedagogies, and Student Learning**		**83**
	Peter Sullivan, Libby Knott, and Yudong Yang		
	3.1	Introduction	83
	3.2	Factors Influencing Task Design and Pedagogies	84
		3.2.1 Knowledge for Teaching Mathematics that Informs Task Design	85
		3.2.2 Anticipatory Pedagogical Decision-Making	86
	3.3	Some Tasks Presented to Inform Subsequent Discussions	87
		3.3.1 The L-Shaped Area: A Lesson from Japan	87
		3.3.2 A Set of Worded Questions	88
		3.3.3 Shopping for Shoes	90
		3.3.4 Determining the Age Range of Students for Which These Tasks Are Appropriately Posed	90
	3.4	Design Elements of Tasks	91
		3.4.1 Five Dilemmas	91
		3.4.1.1 Context as a Dilemma	91
		3.4.1.2 Language as a Dilemma	92
		3.4.1.3 Structure as a Dilemma	93
		3.4.1.4 Distribution as a Dilemma	93
		3.4.1.5 Levels of Interactions as a Dilemma	93
		3.4.2 Task Suitability Criteria	94

	3.5	The Nature of the Mathematics that is the Focus of the Tasks	96
		3.5.1 Tasks that Address Specialized Mathematical Goals	97
		3.5.2 Designing Tasks that Address a Practical Perspective	98
	3.6	Task Design Processes	100
		3.6.1 The Role of the Authority and Autonomy of the Teacher in Designing and Implementing Tasks	102
		3.6.2 Problematic Aspects of Converting Tasks from One Culture to Another	103
		3.6.3 Classroom Culture and Anticipated Pedagogies	105
	3.7	Considering the Students' Responses in Anticipating the Pedagogies	108
		3.7.1 Student Motivation	108
		3.7.2 Introducing the Task to the Students	109
		3.7.3 Access to Tasks by All Students	110
	3.8	Summary and Conclusion	111
	References		112
4	**Accounting for Student Perspectives in Task Design**		**115**
	Janet Ainley and Claire Margolinas		
	4.1	Introduction	115
	4.2	Articulating the Gap Between Teachers' Intentions and Students' Perceptions and Responses	116
		4.2.1 Students' Responses to Word Problems	116
		4.2.2 The Student's Situation	118
		4.2.3 When Student's Milieu and Teacher's Planned Milieu Are Not the Same: An Example	119
		4.2.4 Influence of the Didactical Contract on the Definition of Task	122
		4.2.5 Student's Perception of the Meaning and Purpose of the Task	124
	4.3	Taking Account of Student Perspectives: How Task Design Might Reduce the Gap	126
		4.3.1 Students' Expectations	127
		4.3.2 Reflective Task Design	130
		4.3.3 Emergent Task Design	133
		4.3.4 Open Tasks: Voice and Agency	134
	4.4	Conclusion: Some Topics for Future Research	136
	References		138
5	**Design Issues Related to Text-Based Tasks**		**143**
	Anne Watson and Denisse R. Thompson		
	5.1	Introduction	143
		5.1.1 Research on *Text-Based Tasks* Within Textbooks	144
		5.1.2 Shape of the Chapter	145

	5.2	Nature and Structure of Tasks	146
		5.2.1 Different Kinds of Text Materials	146
		5.2.2 Authorship, Authority, and Voice	148
		5.2.2.1 Authorship	148
		5.2.2.2 Authority for Mathematics	149
		5.2.2.3 Voice of a Task	151
		5.2.3 Nature of Mathematics	153
		5.2.4 Summary	156
	5.3	Pedagogic/Didactic Purpose of Text-Based Tasks	156
		5.3.1 Cultural Differences in Purpose	157
		5.3.2 How Purpose Is Presented to Learners	159
		5.3.3 Coherent Purpose in Collections of Tasks	163
		5.3.4 Embedding New Knowledge with Existing Knowledge	167
		5.3.5 Summary	169
	5.4	Intended/Implemented Mathematical Activity	169
		5.4.1 Principles About Learning	169
		5.4.2 Aims Enacted Through Individual Tasks	171
		5.4.3 Complex Aims Enacted Through Large-Grain-Size Tasks	173
	5.5	Visual Features of Text-Based Tasks	176
	5.6	Teachers' Use of Tasks	181
	5.7	Conclusion: What Text-Based Tasks Can and Cannot Do	183
		5.7.1 A Potential Research Agenda	185
	References		186
6	**Designing Mathematics Tasks: The Role of Tools**		**191**
	Allen Leung and Janete Bolite-Frant		
	6.1	Introduction	191
	6.2	Considerations in Designing Tasks that Make Use of Tools	194
		6.2.1 Epistemological and Mathematical Considerations	194
		6.2.2 Tool-Representational Considerations	195
		6.2.3 Pedagogical Considerations	196
		6.2.4 Discursive Considerations	198
	6.3	Theoretical Frames for Designing Tool-Based Tasks	198
		6.3.1 Didactical Theories	198
		6.3.2 Instrumentation Approach to Tool Use	199
		6.3.3 Cultural Semiotic Frame	200
		6.3.4 Activity Theory	202
		6.3.5 TELMA and ReMath	203
		6.3.6 Other Theories Relevant to Tool-Based Task Design	204
	6.4	Further Design Considerations and Heuristics	207
		6.4.1 Dialectic Between Pragmatic and Epistemic Values	207
		6.4.2 Mediation and Feedback	209
		6.4.3 Discrepancy Potential	212
		6.4.4 Conceptual Blending and Multiplicity	215

	6.5	Synthesis	218
		6.5.1 Strategic Feedback and Mediation	219
		6.5.2 The Instrumental/Pragmatic-Semiotic/Epistemic Continuum	220
		6.5.3 Boundary Between Mathematical and Pedagogical Fidelity	220
		6.5.4 Discrepancy Potential	221
		6.5.5 Multiplicity	221
	References		222

Part III Plenary Presentations

7 E-Textbooks for Mathematical Guided Inquiry: Design of Tasks and Task Sequences 229
Michal Yerushalmy

- 7.1 Textbook Culture: Traditions and Challenges 229
- 7.2 The Design of an Interactive Unit 232
  - 7.2.1 Interactive Exposition 233
  - 7.2.2 Toolbox and Unit Tools 235
  - 7.2.3 Tasks or Problems and Exercises 236
  - 7.2.4 Write an Essay Task 240
  - 7.2.5 Restructuring a Unit Along the Space of Interactive Elements 241
- 7.3 Sequencing in Non-sequential Textbooks 242
- 7.4 Summary: Concept-Driven Navigation in the Space of Interactive Tasks 245

References 246

8 Didactic Engineering as a Research Methodology: From Fundamental Situations to Study and Research Paths 249
Berta Barquero and Marianna Bosch

- 8.1 Didactic Engineering as a Research Methodology 249
- 8.2 Didactic Engineering Within the Theory of Didactic Situations 252
  - 8.2.1 An Example: Measuring Quantities at Primary School 252
    - 8.2.1.1 Preliminary Analysis 253
    - 8.2.1.2 A Priori Analysis: Design of Mathematical and Didactic Situations 254
    - 8.2.1.3 Implementation, Observation, and Data Collection 255
    - 8.2.1.4 A Posteriori Analysis: Results, New Phenomena, New Research Questions 257
  - 8.2.2 Didactic Engineering in the Science of Didactics 258
    - 8.2.2.1 The COREM as a *Didactron* 258
    - 8.2.2.2 An Experimental Epistemology 259

	8.3	Didactic Engineering Within the Anthropological Theory of the Didactic.. 260
		8.3.1 From Situations to Study and Research Paths 260
		8.3.2 An Example: Teaching Modeling at University Level 262
		8.3.2.1 Preliminary Analysis 263
		8.3.2.2 A Priori Analysis: Mathematical and Didactic Design.. 264
		8.3.2.3 Implementation and In Vivo Analysis 266
		8.3.2.4 A Posteriori Analysis and Ecology................... 267
	8.4	Open Questions... 269
	References.. 270

9 The Critical Role of Task Design in Lesson Study............................... 273
Toshiakira Fujii
 9.1 Introduction... 273
 9.2 Japanese Lesson Study... 274
  9.2.1 The Detailed Lesson Proposal 275
  9.2.2 Structured Problem Solving... 276
  9.2.3 Designing the Task as Part of *Kyozaikenkyu*.................... 277
   9.2.3.1 The Meaning of *Kyozaikenkyu* 277
   9.2.3.2 Task Design Principles for Structured Problem-Solving Lessons 278
 9.3 Designing Tasks Using *Kyozaikenkyu* in Lesson Study................ 279
  9.3.1 The Topic: Subtraction with Regrouping.......................... 279
   9.3.1.1 Teachers Know There Are Reasons for the Numbers Used 279
  9.3.2 Why Teachers Begin *Kyozaikenkyu* with Textbooks........ 281
  9.3.3 Exploring Possible Manipulatives 281
  9.3.4 Evaluating the Task in Action 281
 9.4 Discussion.. 283
  9.4.1 Task Design in Lesson Study Always Involves Anticipating Students' Solutions....................................... 283
  9.4.2 Task Design in Lesson Study Goes with Task Evaluation.. 284
  9.4.3 Conclusion 285
 References.. 286

10 There Is, Probably, No Need for This Presentation 287
Jan de Lange
 10.1 Introduction... 287
  10.1.1 Educational Design: Is There a (Need for a) System?..... 288
 10.2 Personal Reflections on the Design Process 289
  10.2.1 Slow Design 289
   10.2.1.1 Slow Design: The Principles 289
   10.2.1.2 Slow Design: The Process................................ 290

		10.2.1.3	Meet the Real World	291
		10.2.1.4	An Example of Slow Design	292
		10.2.1.5	Conditions for Slow Design	294
	10.2.2	Fast Design		294
10.3	Design as Art			295
	10.3.1	The Role of Intuition		295
10.4	Design Examples			296
	10.4.1	Central Concept Design		296
	10.4.2	Central Context Design		296
	10.4.3	Design over Time		299
	10.4.4	Experimental *Free* Design		301
10.5	Restrictions, Conditions, and Challenges			304
10.6	Task Design and Curricula Design			305
	10.6.1	Left Out in the Cold		305
10.7	Summary			306
References				307

Part IV Commentaries

11 Taking Design to Task: A Critical Appreciation 311
Kenneth Ruthven
- 11.1 The Evolution of Task 311
- 11.2 A Scheme for Design 313
  - 11.2.1 A Template for Phasing Task Activity 314
  - 11.2.2 Criteria for Devising a Productive Task 314
  - 11.2.3 Organization of the Task Environment 315
  - 11.2.4 Management of Crucial Task Variables 315
- 11.3 Design Continues in Use 316
- 11.4 Scope for User Agency 318
- 11.5 An Apparatus for Design 319
- References 320

12 Some Reflections on ICMI Study 22 321
Michèle Artigue
- 12.1 Introduction 321
- 12.2 A Subjective Reading 322
- 12.3 Frameworks and Principles for Task Design 325
- 12.4 Task Design Through Text-Based Tasks 328
- 12.5 Task Design, Tools, and Technology 331
- 12.6 Conclusion 333
- References 334

Correction to: Task design in Mathematics Education C1

Index 337

Abbreviations

AMM	American Mathematical Monthly
ATD	Anthropological Theory of Didactics
ATM	Association of Teachers of Mathematics
CAS	Computer algebra system
CCT	Conceptual Change Theory
CGI	Cognitively Guided Instruction
COMPASS	Common Problem Solving Strategies as links between Mathematics and Science
COREM	Centre d'Observations et de Recherches pour l'Enseignement des Mathématiques
DE	Didactical engineering
DGE	Dynamic geometry environment
DME	Digital mathematics environment
ESM	Educational Studies in Mathematics (journal)
GC	Geometric Constructer
GEMAD	Masters in Didactical Analysis
GID	Guiding diagram
GSP	Geometers Sketchpad™
HLT	Hypothetical learning trajectory
ICME	International Congress on Mathematical Education
ICMI	International Commission on Mathematical Instruction
ICMT	International Conference on Multimedia Technology
ICT	Information and communications technology
ID	Interactive diagram
IOWO	Institute for the Development of Mathematical Education, Utrecht
ISDDE	International Society for Design and Development in Education
KOSIMA	Kontexte für sinnstiftendes Mathematiklernen
LS	Lesson study
MARS	Mathematics Assessment Resource Service
MRT	Multiple representational technological environment

PME	The International Group for the Psychology of Mathematics Education
ReMath	Representing Mathematics with Digital Media
RLDU	Resources for Learning and Development Unit
RME	Realistic Mathematics Education
SES	Socio-economic status
SMILE	Secondary Mathematics Individualised Learning Experiment
SMP	School Mathematics Project
SRP	Study and research paths
TDS	Theory of Didactical Situations
TELMA	Technology Enhanced Learning in Mathematics
TIMSS	Trends in International Mathematics and Science Study
VT	Variation theory
ZFM	Zone of free movement
ZPA	Zone of promoted action
ZPD	Zone of proximal development

This original version of this book was published with copyright Springer-Verlag Berlin Heidelberg. This book has now been made open access under a CC BY-NC-ND 4.0 license. For details on rights and licenses please read the Correction https://doi.org/10.1007/978-3-319-09629-2_13.

Part I
Introduction

Chapter 1
Themes and Issues in Mathematics Education Concerning Task Design: Editorial Introduction

Anne Watson and Minoru Ohtani

1.1 Rationale

This study was initiated to produce an up-to-date summary of relevant research about task design in mathematics education and to develop new insights and new areas of relevant knowledge and study. Attention to task design is important from several perspectives in mathematics education research and practice. From a cognitive perspective, the detail and content of tasks have a significant effect on learning; from a cultural perspective, tasks shape the learners' experience of the subject and their understanding of the nature of mathematical activity; from a practical perspective, tasks are the bedrock of classroom life, the "things to do." Recently, there has been growth of research and publication activity about the work of designers in mathematics education: some of it oriented around teams that work globally; some of it focusing on the affordances of digital technologies. There has also been a growth of research activity arising from international comparisons of classroom characteristics, including tasks and task adaptation, and from comparisons of textbooks. It is interesting that globalization in mathematics education research, practice, and policymaking has led us to focus more closely on the minutiae of tasks in mathematics teaching, as well as on the more predictable issues to do with culture, local policy, and technological advances. Task design is also a core issue in research about learning; whether this research takes place through clinical interviews or authentic classroom practice, the detail of the tasks and the way they are presented

---

A. Watson (✉)
University of Oxford, 27 Elms Road, Oxford OX2 9JZ, UK
e-mail: anne.watson@education.ox.ac.uk

M. Ohtani
Kanazawa University, Kakuma, Kanazawa, Ishikawa 920-1192, Japan
e-mail: mohtani@ed.kanazawa-u.ac.jp

This chapter has been made open access under a CC BY-NC-ND 4.0 license. For details on rights and licenses please read the Correction https://doi.org/10.1007/978-3-319-09629-2_13

© The Author(s) 2021
A. Watson, M. Ohtani (eds.), *Task Design In Mathematics Education*,
New ICMI Study Series, https://doi.org/10.1007/978-3-319-09629-2_1

is often reported sketchily, and without full justification, yet tasks have a major influence on assumed findings about capability.

The state of play before this study conference included the following strands of activity:

- Effects of task design on learning and assessment (e.g. Anderson & Schunn, 2000; Runesson, 2005)
- Improvement of communication between designers and researchers, with more exchange about research and principles and practice (e.g. Schoenfeld, 2009; International Society for Design and Development in Education (ISDDE) http://www.isdde.org/isdde/index.htm)
- Inclusion of Topic Study Groups in task design as a regular feature of ICME conferences (Mexico, 2008 http://tsg.icme11.org/tsg/show/35; Korea, 2012; Germany, 2016)
- Publication of tasks, principles of design, and research on effects and implementation by long-standing design teams (e.g. Shell Centre, UK; Freudenthal Institute, the Netherlands; QUASAR, USA; Connected Mathematics, USA)
- Changes in task design at implementation stage (e.g. PME research forum; Tzur, Sullivan, & Zaslavsky, 2008)
- The process of didactic engineering and the influence of tasks on teaching (e.g. Margolinas et al., 2011)
- International textbook comparisons that draw attention to differences in task design (e.g. Valverde, Bianchi, Wolfe, Schmidt, & Houang, 2002)
- Tasks in teacher education (e.g. Tirosh & Wood, 2009; Zaslavsky & Sullivan, 2011; *Journal of Mathematics Teacher Education*, volume 10 (4–6))

However, we recognize that these represent only what came to the attention of the International Programme Committee. Hence, they are restricted to what is available internationally and mainly in English and cannot reflect the working practices of myriad groups of teachers and textbook writers worldwide.

1.2 Structure of the Book

The chapters of the book are organized in four parts. The first part consists of this introductory editorial. In the second part are five chapters which reflect the five organizing themes of the ICMI study conference. The initial themes were identified from a reading of existing research:

- Theme A: Tools and representations
- Theme B: Accounting for student perspectives in task design
- Theme C: Design and use of text-based resources
- Theme D: Principles and frameworks for task design within and across design communities
- Theme E: Features of task design informing teachers' decisions about goals and pedagogies

Reflective work by the International Programme Committee led to new titles and a new sequence for this book that represents more closely the scholarly work undertaken at the conference and subsequently.

The third part of the book consists of chapters by four invited plenary speakers who provide examples of the design process relating to underlying principles and practices. These processes vary widely not only in the ways in which individuals describe their work but also in theoretical perspectives relevant to their working context.

The final part of the book consists of two commentaries, one from Michèle Artigue and one from Ken Ruthven. We invited them to comment as senior scholars in mathematics education who have themselves been intimately involved in the processes of task design and implementation.

1.3 Editorial Overview of Chapters from the Thematic Groups of the Study

1.3.1 *Frameworks and Principles for Task Design*

Chapter 2 is the longest in the book because it presents a way of thinking about the multiple frameworks and sets of principles that arise in the literature on task design. We are deeply grateful to the participants and authors who contributed and hope that future researchers, research students, and designers who wish to publish their practices in scholarly journals find it useful to focus or structure their work according to the ideas in this chapter. It offers a significant theoretical step forward in the field; to a great extent, Chaps. 3–6 depend on Chap. 2 for their theoretical background.

Frameworks and principles for task design are identified as addressing three theoretical *grain sizes*, although a specific set of principles might incorporate different sizes. *Grain size* descriptions are intended to be descriptive tools for thinking in a structured way about task design, rather than being prescriptive. The grain sizes identified are *grand frames*, *intermediate frames*, and *domain-specific frames*. Grand frames present theories about learning in and out of educational settings at a general level. Intermediate frames present the complex interactions between task, teacher, teaching methods, educational environment, mathematical knowledge, and learning so that the purposes and implications for task design are always understood within the total structure of practice. Intermediate theories take time to develop, and applications of them can lead towards elaboration of the theory as well as developments in the practices of both teachers and designers. Communities can develop around both grand and intermediate theories in which there are shared language, shared materials and resources, and shared research studies and conferences. Presentation of work developed within intermediate theories to the outside world cannot always restate the complex background principles, so some background work has to be expected of readers and reviewers. An extra dimension of intermediate frames is that they are based in teaching as craft knowledge and arise from teachers' actions and interactions. Domain-specific frames focus on particular areas of mathematical knowledge or activity and may not be generalizable across mathematics.

A particularly useful contribution from this chapter is the passage on the history of task design in mathematics education. Very often in our field, people only refer to recent research and recent experiences of practice. Where practice and research are based on a mature accumulated body of knowledge, this is not a problem. However, because much of our practice is influenced by policy, ideology, and (at the time of writing) the international testing regime, it is tempting to refer to those as the basis for critical academic work rather than our own past research. It can also be quite difficult for researchers to access past work if it has been locked into the relationships between design, curriculum, and teaching and has not been specifically researched and reported. Globalization of the field helps in this process to some extent: for example, some practices in Singapore can be traced back to work in the UK in the 1970s, even though the influence is now in the opposite direction. This is an example of how worthwhile practices might disappear for some time in one part of the world but be alive and well in another part. More effortful attention to the history of ideas in task design would enable consolidation of the field: there is no need to reconstruct from scratch; there is no need to ignore elements that have been important in the past. Often important ideas continue in practice but are not recognized by researchers, and current recognition by a researcher does not mean that an idea is "new."

1.3.2 The Relationship Between Task Design, Anticipated Pedagogies, and Student Learning

Chapters 3 and 4 between them address the relationships between tasks, teaching, and learning. Sometimes people refer to "gaps" between what is intended by the designer and enacted by the teacher or what is intended by the teacher and perceived by the learner. Such gaps can also be seen as "interactions" which are inevitable in the teaching-learning process. These two chapters are like two sides of a coin, represented by the task as stated and presented to students. On the one side are the teacher's decisions about the nature of mathematics, the collection of students who are being taught, and many emanating practical considerations. On the other side, without the intimate feedback that is available in one-to-one clinical situations, the teacher has to have theoretical support to anticipate and understand students' experiences.

Chapter 3 is very practical. The authors posed questions about the influence on teachers' decision-making of task features and had available to them a wide range of reports about how teachers put given tasks, designed by external sources initially, into practice embedded in pedagogical variables. The authors use three tasks from the study conference as exemplars around which the discussions in the chapter are oriented, raising issues which can then be used to think about any task. They elaborate on how to analyse the thinking that went on in the design, decisions about suitability and applicability, and how these are influenced by teachers' views of the nature of mathematics, the prevailing school and classroom culture, and the relative emphases on mathematical content and broader epistemological goals a teacher may be pursuing with the students. Each of the example tasks is presented as a complex situation, in which the dividing line between task design and implementation, and lesson design and implementation, is not easy to draw and may even be unnecessary. The teacher's knowledge of mathematics pedagogy and ability to anticipate students' responses is critical in all these decisions. The authors had originally intended to address related issues about educating teachers in the processes of using tasks, but these vary so widely between cultures and between teacher education programmes that we cannot do it justice in this volume and refer readers instead to the *Journal of Mathematics Teacher Education*, in particular volume 10 (4–6).

Finally, there is discussion about how to present a task to students initially so that they are motivated, have access to the task and can get started, and also understand what the teacher would like them to be thinking about. These final ideas lead naturally on to Chap. 4, in which learners' perceptions are considered from a different point of view, including the notion of *interest-dense* tasks for sustaining both effort and learning. Questions that arise from Chap. 3 also arise from other chapters, so we shall incorporate them in our final remarks below.

1.3.3 Accounting for Student Perspectives in Task Design

Chapter 4 constructs a theoretical view of the interactions between students, teachers, and tasks in the classroom; it is not possible to infer how the learners "see" a task merely from their actions and their written or verbal products. Merely doing what the teacher hopes and expects is evidence for a certain form of compliance, but might not constitute evidence of learning or evidence of understanding the purpose of the task or even evidence of having the same perception of the task as that of the teacher. The question the authors of this chapter wanted to address is how learners answer this question: "what is this task asking me to do?" The authors became aware of a general dearth of research in this area, and yet knowledge of how learners perceive a task is crucial to planning effective lessons as well as to designing effective tasks. *Perception* therefore has to be imagined, and in some cultures the expertise of the teacher is seen in terms of the accuracy of that process of imagination. We welcome the authors' decision to stay firmly with what can be *known* about the learners' perspective and not be deviated into what might be *assumed*.

The chapter firstly reflects on the literature about word problems, which draws attention to differences in students' perceptions of the purpose of the task and, maybe, designers' intentions, although this is sometimes achieved through inferential reasoning from students' productions. These kinds of differences can arise in any mathematics teaching situation, not only with word problems. So, the authors offer the construct of *didactical situation* as a structure within which to consider the *didactical contract* and *milieu* (terms related to theories discussed in Chaps. 2 and 8) of the learner and, hence, how they might be viewing the nature and purpose of a mathematical task. This chapter includes a rare example of phenomenographic research identifying students' perceptions of a statistical task and what this reveals about their understanding of the purpose of the task and of statistics as a field of study.

The second part of the chapter proposes ways in which various educators, teachers, and researchers have sought to reduce any gaps between the teacher's intentions and the learner's perceptions. In doing so, the authors present the importance of the quality of teachers' expectations of students, the importance of reflective redesign, the idea of *emergent task design*, and considerations of openness. Some of these ideas arise also from other perspectives. For example, teachers' expectations are a component of the discussions in Chap. 3 and also a central aspect of teachers' professional learning described in Chap. 9. Reflective redesign is routinely undertaken with others in Japanese Lesson Study and is also a component of a well-wrought design process for teams and individuals. A degree of openness to allow and also encourage student agency is a key aspect of many problem-solving task design initiatives in which learners' contributions are valued and discussed, and to add to this we would draw attention to the idea of listening *to* learners (Davis, 1996) rather than listening *for* particular solutions. Indeed, in some cultures, generating several methods for approaching a question is a key feature of a lesson. Whereas in Western literature some students' responses might be described as *misconceptions*, in other traditions these are *alternative ways of seeing* and are valuable for the learning of both teachers and students. A welcome development of this is given in the chapter, which offers emergent task design as a process arising from ideas of learners, in the course of a lesson, which are made into tasks by teachers in the moment.

Another way to look at the issues in both Chaps. 3 and 4 would be to consider the work of teaching in Valsiner's terms of aligning the zone of free movement (ZFM) (i.e. what is possible in the situation), the zone of promoted action (ZPA) (i.e. how the teacher directs learners towards particular actions), and the zone of proximal development (ZPD) (i.e. the learning a child can be expected to achieve in that educational situation). Ideally, according to Valsiner, the ZPA matches the ZPD for optimal learning (Valsiner, 1997, p. 198). However, teachers rarely know accurately the ZPD of all their students in school situations, so teachers need to engineer a balance between defining the boundaries of the ZFM through task and situation design and providing loose enough boundaries for the ZPA to allow optimal overlap with their students' relevant ZPD.

1.3.4 Design Issues Related to Text-Based Tasks

This theme group set out originally to focus on the design of tasks in textual format, textbooks more generally, downloadable materials, and other forms of text-based communication designed to generate mathematical learning. The group was aware of differences in the order, development, representation, and presentation of content between textbook series and also between countries and cultures. The group hoped to consider how to analyse the content of individual questions or sequences of questions. Another way to look at tasks would be to view them as the shapers of the curriculum rather than merely presenting a given curriculum and hence consider the differences between author and teacher intentions. The questions originally posed for the study conference concentrated on issues of text design and use.

In practice, few papers were submitted to the theme group that addressed these questions. Instead, the theme received many papers about designed tasks or collections of tasks that were based on clearly enunciated principles and showed how these worked out in practice, generally addressing overarching aspects of students' mathematical learning, such as proof, interdisciplinary perspectives, reasoning, problem-solving, and values. Some papers were on specific examples of textbook issues: involvement of teachers as digital authors and the need for ancillary materials such as assessment tasks. A small number addressed specific details: why this diagram, why these numbers, why this questioning sequence, and so on. Meanwhile, the ICMT *International Conference on Mathematics Textbook Research and Development 2014* signalled an increase in international comparison, cooperation, and knowledge exchange about mathematics textbooks, their design, development, use, and analysis—a field of study focusing on textbooks in particular. The expectations for this conference freed the working group to focus on issues raised by the conference papers and others that could not be addressed at the level of textbook production and use.

The chapter offers a triangular, mutually interactive relationship between the nature and structure of the task, the intended mathematical activity, and the pedagogic purpose. It refers throughout to tasks that are free-standing or situated within *learning management systems*, meaning published textbooks, task banks, programmed systems, and so on. The triangular relationship is relevant for free-standing tasks, home-made task banks, and textbooks, whether digitally delivered or paper based; and tasks created during lessons. Discussions led to the formation of a focus on the learners' perspective when presented with a task, as in Chap. 4, and how the task influences their subsequent mathematical activity, their learning, and their view of mathematics. This perspective is never constructed in isolation from their whole mathematics educational experience, which could include textbook design and use, but is also influenced strongly by pedagogy and presentation. One section of the chapter proposes a detailed consideration of visual appearance and layout as influences on learning.

This chapter focuses on tasks without dynamic or interactive content, while the following chapter addresses tool use, which includes the full range of digital tools.

1.3.5 Designing Mathematics Tasks: The Role of Tools

This theme concerns designing teaching-learning tasks that involve the use of tools in the mathematics classroom and consequently how, under such design, tools can represent mathematical knowledge. This aspect of task design research is currently "coming of age" through combinations of various and widely available digital tools and a Vygotskian understanding of relationships, through semiotic mediation, between artefact and learning. The issue for designers is how to relate the tool-specific discourse representation to mathematical knowledge. There has been an international conference relating to digital technologies since 2004, the *International Conference for Technology in Mathematics Education*, journals, and several special issues of mathematics education journals. For our study, the submitted papers were mainly concerned with practical and theoretical issues of task design in dynamic digital environments, but usefully included papers on physical tool use, thus allowing theories developed in digital environments to be expanded for nondigital tools.

The chapter starts by outlining the practical considerations of tool use in the mathematics classroom and then moves to consider relevant theoretical perspectives that connect instruments and didactics. It then presents various ways in which contributors to the study conference had enacted these connections and introduces the idea of *discrepancy potential of tool*, which is the space between the feedback a learner might experience from using the tool and the mathematics concept, combined with the need for the tool user to make decisions. In this space, unanticipated disturbances might take place, but also the teacher can intervene to introduce disturbances.

The chapter closes with a synthesis of the issues that any task design heuristics need to address: complementarity of feedback and mediation, relationships between pragmatic and epistemic considerations, symbiosis of mathematics and pedagogy, multiplicity of tools, and the discrepancy potential previously described. The design of the task—what the learner is supposed to do with the tool—needs to take account of, bridge, and coordinate these aspects of the activity. The final remark is about the importance of the teacher's perception of the nature of mathematics, particularly as tool-based tasks can challenge the nature of mathematical activity, and hence the nature of mathematical knowledge and competence.

1.4 Overview of the Plenary Chapters

Chapters 7–10 are written by the invited plenary speakers at the study conference. These speakers were selected to represent well-formed examples of task design in practice. Michal Yerushalmy opens this section with an example of *domain-specific* design (as defined in Chap. 2) in which she describes the theory and design of a digital resource that focuses on various features of functions for secondary students. One interesting feature of this resource, and the reason she was chosen to give a plenary, is that she embraces the facility of digital technology to provide flexible sequencing of tasks. In order to use the resource, the teacher (or even student) has to make her own decisions about what to do and when. This is not, therefore, a digital learning management system but rather a digital task world, and she describes the structuring of such a world in terms that can be useful for other designers.

Chapter 8 presents two examples to illustrate the manifestation in practice of the *intermediate frame* of didactics founded by Brousseau. The description of this as intermediate arises from Chap. 2, that is, theories which provide methods of application across mathematics. Berta Barquero and Marianna Bosch illustrate how the theory of didactic situations has been used at a primary level to establish the measurement of quantities, and then they demonstrate the more complex world of didactic engineering in an anthropological development of the original theory. There are also *domain-specific* principles in their descriptions and also craft knowledge, and a disturbing account of how the product of careful longitudinal design research can be subverted by practitioners who do not share the theoretical commitment.

In Chap. 9, Toshiakira Fujii describes an aspect of Japanese Lesson Study, *kyozaikenkyu*, that can be overlooked in some Western adaptations of the process. The disciplined process of Japanese Lesson Study can be seen as an example of a craft-based frame, as described in Chap. 2, and this is becoming widely recognized outside Japan. Typically, each lesson is oriented around one task, which may be one calculation (he gives "12-7" as an example) or may be a conceptual problem (e.g. classifying triangles). The selection and design of one task and how to use it is the focus of teachers' regular professional development activity, and this creates a deep repertoire of "good" tasks that are also reflected in the contents of the authorized textbooks and a pedagogic repertoire for teachers. This approach, where task design is the central focus for teachers' planning and development, runs counter to the comments of Wittmann (1995) who argued for task design to be in the hands of specialist designers, and it poses challenges for teacher knowledge and training and also for the value of externally designed tasks. Each lesson study report uses domain-specific frames alongside, or even as a basis for, generic considerations.

The final plenary chapter also presents a contradiction, this time to the whole book in some respects. Jan de Lange is an experienced designer whose work at the Freudenthal Institute has been influential throughout the world, particularly in the Netherlands, South Africa, and the US reform process. His description of the design process is down-to-earth and practical, setting high standards for the use of intuition and insight as starting points, with a focus on students' learning. He claims, and illustrates, an approach he calls *slow design* that arises from knowledge of mathematics, the environment, teaching, classrooms, and children. While he describes the actions necessary for a designer to take in order to test and improve the design (e.g. not relying on the designer's own teaching), he also challenges the academization of the design process. For him, the nature and direction of research about task design is in danger of moving the emphasis away from direct experience of children and mathematics in classrooms and towards theorization. A series of online articles, *A Designer Speaks* published by ISDDE (http://www.isdde.org/isdde/index.htm), is worthy of mention here, giving alternative insights into designers' working practices.

1.5 Final Comments and Recent Developments

As conveners of this study, and editors of this volume, we have been excited by the breadth and diversity of contributions and impressed by the immense work that has gone into the resultant chapters. We are both actively involved in schools and teacher education as well as in mathematics education research, and our multiple perspectives have helped us consider the different roles of theory in relation to task design. While preparing this volume, Minoru has been involved in several lesson studies and has also been a school principal, while Anne has been teaching mathematics in the UK year 7 and leading teacher workshops. Our main concern in leading this study has been to accelerate the growth of attention to task design given by researchers in their work and their written artefacts. We note that several papers that contributed to the conference have now been expanded and developed and published elsewhere and have included these where possible in the relevant reference lists. A volume about task design with digital technologies follows the study (Leung & Baccaglini-Frank: *Digital technologies in designing mathematics education tasks: Potential and pitfalls* forthcoming from Springer) as does a special issue of the *Journal of Mathematics Teacher Education* edited by Keith Jones and Birgit Pepin. A research forum took place in 2014 on the relationships between task and students (Clarke, Strømskag, Johnsen, Bikner-Ahsbahs, & Gardner, 2014).

Research reports rarely give sufficient detail about tasks for them to be used by someone else in the same way and hence build on knowledge by extending the domain of application. Few studies justify task choice or identify what features of a task are essential and what features are irrelevant to the study. In some intervention/treatment comparison studies to investigate cognitive development, the intervention tasks are often vague, as if the reader can infer what the learning environment was like from a few brief indications. Alan Schoenfeld commented similarly some time ago (Schoenfeld, 1980). As an example of how the task can be *invisible* to researchers, we could look at the commentaries about a well-known and widely accessible video of a mathematics lesson for the TIMSS study (http://www.timssvideo.com/67). In the commentaries reported on the website, it is only the teacher who mentions connections between the task design, its presentation, and students' participation; the researcher talks only generally about the social and structural features of the lesson. Yet the task is central to the success of the lesson in terms of lesson structure and learning and has been included in every official Japanese textbook for at least 40 years, and the diagram that goes with it is one the students are already familiar with. These features of the task are, we believe, crucial to understanding the lesson and the students' mathematical responses, but are hardly mentioned.

To some extent, task design issues are addressed in the literature using design research in which the result of the research is a designed product to fulfil a desired role. Task design also emerges in the growing field of international comparison (Shimizu, Kaur, Huang, & Clarke, 2010). Although there have been two recent edited collections of the use of tasks in mathematics teacher education as previously mentioned, we comment that, perhaps strangely, as yet no comparable international "go-to" collections for thinking about task design for mathematics classrooms have been published, although there have been research foci at ICME and PME in the last decade. We hope this volume will go some way towards filling that gap.

Meanwhile, teams of designers who are now well established have been producing and publishing tasks consistently for decades. Tasks initially invented and disseminated by Alan Bell (1993), Hans Freudenthal (1973), and Guy Brousseau (1997) and their colleagues are widely and effectively used throughout the world. The work of designing, trialling, and publishing often took priority over reporting the design research processes in an internationally accessible way, or researching their own practice, and the degree to which they expected teachers to understand their background theoretical justifications varied. Teachers all over the world might be familiar with the task of graphing the heights achieved by filling bottles of various shapes, or the task of estimating the size of the giant given the dimensions of the handprint, or the task of enlarging the drawing of a rectilinear animal. Teachers use these tasks not because they are committed to the precise background theory that led to their invention nor because their use has been researched and theorized in some other classroom or country. Rather, teachers use these tasks because they match the practices involved in local coordination of curriculum demands, classroom practices, intended mathematical outcomes, and anticipated participation of particular individuals and groups of students, using a craft-based frame as described in Chap. 2.

So we ask ourselves to what extent this book provides a go-to place for thinking about task design in mathematics classrooms. Because of our regular school-based experience, we both have the view that theory in task design should be clear and give meaning to phenomena in classrooms while also having practical meaning for teachers and designers. In Chap. 2, the distinction is drawn between theories as resource and theories as product: *theory for* and *theory of*. The intermediate level frames, as categorized in Chap. 2, combine theoretical structures that are well founded in theories of learning and classrooms with the practical, local theorizing that teachers do on a day-to-day basis. The technical terms used in academic writing might be seen as an obstacle (e.g. milieu; didactic contract in Chap. 4), but the underlying ideas would be familiar to many teachers. By contrast, the notion of instrumental genesis (see Chap. 6) is more abstract and less likely to relate to teachers' day-to-day thinking, although they would see evidence of a *utilisation scheme* in practice. Nevertheless, the concept of instrumental genesis has much to offer designers of both tasks and mathematics software, as well as in research. These thoughts are conjectural, but based on our own recent experience of teaching and talking with teachers in English and Japanese cultures of practice.

Returning to Schoenfeld's paper, which is entitled "On useful research reports," we would therefore ask "useful for whom?" We agree with the closing remarks made in Chap. 2 that there is a need for detailed research reports that are not unhelpfully limited in length and can fully report studies of design and use of tasks as well as pedagogy. We agree that details of tasks and the likely effects of task design features, as well as pedagogy, should be included more frequently in research about classrooms and learning. We point also to a need for researchers to distinguish between theories *of* their observations and theories *for* designers and teachers and to consider drawing on teachers' and learners' situated perspectives when theorizing in either case. In these respects, Chaps. 3–6 can all provide starting points. As for theories of task design, evaluations of effectiveness are always going to take place in natural contexts consisting of specific classrooms, teachers, constraints, and cultures, so it is inevitably the case that empirical studies will not be extensively generalizable, but can be illuminative and give rise to conjectures.

Finally, we pose some areas for further research that arose from the study, sometimes from several theme groups:

- How learners/teachers make sense of, and understand the purpose of, different kinds of tasks
- How different design principles reflect or generate different perceptions of mathematical concepts
- How different combinations of tasks and pedagogy influence learners' perceptions and mathematical activity
- How visual features of task presentation affect activity
- The design and implementation of task sequences
- The professional learning of prospective and practising teachers about task design, sequencing, and adaptation
- The role of task design in promoting equity and other values
- Task design and individual learner differences
- The effectiveness of forms of collaboration and communication between task designers, classroom teachers, educators, and policymakers.

References

Anderson, J. R., & Schunn, C. D. (2000). Implications of the ACT-R learning theory: No magic bullets. In R. Glaser (Ed.), *Advances in instructional psychology: Educational design and cognitive science* (Vol. 5, pp. 1–34). Mahwah, NJ: Lawrence Erlbaum Associates.

Bell, A. (1993). Principles for the design of teaching. *Educational Studies in Mathematics, 24*(1), 5–34.

Brousseau, G. (1997). *Theory of didactical situations in mathematics*. Dordrecht: Kluwer.

Clarke, D., Strømskag, H., Johnsen, H. L., Bikner-Ahsbahs, A., & Gardner, K. (2014). Mathematical task and the student. In: C. Nicol, P. Liljedahl, S. Oesterle, & D. Allan (Eds.), *Proceedings of the Joint Meeting of PME 38 and PME-NA 36* (Vol. 1, pp. 117–144). Vancouver, Canada

Davis, B. (1996). *Teaching mathematics: Toward a sound alternative*. New York: Garland.

Freudenthal, H. (1973). *Mathematics as an educational task*. Dordrecht: Kluwer.

Margolinas, C., Abboud-Blanchard, M., Bueno-Ravel, L., Douek, N., Fluckiger, A., Gibel, P., et al. (Eds.). (2011). *En amont et en aval des ingénieries didactiques*. (Ecole d'été de didactique des mathématiques, 2009). Grenoble: La pensée sauvage.

Runesson, U. (2005). Beyond discourse and interaction. Variation: A critical aspect for teaching and learning mathematics. *The Cambridge Journal of Education, 35*(1), 69–87.

Schoenfeld, A. H. (1980). On useful research reports. *Journal for Research in Mathematics Education, 11*(5), 389–391.

Schoenfeld, A. H. (2009). Bridging the cultures of educational research and design. *Educational Designer, 1*(2). http://www.educationaldesigner.org/ed/volume1/issue2/article5

Shimizu, Y., Kaur, B., Huang, R., & Clarke, D. (Eds.). (2010). *Mathematical tasks in classrooms around the world*. Rotterdam: Sense Publishers.

Tirosh, D., & Wood, T. (Eds.). (2009). *The international handbook of mathematics teacher education* (Vol. 2). Rotterdam: Sense Publishers.

Tzur, R., Sullivan, P., & Zaslavsky, O. (2008). Examining teachers' use of (non-routine) mathematical tasks in classrooms from three complementary perspectives: Teacher, teacher educator, researcher. In O. Figueras & A. Sepúlveda (Eds.), *Proceedings of the Joint Meeting of the 32nd Conference of the International Group for the Psychology of Mathematics Education, and the 30th North American Chapter* (Vol. 1, pp. 133–137). México: PME.

Valsiner, J. (1997). *Culture and the development of children's action: A theory of human development*. Hoboken, NJ: Wiley.
Valverde, G. A., Bianchi, L. J., Wolfe, R. G., Schmidt, W. H., & Houang, R. T. (2002). *According to the book. Using TIMSS to investigate the translation of policy into practice through the world of textbooks*. Dordrecht: Kluwer Academic.
Wittmann, E. C. (1995). Mathematics education as a 'design science'. *Educational Studies in Mathematics, 29*(4), 355–374.
Zaslavsky, O., & Sullivan, P. (Eds.). (2011). *Constructing knowledge for teaching: Secondary mathematics tasks to enhance prospective and practicing teacher learning*. New York: Springer.

Part II
Theme Group Chapters

Chapter 2
Frameworks and Principles for Task Design

Carolyn Kieran, Michiel Doorman, and Minoru Ohtani
with *Einav Aizikovitsh-Udi, Hugh Burkhardt, Liping Ding, Fco. Javier García, Keith Jones, Boris Koichu, Kotaro Komatsu, Francesca Morselli, João Pedro da Ponte, Martin Simon, Michelle Stephan, Malcolm Swan, Wim Van Dooren*
and additional contributions from *Didem Akyuz, David Clarke, Lea Dolev, Ana Cláudia Henriques, Sebastian Kuntze, Joana Mata-Pereira, Birgit Pepin, Marisa Quaresma, Luisa Ruiz-Higueras, Yosuke Tsujiyama, Xenia Vamvakoussi, Lieven Verschaffel, Orit Zaslavsky*

> *The natural sciences are concerned with how things are.*
> *Design, on the other hand, is concerned with how things might be.*
>
> (Herbert. A. Simon, 1969)

2.1 Introduction

The opening citation, drawn from *The Sciences of the Artificial* by H.A. Simon (1969), on design being concerned with "how things might be," evokes the idea that the field of mathematics education has been involved in design ever since its beginnings—going back to the time of Euclid and perhaps even Pythagoras, for whom the term *mathema* meant *subject of instruction*. However, as Wittmann (1995) remarked, in a paper titled *Mathematics Education as a Design Science*, the design of teaching units was never a focus of the educational research community until the mid-1970s. Artigue (2009), too, has argued that "didactical design has always played an important role in the field of mathematics education, but it has not always been a major theme of theoretical interest

---

C. Kieran (✉)
Département de mathématiques, Université du Québec à Montréal,
C.P. 8888, succ. Centre-Ville, Montreal, QC, Canada, H3C 3P8
e-mail: kieran.carolyn@uqam.ca

M. Doorman
Freudenthal Institute, Utrecht University, Princetonplein 5, 3584 CC,
Utrecht, The Netherlands
e-mail: m.doorman@uu.nl

M. Ohtani
School of Education, College of Human and Social Sciences, Kanazawa University,
Kanazawa, Ishikawa 920-1192, Japan
e-mail: mohtani@ed.kanazawa-u.ac.jp

This chapter has been made open access under a CC BY-NC-ND 4.0 license. For details on rights and licenses please read the Correction https://doi.org/10.1007/978-3-319-09629-2_13

© The Author(s) 2021
A. Watson, M. Ohtani (eds.), *Task Design In Mathematics Education*,
New ICMI Study Series, https://doi.org/10.1007/978-3-319-09629-2_2

in the community" (p. 7). According to Cobb, Confrey, diSessa, Lehrer, and Schauble (2003), design experiments are conducted to develop theories, not merely to tune empirically *what works*: "a design theory explains why designs work and suggests how they may be adapted to new circumstances" (p. 9). This movement in design within mathematics education from thoughtful tinkering, growing out of intuition and classroom development, to theoretically based research has not only been interlaced with the emergence of an international research community in mathematics education but also been accompanied by additional influences from the discipline of mathematics and, of no less importance, the work of psychologists (Kilpatrick, 1992).

The objective of this chapter is to give an overview of the current state of the art related to frameworks and principles for task design so as to provide a better understanding of the design process and the various interfaces between teaching, researching, and designing. In so doing, it aims at developing new insights and identifying areas related to task design that are in need of further study. The chapter consists of three main sections. The first main section (Sect. 2.2) begins with a historical overview of the emergence of the mathematics education research community, followed by developments within psychology at large that came to bear on design research in mathematics education. This is followed by a discussion of the ways in which mathematics education researchers took up some of these developments and adapted them to fit with a focus on mathematical content, thereby producing their own design frames. After this historical overview, the section offers a conceptualization of current frameworks for task design in mathematics education and describes the characteristics of the design principles/tools/heuristics offered by these frames. The second main section (Sect. 2.3) presents a set of cases that illustrate the relations between frameworks for task design and the nature of the tasks that are designed within a given framework. Because theoretical frameworks and principles do not account for all aspects of the process of task design, the third main section (Sect. 2.4) addresses additional factors that influence task design and the diversity of design approaches across various professional communities in mathematics education. The chapter concludes with a discussion of the progress made in the area of task design within mathematics education over the past several decades and includes some overall recommendations with respect to frameworks and principles for task design and for future design-related research.

2.2 Emergence and Development of Frameworks and Principles for Task Design

2.2.1 *Brief History of the Emergence of Design-Related Work from the 1960s to the 1990s*

When Wittmann (1995) remarked that the design of mathematical teaching units was never a focus of research until the mid-1970s, he was referring indirectly to the fact that it was only at that time that mathematics education research coalesced as a separate field of study. From the early 1900s, psychologists in various countries had been conducting empirical research on how mathematics is learned, while mathematicians and mathematics educators were more interested in focusing on the mathematical content to be taught and learned (Kilpatrick, 1992). Nevertheless, the post-Sputnik wave of mathematics education reform in the late 1950s and 1960s introduced many new task types using research methods and insights from prior psychological research. However, there was no community as such that could be called a mathematics education research community. That changed in the late 1960s and 1970s. In 1969, the first International Congress on Mathematical Education (ICME) took place in Lyon. A round table at that congress set the stage for the formation in 1976 of what was to quickly become the largest association of mathematics education researchers in the world, the International Group for the Psychology of Mathematics Education (PME). The emergence of this community was accompanied by the creation of several research journals, including *Educational Studies in Mathematics* in 1968, *Zentralblatt für Didaktik der Mathematik* in 1969, and *Journal for Research in Mathematics Education* in 1970. In several countries, research institutes were formed, such as the Shell Centres for Mathematical Education at Chelsea College and at the University of Nottingham in 1968, the Instituts de Recherche pour l'Enseignement des Mathématiques in France in 1969, the Netherlands Institute for the Development of Mathematical Education (IOWO) at Utrecht University in 1971, and the Institut für Didaktik der Mathematik in Bielefeld in 1973. With its annual meetings, its journals, and the fertilization made possible by cross-national collaborations, an international community of mathematics education researchers had taken shape. The late 1960s and 1970s thus signaled a huge surge and interest in research in mathematics education.

2.2.1.1 Influences from Psychology at Large

This surge in research in mathematics education had to rely almost exclusively in its early days on psychology as a source of theory (Johnson, 1980). Piaget's (1971) cognitively oriented, genetic epistemology is but one example of the psychological frames adopted by the emerging mathematics education research community in its studies on the learning of mathematics. However, other forces were beginning to be felt during these years—forces related to design that were being conceptualized and developed by psychologists with an interest in education.

In 1965, Robert Gagné published *The Conditions of Learning*. Based on models from behaviorist psychology, Gagné's (1965) nine conditions of learning were viewed as principles for instructional design—*instructional design* being defined in Wikipedia as the "practice of creating instructional experiences which make the acquisition of knowledge and skill more efficient, effective, and appealing." Gagné classified cognitive learning into the three areas of verbal information, cognitive strategies, and intellectual skills but tended to emphasize the learning and automating of procedures.

In parallel with the instructional design approach being developed by Gagné and others, a new field was emerging, that of cognitive science, often referred to as the *information processing* movement. As Anderson (1995/2000) remarked in his book, *Learning and Memory*, "at the height of the behaviourist era, around 1950, learning was perceived as the key issue in psychology; ... [but] learning was pushed somewhat from center stage by the cognitive movement in the 1960s" (p. vii). Advances in design considerations were stimulated by the theorizing of the cognitive scientist and Nobel laureate, H.A. Simon (1969), in *The Sciences of the Artificial*. He advocated the notion that the design process involved generating alternatives and then testing these alternatives against a range of requirements and constraints. Some of H.A. Simon's design ideas were taken up by the educational psychologist Robert Glaser (1976) in his *Components of a Psychology of Instruction: Toward a Science of Design*.

Glaser distinguished, in line with Bruner, between the descriptive nature of theories of learning and what he referred to as the prescriptive nature of theories of instruction. In integrating design considerations into instructional research, he argued:

> Regardless of the descriptive theory with which one works, four components of a prescriptive theory for the design of instructional environments appear to be essential: (a) analysis of the competence, the state of knowledge and skill, to be achieved; (b) description of the initial state with which learning begins; (c) conditions that can be implemented to bring about change from the initial state of the learner to the state described as the competence; and (d) assessment procedures for determining the immediate and long-range outcomes of the conditions that are put into effect to implement change from the initial state of competence to further development. (Glaser, 1976, p. 8)

Glaser emphasized that the structure of the subject-matter discipline may not be the most useful for facilitating the learning of less expert individuals. While reiterating H.A. Simon's notion that the design process involves the generation of alternatives, he did not designate specific principles on which the "generation of alternatives" might be based. Presumably, these would be related to various theories of learning, especially as design was considered to involve the application of a descriptive theory of learning to the generation of a prescriptive theory of instruction—but, according to Glaser, not necessarily foregrounding subject-matter considerations. Thus, mathematics education researchers would need to develop during the years to come their own scientific approaches to designing environments for the learning of mathematics and to generating frameworks for task design in particular.

2.2.1.2 Early Design Initiatives of the Mathematics Education Research Community

During the 1970s, the focus within the mathematics education research community was squarely on the learning of mathematics and the development of models of that learning. For example, the paper that Hans Freudenthal presented at PME3, held in Warwick, UK, in 1979 (one of the 24 research reports presented at PME that year) dealt with the development of reflective thinking (Freudenthal, 1979); Alan Bishop's, with visual abilities and mathematics learning (Bishop, 1979); and

Richard Skemp's, with goals of learning and qualities of understanding (Skemp, 1979). Nevertheless, two of the 1979 PME papers did touch upon issues related to tasks: one by Claude Janvier and the other by Alan Bell.

Janvier argued that, with the discovery learning movement, emphasis had been put on the "notion of appropriate learning environments and on the idea of rich situations likely to bring about discoveries or to encapsulate rich abstract ideas" (1979, p. 135). In his paper, he made use of one of the tasks (the racing car graph) devised for his doctoral research at the University of Nottingham in order to study various issues involved with the use of situations. At the conclusion of his paper, he remarked that his results were in line with Freudenthal's phenomenological approach, which promoted the use of large-scale situations involving weeks of work and stressing the child's point of view more than that of mathematical structures. In an earlier work, *Weeding and Sowing*, Freudenthal (1978) had introduced the approach of didactical phenomenology, which begins with a thorough mathematical analysis of the topic from which are generated hypothesized learning levels—an approach that he referred to as "developmental research" (see Gravemeijer & Cobb, 2006, 2013) and which was further elaborated by Streefland (1990) and Gravemeijer (1998). Freudenthal's (1979) PME3 text, which reflected his ongoing work, sowed the seeds for a mathematical-psychological approach to task design—an approach that was to develop during the late 1980s and 1990s into the instructional theory specific to mathematics education known as Realistic Mathematics Education.

The PME3 paper presented by Alan Bell focused on the learning that develops from different teaching approaches with various curriculum units that had been designed for the South Nottinghamshire project. The teaching methods that were explored included "embodiment, guided discovery approaches, and cognitive conflict" (Bell, 1979, p. 5). In Bell's research, design considerations were thus seen more through the lens of particular teaching methods than as approaches to the design of tasks per se.

In summary, the work of Hans Freudenthal at IOWO, of Alan Bell at Nottingham, and their colleagues during the 1970s reflected the beginnings of the new community of mathematics education researchers' efforts to grapple with the interaction between curriculum materials and the quality of mathematical teaching and learning—a dimension on which curriculum development efforts over the previous several decades had yielded little information. This embryonic work in task design was characterized mainly by reflection on the nature of mathematics, with aspects drawn from the psychologically based learning theories of the day and supported by personal pedagogical experience, coupled with informal observations of children's, students', or teachers' activity. The main aim seemed a combination of desiring to know more about the nature of learning mathematics and/or improving the teaching of mathematics rather than casting light on the nature of the tasks that might support such teaching or learning.

The 1980s within the mathematics education research community brought some integration of aspects of the design theories of H.A. Simon and others. In his 1984 *ESM* paper (a modified version of his opening address at the 14th annual meeting of German mathematics educators in 1981), titled *Teaching units as the integrating core of mathematics education*, Erich Wittmann (1984) argued for tasks displaying the following characteristics: the objectives, the materials, the mathematical problems arising from the context of the unit, and the mostly mathematical, sometimes psychological, background of the unit. He suggested that a teaching unit is not an elaborated plan for a series of lessons but rather it is an idea for a teaching approach that leaves open various ways of realizing the unit. Wittmann viewed the philosophy behind the teaching units as being embedded in Herbert Simon's *Sciences of the Artificial*: teaching units, according to Wittmann, are simply artificial objects constructed by mathematics educators—objects to be investigated within different educational ecologies.

During the years 1985–1988, one of the PME working groups focused on the extent to which its activities had established principles for the design of teaching. In 1988, a collection of papers from this working group was put together by the Shell Centre under the title *The Design of Teaching: Papers from a PME Working Group* and subsequently published in a special issue of *Educational Studies in Mathematics* in 1993. In his editorial for the special issue, Alan Bell wrote:

> Experimental work on the development of understanding in particular mathematical topics is relatively easy to conduct … but studies of the general properties of different *teaching methods* and *materials* are more difficult to set up. … Types of research on teaching which have been found productive, albeit in different ways, are the following: (1) basic psychological studies of aspects of learning …; (2) *developmental activities in which teaching materials are designed on the basis of theory and practical experience and are then taken through several cycles of trial and improvement*, …; and (3) comparative studies in which the same topic is taught to parallel classes by different methods. Examples of each of these types appear in this issue. But the design of teaching is a creative activity, and readers may hope to gain from these articles not only knowledge of some empirically established principles, but also tested ideas for their practical implementation. (Bell, 1993a, pp. 1–2, italics added)

Note in the quote the integration of "teaching methods" and "materials", that is, principles of teaching practice that are in harmony with principles that have been incorporated into the design of the teaching materials—an integration of two types of principles that will be seen to continue to be important in task design within the community over the decades to come. In his introductory article, "Principles for the Design of Teaching," Bell specified the following set of design principles:

> First one chooses a *situation* which embodies, in some *contexts*, the concepts and relations of the conceptual field in which it is desired to work. Within this situation, *tasks* are proposed to the learners which bring into play the concepts and relations. It is necessary that the learner shall know when the task is correctly performed; hence some form of *feedback* is required. When *errors* occur, arising from some *misconception*, it is appropriate to expose the *cognitive conflict* and to help the learner to achieve a resolution. This is one type of intervention which a teacher may make to assist the learning process. (Bell, 1993b, p. 9, italics in the original)

The underlying psychological learning principles supporting this theory of teaching design were said by Bell to include connectedness, structural transfer across contexts, feedback, reflection and review, and intensity.

The late 1970s and 1980s also witnessed Soviet-style teaching experiments, both in individual settings as well as in classrooms (e.g., Kalmykova, 1966; Menchinskaya, 1969), experiments that explored alternate teaching-with-task designs so as to investigate more deeply the learning of various mathematical concepts. The origin of these teaching experiments dated back to the 1920s with the individualized instructional experiments of Vygotsky, who believed that the development of mental abilities was essentially dependent upon instruction.

Another important development during the decade of the 1980s with respect to design was the emergence within France of *didactical engineering* (DE), an exacting theory-based approach to conducting research that had didactical design at its heart (Artigue, 1992). Despite its success as a design-based research practice, certain problems were encountered, according to Artigue (2009), when the rigorous designs were implemented in everyday classroom practice throughout its first decade. It was observed that the original designs went through a certain mutation in practice, leading her to note that "the relationships between theory and practice as regards didactical design are not under theoretical control" (Artigue, 2009, p. 12). This awareness pointed to one of the inherent limitations in theorizing about task design in isolation from considerations regarding instructional practice.

2.2.1.3 The 1990s and Early 2000s: Development of Research Specifically Referred to as *Design Experiments*

The term *design experiment* came into prominence in 1992 with the psychologist Ann Brown's (1992) paper on design experiments (see also Collins, 1992). Brown emphasized that design experiments aim at increasing the relevance of earlier cognitive science laboratory studies to the real activity of classrooms and that this research is designed to inform practice, as well as benefit from the experience of practitioners. This attention to classrooms, teaching, and teaching practice was a reflection of the movement from cognitive to sociocultural perspectives on learning — a movement that had emerged when Vygotsky's works started to become better known in the West toward the end of the 1980s and one that had already begun to take hold within the mathematics education research community.

Brown's paper signaled a kind of tipping point with respect to interest in design in the mathematics education research community (Lesh, 2002). Several factors had fallen into place, including the maturing of the community over a 20-year period and an evolving desire to be able to study within one's research not just learning or not just teaching. Design experiments aimed at taking into account the entire learning picture. As Cobb et al. (2003) pointed out:

> Design experiments ideally result in greater understanding of a learning ecology. ... Elements of a learning ecology typically include the tasks or problems that students are asked to solve, the kinds of discourse that are encouraged, the norms of participation that are established, the tools and related material means provided, and the practical means by which classroom teachers can orchestrate relations among these elements. (p. 9)

Within this conception of design experiments, the task or task sequence is considered but one of a larger set of design considerations involving the entire learning ecology—*task* or *task sequence* (which could take an entire lesson or more) being characterized in the ICMI Study-22 Discussion Document as "anything that a teacher uses to demonstrate mathematics, to pursue interactively with students, or to ask students to do something … also anything that students decide to do for themselves in a particular situation" (Watson et al., 2013, p. 12).

Another central feature of design experiments is, according to Cobb et al. (2003), the role played by theory, as well as the nature of this theory:

> General philosophical orientations to educational matters—such as constructivism—are important to educational practice, but they often fail to provide detailed guidance in organizing instruction. The critical question that must be asked is whether the theory informs prospective design and, if so, in precisely what way? Rather than grand theories of learning that may be difficult to project into particular circumstances, design experiments tend to emphasize an intermediate theoretical scope. (pp. 10–11)

Cobb et al.'s argument that theories of intermediate scope do a better job of informing prospective design leads naturally to the question of the nature of such theories and the ways in which they can inform prospective design. A broad approach to answering this question suggests that it might be helpful to first draw upon an example from outside the field to see what kinds of theories educational designers who specialize in the work of design integrate into their work.

2.2.1.4 An Example of Design Work by Educational Technologists

Jeroen van Merriënboer and his colleagues (van Merriënboer, Clark, & de Croock, 2002) are leading educational technologists who have developed a design model that consists of the following four components of instructional design for complex learning: (1) classes of learning tasks that are ordered and that promote schema construction, along with rule-oriented tasks for routine aspects, (2) supportive bridging information to link with prior knowledge, (3) just-in-time prerequisite information, and (4) part-task practice. The elaboration of these components is supported explicitly by major theoretical foundations in cognitive psychology, accompanied by a mix of different instructional approaches suitable for the different components of the design model. For example, in discussing the ordering of tasks with respect to their complexity, van Merriënboer et al. (2002) refer to cognitive load theory (Sweller, van Merriënboer, & Paas, 1998); in describing the amount and nature of learner support required, they refer to the framework of human problem-solving provided by Newell and Simon (1972); in noting the important role of cognitive feedback, they refer to the cognitive apprenticeship model (Collins, Brown, & Newman, 1989); and for just-in-time information that is best characterized as "how-to instruction or rule-based instruction", they cite Fisk and Gallini (1989). As an example of the kinds of instructional support useful in helping learners identify relevant relationships, van Merriënboer et al. distinguish between the inquiry method (e.g., "ask the learners to present a well-known, familiar example or counterexample for a particular idea") and the expository method (e.g., the instructor "presents a well-known, familiar example or counterexample for a particular idea") and note that "inquiry approaches are time-consuming, but because they directly build on learners' prior knowledge they are very appropriate for interconnecting new information and already existing cognitive schemata" (p. 48).

Van Merriënboer et al. state that their model was not developed for teaching conceptual knowledge or procedural skills per se nor is it very useful for designing very short learning programs that only take an instructional time of hours or a few days—it was generated with the aim of developing solutions to complex problems and has its roots in vocational education. Nevertheless, we consider it useful for illustrating how this team of professional task designers relies on a variety of theories of intermediate scope to underpin their design, as well as for pointing out how their suggested teaching approaches vary according to the nature of the given component of the instructional design.

2.2.1.5 From Early 2000 Onward

Theorizing related to design in mathematics education research, and in educational research more broadly, continued to evolve during the 2000s (Kelly, Lesh, & Baek, 2008). In addition, the term *task design* came to be more clearly present. For example, at the 2005 PME conference, a research forum was dedicated to task design, having as its stated theme, "The significance of task design in mathematics education" (Ainley & Pratt, 2005). One of the presenting teams, Gravemeijer, van Galen, and Keijzer (2005), pointed out that, "in Realistic Mathematics Education, instructional design concerns series of tasks, embedded in a local instruction theory; this local instruction theory enables the teacher to adapt the task to the abilities and interests of the students, while maintaining the original end goals" (pp. 1–108)—a local instructional theory being described by Gravemeijer et al. as the rationale for the instructional sequence, a rationale that evolves over several design experiments that involve testing and revising the sequence. Gravemeijer et al.'s statement suggests another view of theory as one that not only informs prospective design but is also a product of instructional design—an issue to which we will return immediately below. These ideas continued to be explored at ICME-11 in 2008 where the scientific program, for the first time, included a Topic Study Group (TSG) on task design: "Research and development in task design and analysis". The excitement generated regarding this research area was such that a similar TSG was put on the program for ICME-12 in 2012, as well as for ICME-13 in 2016. This interest was further illustrated by the holding of the 2013 ICMI Study-22 Conference on the same theme, a conference whose scientific work and discussions are the subject of this very volume.

2.2.1.6 Two Key Issues

In bringing to a close this first subsection devoted to a historical overview of the emergence of research related to design activity, we emphasize two central issues in need of clarification regarding the place of task design within design research. These issues underpin and run through much of the discussion that follows in the next subsection on frameworks and principles. In a recent article on design tools in didactical research, Ruthven, Laborde, Leach, and Tiberghien (2009) elaborated on the distinction between *design as intention* and *design as implementation* (Collins, Joseph, & Bielaczyc, 2004). *Design as implementation* focuses attention on the process by which a designed sequence is integrated into the classroom environment and subsequently is progressively refined, whereas *design as intention* addresses specifically the initial formulation of the design. While many studies address both, the distinction can be useful for understanding certain nuanced differences between one study and another. Ruthven et al. state that design as intention emphasizes the "original design and the clarity and coherence of the intentions it expresses" (p. 329). Design as intention makes use, in general, of theoretical frames that are well developed so as to provide this clarity and coherence. Although Ruthven et al. add that "the availability of design tools capable of identifying and addressing specific aspects of the situation under design can support both the initial formulation of a design and its subsequent refinement in the light of implementation" (p. 329), their examples cast light in particular on the *design as intention* orientation. In so doing, they illustrate clearly the role that theoretical tools play in the initial design.

In contrast to the front-end importance given to theory-based design tools by Ruthven et al. (2009), Gravemeijer and Cobb (2006) put the focus more toward the *development* of theory and its role as a product of the design research. In their design experiment studies, the initial theoretical base for the study and its accompanying instructional plan undergo successive refinements by means of the implementation process. The description of the entire process constitutes the development of the theory. Because of the centrality of the implementation process in the development of the resulting theory, such studies are characterized as *design as implementation* studies—even if their theoretical starting points could also qualify them as *design as intention* studies. For example, Gravemeijer and Cobb (2006) point out that

> from a design perspective, the goal of the preliminary phase of design research experiments is to formulate a local instruction theory that can be elaborated and refined while conducting the intended design experiment; ... this local instruction theory encompasses both provisional instructional activities, and a conjectured learning process that anticipates how students' thinking and understanding might evolve when the instructional activities are employed in the classroom. (p. 48)

They emphasize that the "products of design experiments typically include sequences of activities and associated resources for supporting a particular form of learning, together with a domain-specific, instructional theory that underpins the instructional sequences and constitutes its rationale; a domain-specific, instructional theory consists of a substantiated learning process that culminates with the achievement of significant learning goals as well as the demonstrated means of supporting that learning process" (Cobb & Gravemeijer, 2008, p. 77).

More precisely, Cobb et al. (2003) insist that "design experiments are conducted to develop theories" (p. 9).

Put another way, theories are both a resource and a product. As a resource, they provide theoretical tools and principles to support the design of a teaching sequence (e.g., Ruthven et al., 2009) and, as a product of design research, theories inform us about both the processes of learning and the means that have been shown to support that learning (Cobb et al., 2003).

A second issue related to the role and nature of theory in design is the significance given to task design itself within the design process. When theory and its design tools are viewed as a front-end resource in the design process, the way in which task design is informed by these theory-based tools moves to center stage (e.g., Ruthven et al., 2009). By way of contrast, when theory development is viewed as the aim of design experiments, task design tends to be less central: "One of the primary aims of this type of research is *not* to develop the instructional sequence as such, but to support the constitution of an empirically grounded local instruction theory that underpins that instructional sequence" (Gravemeijer & Cobb, 2006, p. 77, emphasis added). This is not to say, in the latter case, that task design is unimportant (it clearly is) but rather the design of the teaching/instructional sequence is only one of several all-encompassing considerations within the whole interactive learning ecology. In practice, most design experiments combine both orientations: the design is based on a conceptual framework and upon theoretical propositions, while the successive iterations of implementation and retrospective analysis contribute to further theory building that is central to the research. In fact, both orientations will be seen to be present in most of the design studies exemplified below.

These two issues, that is, (1) *design as intention* and *design as implementation* and (2) the status given to the initial design of the set of tasks, point to central differences in the way in which the roles of theory and task design are considered within the design process in the mathematics education research community. In the presentation that follows—one that focuses on current frameworks and principles for task design—we shall attempt to interweave these distinctions into our discussion of the nature of the frames adopted, adapted, and developed within the activity of design. By so doing, we hope to be able to contribute to clarifying some of the ways in which theory and task design are related.

2.2.2 *A Conceptualization of Current Theoretical Frameworks and Principles for Task Design in Mathematics Education Research*

2.2.2.1 Introduction

Our historical look at the early research efforts related to task design revealed a mix of task and instructional considerations. However, the extent to which instructional aspects are factored into task design is but one of the ways in which design frameworks can vary. Frameworks can also differ according to the degree to which the learning environments are student centered, knowledge centered, or assessment centered (Bransford, Brown, & Cocking, 1999), as well as the manner in which they draw upon cognitive, sociological, sociocultural, discursive, or other theories. In addition, frameworks are distinguishable according to the extent to which they can be related to various task genres, that is, whether the tasks are geared toward (1) the development of mathematical knowledge (such as concepts, procedures, representations; see, e.g., Swan, 2008), (2) the development of the processes of mathematical reasoning (such as conjecturing, generalizing, proving, as well as fostering creativity, argumentation, and critical thinking; see, e.g., Leikin, 2013; Lin, Yang, Lee, Tabach, & Stylianides, 2012; Martinez & Castro Superfine, 2012), (3) the development of modeling and problem-solving activity (e.g., Lesh, Hoover, Hole, Kelly, & Post, 2000; Ponte, Mata-Pereira, Henriques, & Quaresma, 2013; Schoenfeld, 1985), (4) the assessment of mathematical knowledge, processes, and problem-solving (e.g., Swan & Burkhardt, 2012), (5) the context of mathematical team competitions (e.g., Goddijn, 2008), and so on. As well, some frameworks may be more suited to the design of specific tasks, others to the design of lesson flow (e.g., Corey, Peterson, Lewis, & Bukarau, 2010), and still others to the design of sequences involving the integration of particular artifacts (e.g., Kieran & Drijvers, 2006). Because several considerations enter into an overall design—considerations that include the specific genre of the task, its instructional support, the classroom milieu, the tools being used, and so on—each part of the design might call for different theoretical underpinnings. Thus, the resulting design can involve a *networking* of various theoretical frames and principles (Prediger, Bikner-Ahsbahs, & Arzarello, 2008) or a *bricolage* (Gravemeijer, 1994) or a *bridging* (Koedinger, 2002). Furthermore, the nature of the principles or heuristics associated with various frames and the way in which these heuristics are construed—according to whether they are viewed as illuminating, inspiring, guiding, systematizing, or even constraining—all have a part to play (see Sect. 2.4 for discussion of other factors, such as the artistic and value-related aspects of task design). A more holistic way of thinking about frames is to view them as being of different levels (e.g., Goldenberg, 2008) or types, for example, grand frames, intermediate-level frames, domain-specific frames (i.e., frames related to the learning of specific mathematical concepts and reasoning processes), and frames related to particular features of the learning environment (e.g., frames for tool use)—all of them together constituting any one theoretical base for the design of a given study (Gravemeijer & Cobb, 2006). This manner of conceptualizing design frames according to the levels of grand, intermediate, and domain specific (note that tool-related frames are treated in Chap. 6) will now be used as a backdrop for examining the nature of current theoretical frameworks and principles for task design in mathematics education research.

2.2.2.2 Grand Theoretical Frames and Their Affordances

Mathematics education research has tended in large measure to adopt such grand theoretical perspectives as the cognitive-psychological, the constructivist, and the socioconstructivist. However, as pointed out by Lerman, Xu, and Tsatsaroni (2002), these are but three of the vast array of theoretical fields, in addition to those from educational psychology and/or mathematics that have backgrounded mathematics education research. In line with Cobb (2007), who has argued that such grand theories need to be adapted and interpreted in order to serve the needs of design research, our discussion of them will be brief and limited to a few selective aspects.

A cognitive-psychological theoretical perspective has dominated research on the learning of mathematics ever since the days of Piaget—at least up until the late 1980s and early 1990s when Vygotsky's work came to be better known among Western mathematics educators (Bartolini Bussi, 1991). Tightly linked with the cognitive-psychological perspective is the constructivist frame (von Glasersfeld, 1987) that stemmed mainly from Piaget's genetic epistemology (Steffe & Kieren, 1994). Learning came to be widely interpreted as a constructive process, a process in which students actively construct mathematical knowledge. While constructivism has always been with us, perhaps under another guise, its growing acceptance as an educational tenet during the 1980s (Cobb & Steffe, 1983) helped to oust the view of mathematical teaching as the transmission of the teacher's knowledge and mathematical learning as the reception of that knowledge. However, constructivist, cognitively oriented research soon became hard pressed to reconcile the notion that all learning is individually constructed with the evidence of commonalities found across individuals. Constructivists had to admit the social dimensions of learning, thereby paving the way for integrating the Soviet work of Vygotsky and Leont'ev. The view of learning as situated with respect to social and cultural practices, and thus a socioconstructivist frame of reference, soon became widely accepted (Lerman, 1996). This frame directly allowed for a focus on the role of teaching and of classroom interactions in the learning process.

Within the tradition of cognitive psychology, two types of theories have been developed by mathematics education researchers (Cobb, 2007). One concerns theories of learning across mathematics in general (e.g., Pirie & Kieren's, 1994, recursive theory of mathematical understanding; Sfard's, 1991, theory of reification); the other concerns theories of the development of students' learning in specific mathematical areas (e.g., Filloy & Rojano's, 1989, theory of algebraic reasoning; Clements & Battista's, 1992, theory of geometric reasoning). As will be seen below, these theories that have been inspired by and are situated within the grander theories are key components of design in mathematics education research, even if, as Cobb (2007) insists, they are not instructional and require adaptation or combination with other theories, in order to serve the needs of instructional design. Cobb's point of view was also emphasized earlier by Bransford et al. (1999) in their volume *How People Learn*: "Learning theory provides no simple recipe for designing effective learning environments, but it constrains the design of effective ones" (p. xvi).

2.2.2.3 Intermediate-Level Frames

Intermediate-level frames have a more specialized focus than the grand theories of socioconstructivism and the like, yet intermediate-level frames still tend to be situated within the perspective of one or the other of these grand frames. Even if their focus is more specialized, intermediate-level frames have the property that they can be applied across a wide variety of mathematical areas. In brief, intermediate-level frames are located between the grand theories and the more local, domain-specific frames that address particular mathematical concepts, procedures, or processes. A multitude of intermediate-level frames have been developed that are being applied in adaptive ways to design research in mathematics education. They include, for example, Realistic Mathematics Education theory (Treffers, 1987), the Theory of Didactical Situations (Brousseau, 1997), the Anthropological Theory of Didactics (ATD) (Chevallard, 1999), Lesson Study (Lewis, 2002), Cultural-Semiotics theory (Radford, 2003), Commognitive Theory (Sfard, 2008), and so on (see Sect. 2.3 for an elaboration of the ways in which some of the of various intermediate-level frames have been adopted and adapted for use in research on task design).

In general, intermediate-level frames can be characterized by explicit principles/heuristics/tools that can be applied to the design of tasks and task sequences. Because these frames tend to be highly developed, they are often used in *design-as-intention* approaches. In addition, intermediate-level frames can also be characterized according to whether their roots are primarily theoretical or whether they are based to a large extent on deep craft knowledge. The two examples of intermediate-level frames and their accompanying principles for task design that we offer immediately below reflect these two roots. The first is the Theory of Didactical Situations and the second is that of Lesson Study. For both types, we examine the framework and associated principles that support the process of task design.

An Example of a Theory-Based Intermediate-Level Frame: Theory of Didactical Situations

The Theory of Didactical Situations (TDS) is generally associated with Guy Brousseau (1997); however, its development over the years has been contributed to by the French mathematical *didactique* community at large. A central characteristic of TDS research is its framing within a deep a priori analysis of the underlying mathematics of the topic to be learned, integrating the epistemology of the discipline, and supported by cognitive hypotheses related to the learning of the given topic. TDS is said to be an intermediate theory in that it draws upon the grand theory of Piaget's work in cognitive development. According to Ruthven et al. (2009), one of the central design tools provided by TDS is the *adidactical situation*, which mediates the development of students' mathematical knowledge through independent problem-solving. The term *adidactical* within TDS refers specifically to that part of the activity "between the moment the student accepts the problem as if it were her own and the moment when she produces her answer, [a time when] the teacher refrains from interfering and suggesting the knowledge that she wants to see appear" (Brousseau, 1997, p. 30).

A situation includes both the task and the environment that is designed to provide for the adidactical activity of the student. According to the TDS frame, the adidactical situation tool furnishes guidelines as to: "the problem to be posed, the conditions under which it is to be solved, and the expected progression toward a strategy that is both valid and efficient; this includes the process of 'devolution' intended to lead students to directly experience the mathematical problem as such and the creation of a (material and social) 'milieu' that provides students with feedback conducive to the evolution of their strategies" (Ruthven et al., 2009, p. 331). During the early years of the development of the TDS, it was found that the frame needed some modification so as to take into account the necessary role played by the teacher in fostering the later institutionalization of the student's mathematical knowledge acquired during the adidactical phase—the term *institutionalization* referring specifically to the process whereby the teacher gives a certain status to the ideas developed by students by framing and situating them within the concepts and terminology of the broader cultural body of scientific knowledge (see also Chaps. 3 and 5).

Identifying a suitable set of problem situations that can support the development of new mathematical knowledge is absolutely central to the design of a TDS teaching sequence. The adidactical situation must be one for which students have a starting approach but one that turns out to be unsatisfactory. Students must be able to obtain feedback from the milieu that both lets them know that their approach is inappropriate and also provides the means to move forward. The "enlargement of a shape puzzle" (see Ruthven et al., 2009, pp. 332–334) is a fine example of the design of an adidactical situation. When students (who are working in small groups) are asked to make a larger puzzle of the same shape but with the edge whose length is 4 cm being enlarged to 7 cm, it is expected that they would use additive reasoning. But the feedback provided by the attempt to put the enlarged puzzle pieces together lets the students know that their way of solving the enlargement problem is incorrect. Eventually, "intellectual" feedback is provided by the teacher in order to help the students to arrive collectively at a multiplicative model.

In addition to the adidactical situation tool, TDS-based design is also informed by a second design heuristic, that of the *didactical variables* tool. This supplementary design tool allows for choices regarding particular aspects of the main task and how it is to be carried out (e.g., shape and dimensions of the pieces, the ratio of the enlargement, the various pieces of the puzzle being constructed by different students), aspects that are subject to modification as a result of successive cycles of the teaching sequence. Although certain modifications are made to those aspects of the task that are found to improve the learning potential of the situation (i.e., that students are more likely to learn what is intended), the initial design of the task is absolutely central to the TDS-framed *design-as-intention* process.

An Example of a Craft-Based Intermediate-Level Frame: Lesson Study

Lesson Study is typically associated with Japanese education where its roots can be traced back to the early 1900s (Fernandez & Yoshida, 2004). However, variants of Lesson Study have been developed in China (Huang & Bao, 2006; Yang & Ricks, 2013), as well as in other countries (Hart, Alston, & Murata, 2011). For example, in the Chinese version, according to Ding, Jones, and Pepin (2013), the role of the expert in the development and refinement of a lesson plan is of critical importance. This role consists of contributions that are said to go beyond that of deep craft knowledge—contributions that Ding, Jones, Pepin, and Sikko (2014) describe as consisting of a complex combination of considerable knowledge of mathematical didactics and general theories of learning and of students, as well as the "accumulated wisdom of practice." Thus, the distinction we are proposing between craft-based and theory-based intermediate-level frames may be rather blurry for certain versions of Lesson Study. Even within Japan, various types of Lesson Study exist: at the school level, at the local and prefectural level, and at the national level. Nevertheless, in that the majority of Lesson Studies in Japan occur at the school level and that school-based Lesson Study in Japan tends to be considered around the world as prototypical of Lesson Study practice, it is this latter version of Lesson Study that is the focus here.

Lesson Study is a culturally situated, collaborative, approach to design—one where teachers with their deep, craft-based knowledge are central to the process and which at the same time constitutes a form of professional development (Krainer, 2011; Ohtani, 2011). Fundamental to Japanese teachers' ability to design and implement high-quality mathematics lessons that are centered on high-quality mathematical tasks is a detailed, widely shared conception of what constitutes effective mathematics pedagogy (Jacobs & Morita, 2002). Thus, Lesson Study, with its cultural and collaborative foundations, could be said to be situated within the grand theory of socioculturalism.

Lesson Study consists of the following phases: (1) collaboratively planning a research lesson, (2) seeing the research lesson in action, (3) discussing the research lesson, (4) revising the lesson (optional), (5) teaching the new version of the lesson, and (6) sharing reflections on the new version of the lesson. Phases 4–6 are sometimes replaced with a single phase of "consolidating and reporting". In any case, it is the research lesson—its planning, its implementation, and its evaluation, but especially its planning—that is the focus here. As will be seen, Lesson Study is a frame devoted as much to *design as intention* as it is to *design as implementation*.

A typical lesson plan proposal contains the following seven items (Lewis, 2002):

1. Name of the unit
2. Unit objectives
3. Research theme
4. Current characteristics of students
5. Learning plan for the unit, which includes the sequence of lessons in the unit and the tasks for each lesson
6. Plan for the research lesson, which includes:
    - Aims of the lesson,
    - Teacher activities
    - Anticipated student thinking and activities
    - Points to notice and evaluate
    - Materials

- Strategies
- Major points to be evaluated
- Copies of lesson materials (e.g., blackboard plan, student handouts, visual aids)

7. Background information and data collection forms for observers (e.g., a seating chart).

While the unfolding of the research lesson and its evaluation takes only 1 day, its planning can occupy anywhere from 1 to 2 months. As can be seen from Lewis's (2002) list of seven items, the plan of the lesson includes not only the detailed design of the task itself, which constitutes the essence of the research lesson, but also the links with other tasks in the larger unit. Central to this planning is the process of *kyozaikenkyu*. *Kyozaikenkyu* means literally "instructional materials research" and constitutes a first principle for task design. The study of instructional materials goes beyond the textbook series being used in the classroom. As pointed out by Fujii (2013), *kyozaikenkyu* involves examining teaching materials and tasks from a mathematical point of view (mathematical content analysis), an educational point of view (considering broader values such as "skills for living"), as well as from the students' point of view (readiness, what students know, anticipated students' thinking and misconceptions, etc.). It includes studying other textbook series treating the same topic, thinking about the manipulatives being used, and analyzing what the curriculum standards and research have to say about the topic and its teaching and learning. If the decision is ultimately made to modify an existing textbook task, that decision is made with great care because the teachers know that the textbook task was designed with considerable thoughtfulness. Tasks that will lead to multiple strategies are crucial to the task design process of Lesson Study (for details, see the Lesson Study case that is illustrated in Sect. 2.3)—strategies that will ultimately comprise the basis for the classroom discussion phase of the research lesson.

Consequently, a second design principle concerns the actual form that the research lesson takes. Referred to as *structured problem-solving* by Stigler and Hiebert (1999), the research lesson involves a single task and the following four specific phases: (1) teacher presenting the problem (*donyu*, 5–10 min), (2) students working at solving the problem without the teacher's help (*jiriki-kaiketsu*, 10–20 min), (3) comparing and discussing solution approaches (*neriage*, 10–20 min), and (4) summing up by the teacher (*matome*, 5 min). During students' independent working, the teacher walks between the desks (*kikan-junshi*) and silently assesses students' work; she is in the process of making a provisional plan as to which student contribution should be presented first in order to make clear the progress and elaboration from simple idea to sophisticated one: this is the core of *neriage*, a phase during which students' shared ideas are analyzed, compared, and contrasted. During the fourth phase of the research lesson (*matome*), the teacher will usually comment as to the more efficient of the discussed strategies, as well as the task's and the lesson's mathematical and educational values. As an aside, it is noted that Japanese teachers use these specific didactical terms to discuss their teaching and that such didactical terms not only mediate the activity of the various participants involved in Lesson Study but also lead to the co-construction of deep craft knowledge.

After the research lesson has been observed by other teachers, school administrators, and sometimes by an outside expert, it is then discussed and evaluated in relation to its overall goals. This process of lesson evaluation, and in particular *task evaluation*, is considered a third design principle. The post-lesson discussion focuses to a large extent on the effects of the initial task design with respect to student thinking and learning. The teacher's thought-out key questioning receives much attention. Another of the main aspects discussed is whether the anticipated student solutions were in fact evoked by the task and its accompanying manipulative materials or whether improvements in specific parts of the task design are warranted.

Of the three design principles that are the core of the research lesson of Lesson Study, *kyozaikenkyu* is the most all encompassing. However, it is one that is often underrepresented or even overlooked in Lesson Study practice in other countries (Doig, Groves, & Fujii, 2011). *Kyozaikenkyu* is, in fact, central to Japanese teachers' everyday practice. As such, it is a key component of the Japanese Lesson Study example of craft-based frames for task design.

2.2.2.4 Domain-Specific Frames

In contrast to intermediate-level frames whose characterizations do not specify any particular mathematical reasoning process or any particular mathematical content area, domain-specific frames for the design of tasks or task sequences do specify particular reasoning processes (e.g., conjecturing, arguing, proving) or particular content (e.g., geometry, integer numbers, numerical concepts, algebraic techniques). Task design research involving domain-specific frames typically draws upon past research findings in a given area, in addition to being situated within more general, intermediate-level, and grand-level frameworks. As such, domain-specific frames for task design research tend to be more eclectic than their intermediate-level counterparts. As an aside, note that Realistic Mathematics Education theory has at times been referred to by its adherents as a domain-specific instructional theory in that it is an instructional theory for the domain of mathematics education; however, in this chapter we reserve the term *domain specific* for frames dealing with specific mathematical content areas or reasoning processes. Some researchers use the term "local theories" or "local frames" for what we are referring to as domain-specific frames. In general, domain-specific frames are associated with *design as implementation* in that the main aim of the research is the further development of the domain-specific frame by means of the implementation process. However, this is not a hard distinction. As will be seen, for some examples of design research studies that make use of and develop domain-specific frames, the approach is as much *design as intention* as it is *design as implementation*.

A Domain-Specific Frame for Fostering Mathematical Argumentation Within Geometric Problem-Solving

In a recent article, Prusak, Hershkowitz, and Schwarz (2013) reported on a yearlong, design research-based course with third graders in mathematical problem-solving that aimed at instilling inquiry learning and argumentative norms. The researchers investigated if, and in which ways, principled design is effective in promoting a problem-solving culture, mathematical reasoning, and conceptual learning. Their design was situated in a multifaceted framework that drew upon principles from the intermediate-level, educational theory of Cognitive Apprenticeship, as described in Schoenfeld (1994), and from domain-specific research in geometric reasoning (Hershkowitz, 1990) and argumentation (Arzarello & Sabena, 2011; Duval, 2006), as well as from multiple studies with a sociocultural orientation. The Prusak et al. study was, in fact, one that articulated explicitly two design components: one for the task and one for the learning environment.

The task that Prusak et al. (2013) discuss in their paper is the *sharing a cake* task:

> Yael, Nadav, and their friends Itai and Michele came home from school very, very hungry. On the kitchen table was a nice square piece of cake, leftover from their birthday. They wanted to be fair and divide the square into four equal pieces so that everyone would get a fourth (1/4) of the leftover cake. Suggest different ways in which the children can cut up and divide the square piece of cake. For each suggestion, explain why this would give each child exactly a fourth of the leftover cake. (p. 6)

Accompanying the text on the task worksheet was a set of nine square grids upon which the students, who worked first individually and then in groups, could draw their suggested cuttings of the cake and several blank lines per grid where they were to explain their thinking. Prusak et al. state that the design of this task, as well as that of the others used within their yearlong study, relied on the following five principles:

- Encourage the production of multiple solutions (Levav-Waynberg & Leikin, 2009).
- Create collaborative situations (Arcavi, Kessel, Meira, & Smith, 1998).
- Engage in socio-cognitive conflicts (Limón, 2001).
- Provide tools for checking hypotheses (Hadas, Hershkowitz, & Schwarz, 2001).
- Invite students to reflect on solutions (Pólya, 1945/1957).

Setting up a problem-solving culture in the classroom was an integral part of the Prusak et al. design study. More specifically, they brought into play Schoenfeld's (1994) use of the Cognitive Apprenticeship model by which he scaffolded students' problem-solving in a classroom culture that emphasized communication, reflective mathematical practice, and reasoning rather than results. In line with Schoenfeld, the following instructional-practice principles constituted a second overall design frame for the Prusak et al. study:

- Emphasize processes rather than solely results.
- Use a variety of social settings (individual, small group, and whole class).

- Develop a critical attitude toward mathematical arguments using prompts like, "Does it convince me?"
- Encourage students to listen and try to persuade each other and, thus, to develop ideas together.
- Have students learn to report on what they do, first verbally, then in written form, explaining their solutions to their teammates or to the entire class.

The authors argue that the findings of their study provided evidence that

> the meticulous design as well as the problem-solving culture triggered a general process according to which students capitalised on problem-solving heuristics and engaged in multimodal argumentation, subsequently reaching deep understanding of a geometrical property (the fact that non-congruent shapes may have equal areas). ... The activity we described encourages the production of multiple solutions, which is an explicit instruction in the task. Also, students were arranged in small groups, and were asked to collaborate. Collaboration led students to compare solutions. Since they were asked to justify their solutions, these justifications naturally created socio-cognitive conflicts. The nine grids in the task provide a tool for checking hypotheses. (pp. 16–17)

The authors concluded their paper with a theoretical model for learning early geometry through multimodal argumentation in a problem-solving context—a model that includes the description of the learning process and the demonstrated means of supporting that learning process. They emphasize that the designed task served as a principle-based research tool, one that was central to the elaboration of their domain-specific model.

The Prusak et al. study presents an example of the use of well-defined, even if quite general, principles as a front-end resource for the design of the tasks. A second set of principles provided the frame for the design of the learning culture in which the tasks would unfold. Both sets of principles make their study one that could be described as *design as intention*. The empirical evidence that the initial design was effective in eliciting the aimed-for learning of specific geometrical notions through argumentation within a problem-solving setting led to the theoretical elaboration of a domain-specific model. In this sense, the study could be said to be also an example of *design as implementation*. Additionally, and of pivotal importance for design in mathematics education research, the design of the task activities was supported by the accompanying design of an instructional environment involving specific teaching practices that would nurture a collaborative problem-solving culture. This emphasizes the crucial interactive relation between the design of a task or task sequence and the design of the instructional culture in which the task is to be integrated—an emphasis that is also seen in design research involving intermediate-level frames.

A Domain-Specific Frame for Proof Problems with Diagrams

The frame used by Komatsu and Tsujiyama (2013) in their design research, which centered on eighth grade proof and proving, was inspired by the notion of deductive guessing—a notion formulated by Lakatos (1976) as a heuristic rule for coping with counterexamples. In deductive guessing, after one proves conjectures and then faces their counterexamples or non-examples, one invents deductively more general conjectures that hold true even for these examples. Because deductive guessing is a mathematical notion, some adaptation with respect to pedagogical perspectives was necessary so as to use deductive guessing as a frame for task design. Its adaptation yielded the *proof problems with diagrams* frame—a proof problem with diagrams being a problem in which a statement is described with reference to particular diagrams with symbols (one diagram in most cases) and solvers are required to prove the statement and then to deal with related diagrams involving counterexamples and non-examples. The frame was also informed by the earlier research of Shimizu (1981) who had argued that, after students solve proof problems with diagrams, it is important for them to further inquire "of what (mathematical) relations the given diagram is a representative special case" (p. 36) by utilizing the already obtained proof.

As is the case with much of the current task design research in the field, Komatsu and Tsujiyama (2013) point out that, because "it is unrealistic to expect that only posing the designed problems will facilitate students' activities and mathematical learning, task design involves not only selection or development of problems but also teachers' instructional guidance related to the problems" (p. 472). In line with (a) deductive guessing in Lakatos's work, (b) the nature of proof problems with diagrams, and (c) the instructional guidance to be provided by the teacher, the researchers derived the following three task design principles:

- Educators and teachers should select or develop certain kinds of proof problems with diagrams where students can find counterexamples or non-examples and engage in deductive guessing through changing the attached diagrams.
- Teachers should encourage their students to change the attached diagrams while keeping the conditions of the statements, so that they find counterexamples or non-examples of the statements.
- After students face the counterexamples or non-examples, teachers should plan their instructional guidance by which students can utilize their proofs of initial problems to invent more general statements that hold true for these examples.

Komatsu and Tsujiyama illustrate their principles for task design by means of a problem involving parallelograms, drawn from Okamoto, Koseki, Morisugi, and Sasaki et al. (2012) (see also Komatsu, Tsujiyama, Sakamaki, & Koike, 2014). Their principle-based description of the design of the parallelogram task, accompanied by suggestions related to specific instructional guidance (see Komatsu & Tsujiyama, 2013, pp. 476–477), provides a detailed plan for the teaching of proof problems with diagrams, one that will eventually be subjected to further classroom implementation and possible revisions. Thus, the domain-specific frame crafted by Komatsu and Tsujiyama yielded, at this stage of their research, a primarily *design-as-intention* tool—a tool for task design that integrated earlier research on proof problems with diagrams, a novel theoretical frame based on Lakatosian deductive guessing, and a cultural tradition involving the role of the teacher.

A Domain-Specific Frame for the Learning of Integer Concepts and Operations

The design research of Stephan and Akyuz (2013) involved creating and implementing a hypothetical learning trajectory (HLT) and associated sequence of instructional tasks for teaching integers in a middle-grade classroom over a 5-week period. Grounded in the researchers' deep knowledge of past research on the learning of integers and integer operations, the design of their instructional sequence was underpinned by the following three heuristics of the intermediate-level frame of Realistic Mathematics Education (RME):

- Guided reinvention—"To start developing an instructional sequence, the designer first engages in a thought experiment to envision a learning route the class might invent with guidance of a teacher" (p. 510).
- Sequences experientially real for students—"Instructional tasks draw on realistic situations as a semantic grounding for students' mathematizations" (p. 510).
- Emergent models—"Instructional activities should encourage students to transition from reasoning with models of their informal mathematical activity to modeling their formal mathematical activity, also called *emergent modeling* (Gravemeijer & Stephan, 2002)" (p. 510).

The anticipated learning path (HLT) led to the generation of a six-phase instructional sequence involving various mathematical tools, which was then implemented in the classroom. The authors used a version of social constructivism, called the *emergent perspective* (Cobb & Yackel, 1996), to situate their interpretation of classroom events. In the emergent perspective, learning is considered both an individual, psychological process and a social process. Thus, two frames were used by Stephan and Akyuz to analyze their classroom data: (a) a framework for interpreting the evolving classroom learning environment, that is, the emergent perspective, and (b) a framework for interpreting student mathematical reasoning and learning of integer concepts, that is, a frame based on the instructional theory for Realistic Mathematics Education. After implementation and analysis of the collective learning of the class, the authors considered various possible revisions to the instructional sequence. The details of the design of the instructional sequence, its implementation, classroom analysis, suggested revisions, and reflective theoretical discussion can be found in Stephan and Akyuz (2012).

The description of the entire process, which constitutes an empirically sustained, domain-specific theoretical model for the teaching of integers and integer operations, is a classic example of *design as implementation*. In the spirit of Cobb and Gravemeijer (2008), Stephan and Akyuz generated a domain-specific, instructional theory that embodied the classroom-based, activity-oriented process of learning a specific mathematical content and which included a very detailed description of the representational tools, classroom interactions, and teacher interventions that sustained this learning. The elaboration of their domain-specific theoretical frame was supported explicitly in its design, implementation, and analysis by the two frames of Realistic Mathematics Education and the emergent perspective and implicitly by its reliance upon prior research and previous domain-specific design work on the learning of integers.

In describing their research, Stephan and Akyuz stress students' engagement with tasks: "In RME, … tasks are defined as problematic situations that are experientially real for students" (Stephan & Akyuz, 2013, p. 509), a perspective based on Freudenthal's assertion that "people need to see mathematics not as a closed system, but rather as an activity, the process of mathematizing reality and if possible even that of mathematizing mathematics" (Stephan & Akyuz, 2012, p. 432). This emphasis on the way in which students engage with tasks, and the way in which teachers actually facilitate that activity, is central to *design as implementation*. It also helps to shed an explanatory light on Gravemeijer and Cobb's (2006) earlier statement that one of the primary aims of design research is *not* to develop an instructional sequence as such. More precisely, the description of the entire design process (including initial design, implementation, and revision) is intended to foster an understanding of why and how the final sequence is supposed to promote learning. The whole description supports others in implementing the sequence in other contexts and as such constitutes its theoretical role: that of a local instructional theory for a specific mathematical domain.

2.2.2.5 In Drawing This Section to a Close

A main objective of this section of the chapter has been to examine the nature and roles of frameworks and principles in the design process and, at the same time, to draw out the relative centrality given to the design of the task or task sequence itself. Building upon the pioneering work of scholars during the early years of the growth of the mathematics education research community and its evolution through to design experiments and beyond, a double lens was used to explore the nature of current theoretical frameworks and principles for task design: (1) an analysis according to *levels of frames* that focused particular attention upon both intermediate and domain-specific frames and (2) a consideration of the constructs of *design as intention* and *design as implementation* within the design process. The lenses that were used, accompanied by a sampling of examples drawn from the international body of research literature related to design in mathematics education, helped to clarify some of the ways in which theory and task design are related. Among the relationships that emerged from the analysis of frames and their roles in design in general and task design in particular, one was particularly salient: it was the design consideration related to instructional support that was common to all the examples and central to each.

The examples all included attention to instructional support, some in the form of quite explicit principles. For instance, in the Prusak et al. (2013) example, a separate list of specific principles related to the design of the instructional environment was provided—principles that delineated a clear set of indices related to the way in which the instructional environment and the designed task were to mutually support each other. This example offers a viable model for further productive work in design in mathematics education and for its reporting. In fact, the way in which instructional principles were incorporated into the design of the studies exemplified so far in this chapter leads to suggesting that it might be more appropriate, terminology-wise, to refer to this field as *task design for instruction*. This more precise terminology could thereby give weight to the notion that the initial formulation of the design and its description include principles related to the design of the instruction and instructional environment, as well as to the design of the task. This terminology would also capture the spirit of the early research efforts in this area by Freudenthal, Bell, Wittman, and others. But even more importantly, integrating the terms *task design* and *for instruction* would allow us to emphasize that which would appear to be fundamental to design in mathematics education, a fundamental that was well expressed by Komatsu and Tsujiyama (2013, p. 472), namely: "It is unrealistic to expect that only posing the designed problems will facilitate students' activities and mathematical learning; task design involves not only selection or development of problems but also teachers' instructional guidance related to the problems."

2.3 Case Studies Illustrating the Relation Between Frameworks and Task Design

2.3.1 *Introduction*

The cases within this section illustrate the variety in types of frameworks for task design and the variety in relationships among the frameworks, design principles, and the actual design process. The heart of the discussion of each of the prototypical cases is guided by two questions:

(a) What do tasks look like when designed within a given theoretical frame or according to given design principles?
(b) Why do they look the way they do?

Using these questions as a lens, this section goes into some detail with respect to each example and thus extends the discussion that was initiated in the previous section. Seven cases are herein presented, most of them drawing upon aspects of grand theories and illustrating the use of intermediate-level theories. The cases are based on the contributions of the participants of Theme Group D (Frameworks and Principles for Task Design) of the ICMI Study-22 Conference (Margolinas, 2013). They are not intended to represent a sample of all possible design principles and frameworks that are currently used or investigated all over the world. The cases reflect the different levels of frames discussed in the previous section and illustrate how these frameworks can be applied across a variety of mathematical domains, as well as offer design approaches related to particular mathematical understandings. They also include cases that exemplify principles from design frames based on deep craft knowledge and from design related to various task genres, such as concept development and assessment. In sum, the seven cases being discussed, often too briefly to do justice to the richness of the underlying theory and the design, are intended to provide insight into the current state of the art of task design in mathematics education.

2.3.2 Cases

2.3.2.1 Case 1: Anthropological Theory of Didactics

Within the ATD, mathematics is conceived as a human activity, institutionally situated, and modeled in terms of practices that go beyond learning "concepts" or "processes." This results in the need for a renewed paradigm of learning mathematics in school (Chevallard, 2012). The paradigm thus changes from *visiting mathematical concepts and skills* to *questioning the world* (motivated, functional encounters). This elaboration of the ATD has its roots in Chevallard's earlier ATD work of the 1990s, as well as in his collaborative research with Brousseau on the notions of didactic engineering, the didactic transposition, and the Theory of Didactical Situations.

The following example focuses on the application of an intermediate-level frame to the design of a mathematical activity involving young children. It illustrates design principles that are related to the previously mentioned paradigm shift. These principles were not extracted from this particular case but result from a collection of ATD study and research paths that have been designed in the last 10 years (e.g., Barquero & Bosch, 2015).

The aim of the task was to embed the emergence and use of numbers and addition in the study of a system that is real and that gives rise to a meaningful mathematical activity for (preschool) students (García & Ruiz-Higueras, 2013). The initial question for the students was *if we've got a box with silkworms, how many leaves do we need to feed them?*

Firstly, students would collect leaves by themselves. But after a few days, they would ask the gardener to collect the leaves for them, using a written message. That would provoke the need of being aware of quantities, as well as using codes to express them. Next, the biological system would start to evolve: silkworms turn into cocoons, then moths arise, and finally, they die. Students would have to control a heterogeneous collection made of silkworms, cocoons, and moths. As change happened, they would need techniques to record the evolution of the system. The teacher would prepare different tables to record and control the evolution of the system. She would introduce this tool so that students could take control of the evolution of the system under their own responsibility. This would widen students' activity, particularly toward addition, time control, and recording. At the end, when all the moths would have died, the system would disappear. However, students would have lots of information (models) about its evolution. Through the interpretation of these models, pupils would carry out the final task: reconstructing the system and its evolution. Figure 2.1 illustrates the unfolding of part of the task activity in class.

Designing tasks for a renewed paradigm of learning mathematics, from visiting mathematical works to questioning the world, is operationalized within the ATD by design principles for creating research paths for students (Table 2.1). The whole task, called a *study and research path* (SRP), is linked within the ATD to an epistemological conception of mathematics as a human activity and modeled in terms of practices.

Context: Taking care of our silkworms
Characteristics of this rich context:
• Dynamic system (evolving over time)
• Many different quantities to be measured
• Communicative tasks can be naturally formulated (representing quantities with numbers & numerals)
• Increasing complexity

Fig. 2.1 Children taking care of silkworms (García & Ruiz-Higueras, 2013)

Table 2.1 Design principles for a *study and research path* in ATD

Design principle	Illustrated by the case
Develop an epistemological reference model for the mathematical activity the task is aiming at. Investigate how the mathematical objects of study are related, how they are articulated and used in specific (out-of-school) practices, and how these can be transposed into the educational system.	Numbers (as mathematical objects) and codes to express them (numerals) emerge in communicative situations where the aim is not just to measure a discrete set but to communicate about it so that another person can understand the evolution of the system without having access to it (neither visually nor manipulatively).
Look for generating questions beyond school mathematics that are crucial and alive for the students, connected with society and its problems (questioning the world).	"When we've got a box of silkworms, how many leaves do we need to feed them?"
Generate questions that do not lead the study process to a dead end but that give rise to new questions that could expand it.	How to communicate the number of leaves needed? How to keep track of the number of silkworms, cocoons, and moths?
Create a collaborative and shared study process with shared responsibilities and shared norms for justification.	The teacher introduced tools so that students could take control of the evolution of the system under their own responsibility.
Support the search for answers by stimulating the study of (extra) mathematical works or consulting other communities.	Students were stimulated to ask parents about the time needed for the cocoon phase.

For tasks like these, designers need to leave the school, step out of traditional school mathematics, and question the meaning of the objects they want students to work with (their origin, evolution, and purpose in current society). This leads to a reference model for the design of a study path and will inform possible overarching generative questions. Piloting is an essential phase in the design process for checking conjectured teaching and learning processes and for improving the ecological and economic robustness of the task.

2.3.2.2 Case 2: Variation Theory

Variation Theory (VT) focuses task designers on what varies and what remains invariant in a series of tasks in order to enable learners to experience and grasp the intended object of learning (Runesson, 2005). The learners' experiences depend on the critical features of the object to which their awareness is directed. Consequently, designing task sequences requires an analysis of possible variations so that learners "might observe regularities and differences, develop expectations, make comparisons, have surprises, test, adapt and confirm their conjectures within the exercise" (Watson & Mason, 2006, p. 109). Analyses of variation space, patterns in learners' experiences, and how these patterns are compatible with the intended object of learning are key elements in the intermediate-level frame of VT.

Three successive versions of a task for teachers illustrate how VT guided a cyclic process of task design, analysis, and redesign. The learning objective of the task was to facilitate the teachers' awareness of mathematics as a connected field of study by directing their attention to structural similarities and differences among the basic concepts of analytical geometry and loci of points (Koichu, Zaslavsky, & Dolev, 2013).

The first version of the task consisted of 24 representations of loci of points that had to be sorted by the teachers by creating groups of similar loci (see Fig. 2.2). It was created so that three types of controlled variation would be maximized:

- The first type of variation was related to the mathematical objects described in the cards for sorting (e.g., a straight line, circles, parabolas, ellipses, hyperbolas).
- The second type of variation was related to the type of representation (symbolic, graphical, and verbal).
- The third type of variation was related to the type of experience needed to handle the task (prior knowledge, information provided with the task).

During the trial of this first version of the task, it was found that a lot of time was devoted to technical work and to classifying the items by surface features. To reduce the amount of time and the attraction of surface features, the second version consisted of 18 items. The items that were approached in all the groups only algebraically were excluded (items 8–10, 15, and 19–21). In spite of a smaller intended variation space, it appeared that the enacted variation space became richer and the teachers more engaged. However, the presence of the well-familiar graphical and symbolic representations in the task postponed, and likely hindered, the learning experiences offered by the verbal items. For this reason, pictorial representations were eliminated and the third version of the task contained only 11 verbal items

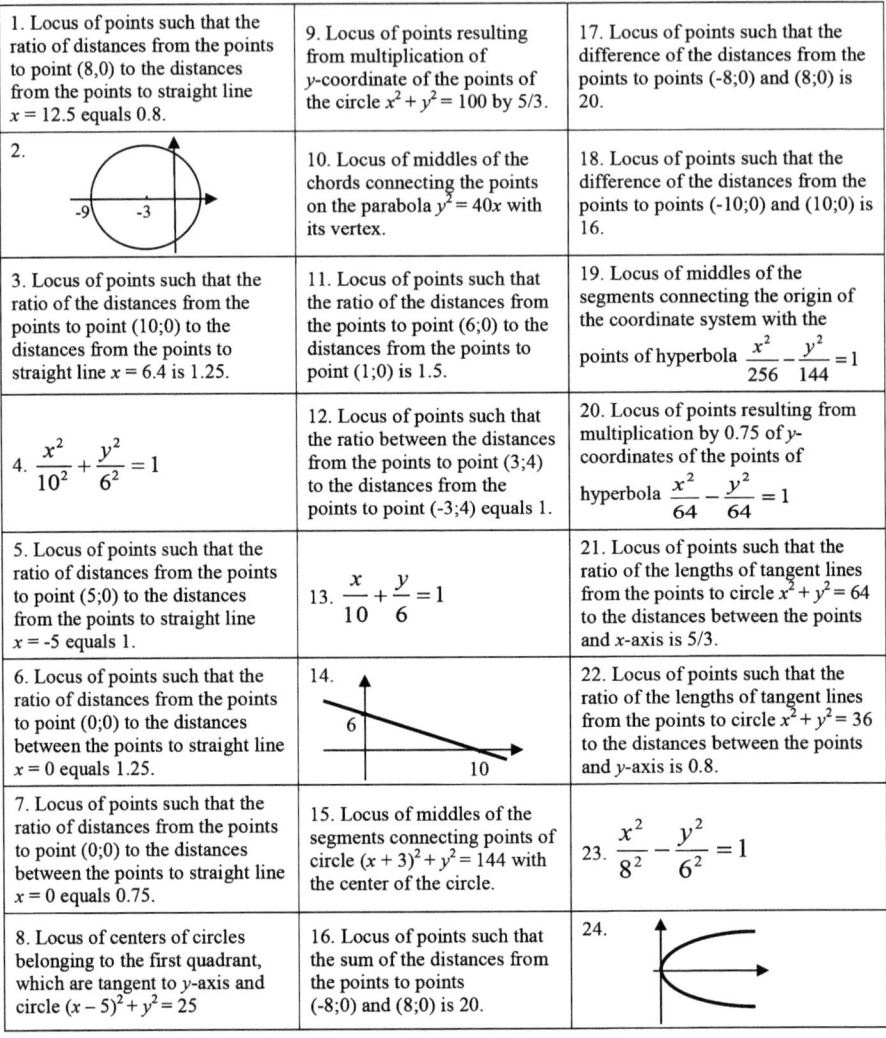

Fig. 2.2 The first version of the sorting task (Koichu et al., 2013, edu.technion.ac.il/docs/KoichuZaslavskyDolevThemeA_Supplementary_material.pdf)

[i.e., items 1, 3, 5–7, 11, 12, 16–18, and a new item 25: "Locus of points such that the distance from them to point (−3, 0) is 6."].

The intention of the third version of the task was to suppress the affordance of using sorting criteria based upon surface features, in favor of criteria related to the identification of structural similarities and differences. The experiences showed that the two main enacted subcategories of the by-keywords criterion in the third version of the task (i.e., by main operation and by main generating elements) were remarkably close to one of the intended types of variation of the task.

2 Frameworks and Principles for Task Design

Table 2.2 Design principles underlying a VT example

Design principle	Illustrated by the case
Identify and analyze the object of learning and its critical features that constitute a variation space (Marton, Runesson, & Tsui, 2004).	The object of learning for the teachers is to facilitate their awareness of structural similarities and differences among the basic mathematical concepts of analytical geometry. A critical feature is to classify conics by names of loci of points, because this requires an understanding of structural similarities and differences. The variation space consisted of the mathematical objects, their various representations, and the types of prior experience needed to handle the task.
Create task(s) so as to have the learners discern critical aspects of the intended learning object and aim for coinciding the intended and enacted variation space.	Map the types of variation in the sorting task and connect them to the intended object of learning. The teacher-awareness facet of the study prompted a first version of the task where the space of variation was maximized.
Focus on the central role of the main intended activity (be careful with including mathematically challenging items and affording complementary mathematical techniques).	The central activity was discovering structural similarities and differences among the basic concepts of analytical geometry, but this was obscured by technical manipulations evoked by the first version of the task.
Carefully analyze whether the variation space of a task can be improved toward the intended object of learning.	The final version had a reduced variation space that was more engaging and resulted in richer learners' experiences. This version succeeded in suppressing sorting criteria by surface features, in favor of criteria related to the identification of structural similarities and differences.

This example illustrates the process of task design guided by the interplay between analyzing and providing variation space and observing patterns in learners' experiences. Design principles drawn from this example in connection with VT orient the designer to what varies and what remains invariant in a series of tasks (Table 2.2).

This case, which exemplifies the use of the VT frame in task design, shows that design decisions can easily hinder or support affordances of a task with respect to the intended object of learning. The challenge for task design is to anticipate and organize learners' experiences so that they serve as reference points to more meaningful decisions.

2.3.2.3 Case 3: Conceptual Change Theory

A particular issue for task design is the teaching of concepts that are known to be difficult for students because prior knowledge is in conflict with what is to be learned. Conceptual Change Theory (CCT) is an intermediate-level frame that allows researchers to specifically investigate this issue. The case, which is drawn from research into students' learning of nonnatural numbers (Van Dooren, Vamvakoussi, & Verschaffel, 2013), illustrates design principles that are derived from CCT, as well as from existing domain-specific research related to the learning of rational number.

Many difficulties that students have with nonnatural numbers are rooted in prior knowledge about whole (natural) numbers. The conceptual change perspective provides an explanatory framework for these difficulties as it analyzes them in terms of students' initial, intuitive theories that shape their predictions and explanations in a coherent way (Vosniadou, Vamvakoussi, & Skopeliti, 2008). This results in the following starting point for task design: How to deal with an incompatibility between students' initial theories and intended mathematical development that unavoidably will occur? The initial theories students rely on when encountering the ideas of nonnatural numbers are related to their understanding of whole numbers. Consequently, students see numbers as being discrete, used for counting, and grounded in additive reasoning (Ni & Zhou, 2005; Vamvakoussi & Vosniadou, 2012). These initial theories easily lead to typical misconceptions like longer decimals are larger, for example, $2.12 > 2.2$; a fraction gets bigger when one of its parts is larger, for example, $2/5 < 2/7$; and the density misconception that between two non-equal numbers, there is a finite number of other numbers. Furthermore, students are often unaware of the background assumptions of their reasoning.

The sequence of tasks in this example takes these background assumptions into account and supports overcoming the incompatibility between the discreteness of whole numbers and the density of nonnatural numbers (see Table 2.3). The tasks were accompanied by the introduction of a tool-like representation that fostered

Table 2.3 A sequence of tasks supporting conceptual change

Task	Goal
1. What do you know about the number line? Describe it as well as you can. Read and comment upon the answers of your fellow students.	Express prior knowledge about the number line.
2. We often use the term "the set of all numbers". Suppose someone tries to understand what we mean by that. Could you draw a picture to help him/her understand?	Construct a representation for all numbers.
3. Imagine the number line as a rubber band that can be stretched. Position 0 and 1 on the band and place numbers between them until it looks like you have used all the available points. If you stretch the rubber band, then you will find out there are more points, corresponding to more numbers. This procedure can be repeated infinitely many times—your imaginary rubber band never breaks!	Construct the imaginary rubber band as a representation for all numbers.
4. We have been talking about two different representations of numbers: A "formal" one, which we usually use at school, and a second one, which was proposed in our discussion and you seem to find adequate. Could you find a solid reason why we should prefer one over the other?	Compare two different representations.
5. Imagine that you can become as small as a point of the number line. Then you could see other points up close. Suppose that you are on the point that stands for the number 2.3. Can you define what point is the one closest to you? Describe in words or by drawing a picture.	Reason about density with the number line.

Table 2.4 Design principles for a Conceptual Change example

Design principle	Illustrated by the case
Take students' prior knowledge and potential initial understandings into consideration (explore existing literature).	Build on students' prior knowledge of differences and similarities between natural and nonnatural numbers by explicitly addressing the number line (tasks 1 and 2).
Facilitate students' awareness of their background assumptions by creating opportunities for them to externalize their ideas, to compare them with peers' ideas, and to reflect on them.	Let students compare and discuss representations for all numbers (task 2).
Use models and external representations, know their power and their limitations.	The rubber band was introduced to prevent the number line from continuing to be interpreted as a ruler with a finite number of points (task 3, also task 4).
Foster analogical reasoning that supports conceptual restructuring.	The rubber band is a bridging analogy that fosters students' comparison between a continuous geometrical object and the real number line (task 5).

reasoning with numbers in a geometrical analogy: an imaginary rubber band (Vamvakoussi & Vosniadou, 2012).

The sequence of tasks in this example illustrates how initial, usually largely unconscious assumptions can be elicited and made explicit. It also shows how cross-domain mapping between continuous magnitudes (points on the rubber band) and the set of numbers can be fostered (see Table 2.3). Design principles drawn from this conceptual change example are listed in Table 2.4.

CCT is primarily a cognitively oriented theory and therefore does not encompass all aspects related to instructional design. The design principles emerging in this case are intended as instructional tools to change, to move forward, students' cognitions. As such, this theory offers added value for task design when dealing with difficult concepts. The sequence of tasks in this example was designed on the basis of specific theoretical principles, was empirically tested, and appeared useful for teachers as well as for students. The resulting design principles have a wide field of application in that instructional design has to cope with similar prevailing misconceptions in many domains of mathematics and beyond.

2.3.2.4 Case 4: Conceptual Learning Through Reflective Abstraction

This case (M. Simon, 2013) is derived from a one-on-one teaching experiment (M. Simon et al., 2010) with a prospective primary school teacher, Erin. The teaching experiment focused on developing a common-denominator algorithm for the division of fractions with conceptual understanding. Conceptual learning in this case is understood as the process of developing new and more powerful abstractions through activity. The approach draws task design principles from the grand-level frame of Piaget's construct of reflective abstraction.

Solving Word Problems Using Rectangular Drawings
1. I have seven-eighths of a gallon of ice cream, and I want to give each of my friends a one eighth portion. How many friends can I give ice cream to?
2. A scuba diver has two hours worth of air in her tank. If each dive to the bottom of the bay takes three-eighths of an hour, how many dives can she make with the air she has?
3. Each ticket at an amusement park in France is worth four fifths of a Euro. If a pack of tickets costs four Euros, how many tickets are in a pack?

Solving Context-Free Problems Using Rectangular Drawings
4. $3/4 \div 1/4 =$
5. $7/3 \div 2/3 =$
6. $8/5 \div 3/5 =$
7. $5/6 \div 4/6 =$

Solving Context-Free Problems Using Mental Runs of Rectangular Drawings
8. $23/25 \div 7/25 =$
9. $7/167 \div 2/167 =$
10. $7/103 \div 2/103 =$

Fig. 2.3 A task sequence for learning to solve division problems with common denominators (M. Simon, 2013, p. 508)

The researcher engaged Erin in a sequence of tasks, probed her thinking, and allowed Erin to develop her understanding without input from the researcher. The task sequence began with division-of-fraction word problems whose dividend and divisors had common denominators. Erin was asked to solve them by drawing a diagram. She was able to solve the first task without difficulty ("I have $\frac{7}{8}$ of a gallon of ice cream and I want to give each of my friends a $\frac{1}{8}$-gallon portion. To how many friends can I give ice cream?"). The task sequence progressed to word problems in which the dividend and the divisor still had common denominators, but the divisor did not divide the dividend equally, and then to similar tasks presented using only number expressions (e.g., $\frac{8}{5} \div \frac{3}{5} =$). Erin still drew rectangles for solving the problem. For $\frac{8}{5} \div \frac{3}{5}$, she first drew two whole rectangles divided into fifths. Next, she shaded $\frac{2}{5}$ of one rectangle leaving $\frac{8}{5}$ unshaded. She circled each $\frac{3}{5}$, counted 2 groups, and was able to deduce that the remainder $\frac{2}{5}$ is $\frac{2}{3}$ of $\frac{3}{5}$, thereby finding the solution $2\frac{2}{3}$.

Next, Erin was asked to solve a more complex fraction division task $\frac{23}{25} \div \frac{7}{25}$ (see Fig. 2.3).

Erin made clear that she did not know the answer and the researcher encouraged her to talk through a diagram solution without actually drawing. Erin described the diagram process she would use and the result she would get. Erin easily solved the next task, $\frac{7}{167} \div \frac{2}{167}$, using the same approach, that is, narration of a diagram solution. However, when that task was followed by the task, $\frac{7}{103} \div \frac{2}{103}$, Erin gave the answer immediately. She realized that the change in the fractional units would not affect the quotient. Further, she was able to explain the invariance of the quotient across a range of denominators by creating a general diagram. Erin had made an

2 Frameworks and Principles for Task Design

Table 2.5 Design principles for Conceptual Learning by Reflective Abstraction

Design principle	Illustrated by the case
Identify a potential activity that is already available to the learner and that can be the basis for the intended abstraction (the identified learning goal).	The student's informal diagram solutions supported anticipations toward a common-denominator algorithm for the division of fractions. This learning goal affected the identification of the solution strategy and the strategy affected the specific goal toward which the design was oriented.
Design tasks to elicit the available activity and to promote reflective abstraction (a learned anticipation supported by a shift from activities with external representations to mental runs).	The task sequence starts with word problems and context-free tasks to elicit and reinforce the diagram-drawing strategy. Once the student is using the intended strategy, the task sequence provokes the anticipated abstraction. For this purpose, larger numbers for the denominators and invited mental runs of diagram drawings were used.

abstraction as a result of this task sequence. She perceived a commonality in her activity involving these mental diagram solutions.

In this example, two design principles fostering Conceptual Learning by Reflective Abstraction can be recognized (see Table 2.5).

When Erin was faced with a second task with the same pair of numerators and different common denominators, she realized that she was about to enact the same activity as in the previous task. At that moment, she also realized why the size of the common denominators did not change the quotient. This was an example of Erin's reflection on her (mental) activity. That is, she perceived the commonality in her activity in the two cases that led to an abstraction. These tasks helped her to foreground key quantitative relationships and to create a need to invoke a new concept and mental operations that are critical to the concept being developed.

This is an example of task design for concept development that does not depend on students making a leap through problem-solving. Rather, the task sequence affords them the opportunity to build an abstraction from already available activity. In this case, the abstraction was built from the activity of creating informal diagram solutions for solving simple sharing tasks. The approach illustrated by this example can serve to inform ongoing and future research work on the crafting of domain-specific frames for task design related to the process of mathematical abstraction.

2.3.2.5 Case 5: Realistic Mathematics Education

Realistic Mathematics Education (RME) is an intermediate-level frame that has been developed in the Netherlands (see Van den Heuvel-Panhuizen & Drijvers, 2013). RME is rooted in the work of Freudenthal (1973, 1991) who argued for teaching mathematics that is relevant for students and instigated research in how students can be offered opportunities for guided reinvention of mathematics. This example illustrates design principles drawn from RME by presenting one task from a longitudinal sequence on the topic of percentage (Van den Heuvel-Panhuizen, 2003).

Three performances will take place in the school theater. How busy will the theater be during each performance? Color the part of the hall that is occupied and write down the percentage of the seats that are occupied.

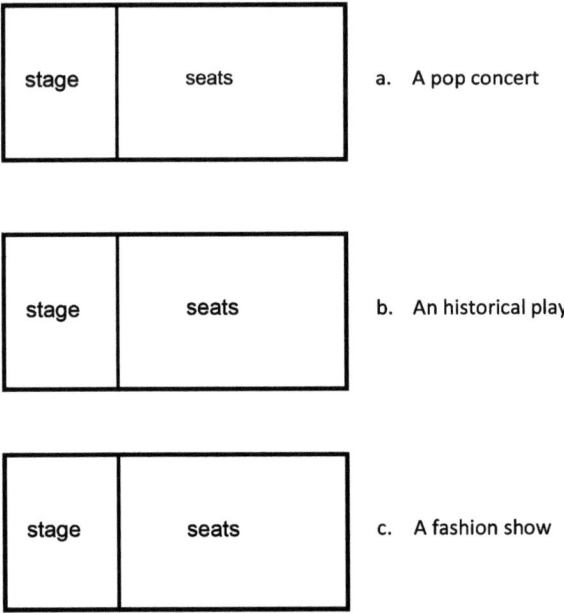

Fig. 2.4 Percentage of occupied seats in a school theater (adapted from Van den Heuvel-Panhuizen, 2003, p. 19)

The learning of percentage is embedded within the domain of rational numbers and is strongly intertwined with learning fractions, decimals, and ratios. The example is taken from a sequence that starts with a qualitative informal introduction to percentages before it proceeds toward quantitative formal procedures. The underlying notion is that you first need to know what the procedures are about before you can perform and practice them. The introduction attempts to evoke the use of so-many-out-of-so-many reasoning in everyday situations. The design question from the RME perspective is *How to evoke and build on informal and outside-school knowledge of students when aiming at having them make sense of percentage?*

The introductory exploratory activities of the sequence are designed to make students aware of the daily life use of percentages, to evoke tentative representations, and to prepare for model building. The activities are cast in problem situations that "beg to be organized" by means of the mathematics under study (Freudenthal, 1983, p. 32). Some of these initial tasks are based on a school theater scenario. The students are asked to indicate for different performances how busy the theater will be. They can do this by coloring in the part that is occupied and then writing down the related percentage (see Fig. 2.4).

Table 2.6 Design principles for an RME example

Design principle	Illustrated by the case
Identify the fundamental concept, potential starting points, and models that support the learning of mathematics through a phenomenological didactical analysis, thought experiments, discussions with teachers, and working with students.	Starting points for the design of the sequence are the relative character in percentages (so many out of so many), the use of contexts like comparing the occupation of a school theater for various performances, and the bar model that supports the shift from intuitions to mathematical reasoning.
Model-eliciting activities are at the heart of an instructional sequence. They are cast in contexts that are familiar for students and provide relevant and challenging elements that need to be organized or schematized mathematically so as to have the potential to evoke their (informal) knowledge.	The theater context offers (limited) opportunities to be mathematically creative, to learn to solve problems for which the students do not have a standard solution procedure yet, and, at the same time, to learn about percentage.
A task sequence guides students from informal to formal mathematical reasoning. Models play a key role by shifting from a "model of" a particular situation to a "model for" mathematical reasoning (Streefland, 1993).	The drawings in the theater are expected to first become a "bar model of" so-many-out-of-so-many situations and later turn into a "model for" mathematical reasoning about percentages and fractions.
Take into account the design of skill development and connections with related mathematical topics to develop strong structures and procedures.	The notion of percentage is being taught in close connection with fractions, decimals, and ratios. A qualitative understanding precedes the development of quantitative skills.
Design whole-class and peer-to-peer interaction.	Whole-class discussion of students' answers, their drawings, and estimated percentages is essential for the progress of the teaching process (not included in the example task).

This task is an example of an exploratory activity to support students in building models (i.e., the bar model) based upon their prior ideas and experiences. For the students, the coloring of theater halls is intended to lead to a way to express so-many-out-of-so-many situations. Furthermore, it is expected that students will spontaneously use fractions to "explain" the percentage of fullness. With a system of tasks, including more closed practicing tasks, students are guided to reinvent the mathematics of percentages.

This example illustrates core principles of RME that were articulated originally by Treffers (1987) but were reformulated over the years (see Table 2.6).

In recent years, new aspects, like mathematics in vocational education, in special education, and in linguistically diverse classrooms, have also been approached from an RME perspective. These projects enrich RME and enhance the robustness of the research that accompanies its further development.

2.3.2.6 Case 6: Formative Assessment for Developing Problem-Solving Strategies

This case is drawn from the work at the Shell Centre at Nottingham University (UK) (i.e., Burkhardt & Swan, 2013; Mathematics Assessment Project[1]). The case illustrates how formative assessment can support the development of problem-solving strategies in mathematics. The power of formative assessment for enhancing learning in mathematics classrooms is well known (Black, Harrison, Lee, Marshall, & Wiliam, 2003; Black & Wiliam, 1998).

Formative assessment includes "all those activities undertaken by teachers, and by their students in assessing themselves, which provide information to be used as feedback to modify the teaching and learning activities in which they are engaged; such assessment becomes 'formative assessment' when the evidence is actually used to adapt the teaching work to meet the needs" (Black & Wiliam, 1998, p. 140). Herein lies the real challenge: For assessment to be formative, the teacher must develop expertise in becoming aware of and *adapting to* the learning needs of students, both in planning lessons and in the moment-by-moment of the classroom.

These problem-solving lessons are not about developing understanding of new mathematical concepts but rather about students developing and comparing alternative approaches to nonroutine tasks. The structure of a typical lesson is illustrated with the *Counting Trees* task (Fig. 2.5). This task is intended to assess how well students are able to select an appropriate sampling method and use it, together with mathematical concepts such as area and proportion, to solve an unfamiliar problem.

In a preliminary lesson, students are invited to tackle the problem individually. They are told not to worry if they don't find an answer, that there are many ways to

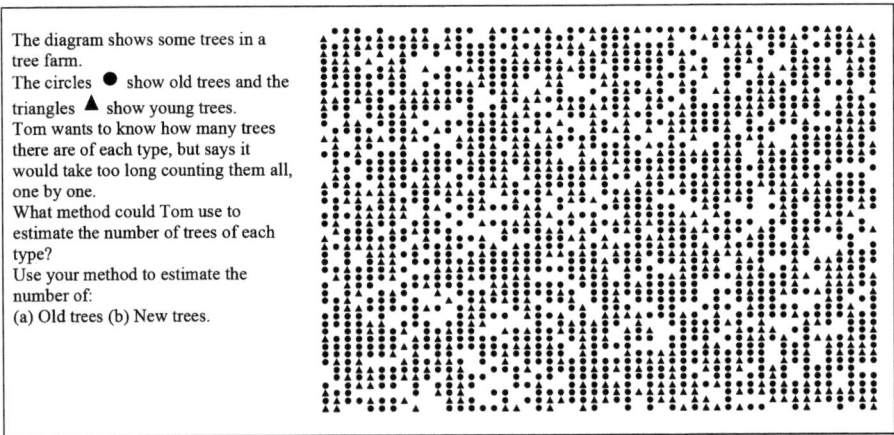

Fig. 2.5 The counting trees task (adapted from MARS, 2012)

---

[1] http://map.mathshell.org

tackle the problem, and that there may be more than one correct answer. The task is used to expose students' different intuitive approaches to the problem. Students' responses are collected by the teacher and analyzed *before* the actual lesson. This gives the teacher time to plan well-considered responses to students.

The lesson itself begins with the teacher returning students' attempts along with questions (not explanations) that are intended to move their thinking forward. This role shift for students encourages them to reflect on their own strategy and to consider alternative methods. Instead of using the work of fellow students, the teacher introduces sample student work from materials provided. These samples are carefully chosen to highlight different approaches and common mistakes. Each piece of work is annotated with questions to focus students' attention. Figure 2.6 shows two examples of this work. The first (from Laura) contains some common mistakes that students make (ignoring gaps, assuming that there are an equal number of old and new trees), while the second (from Amber) introduces students to a sampling method they may not have considered. Introducing work from outside the classroom is helpful in that (1) students are able to critique it freely without fear of other students being hurt by criticism and (2) handwritten "student" work carries less status than printed or teacher-produced work and it is thus easier for students to challenge, extend, and adapt.

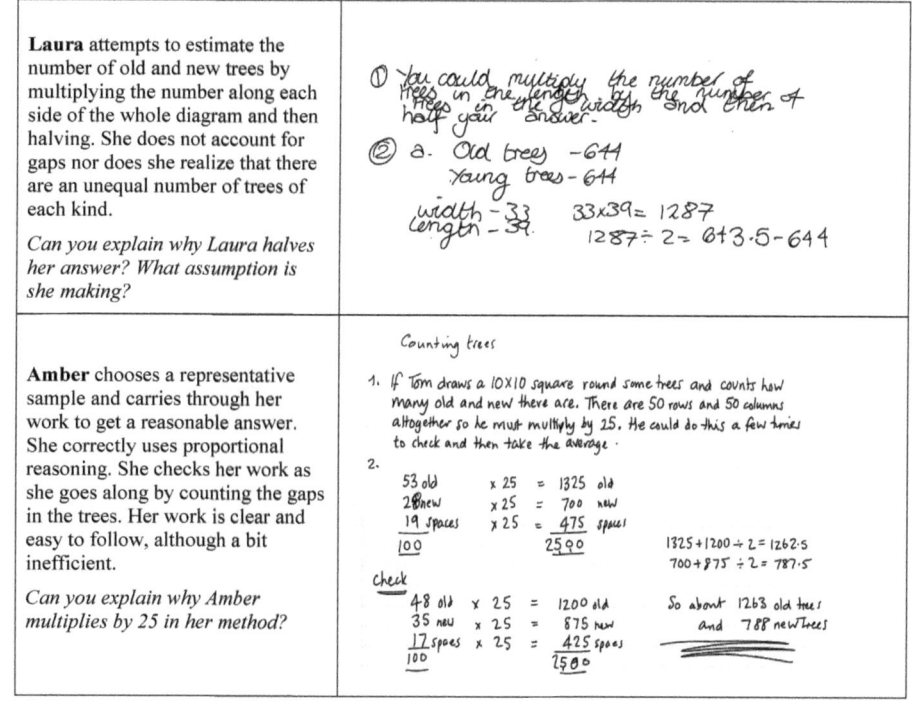

Fig. 2.6 Sample student work with commentary for discussion (MARS, 2012© 2007–2012 Mathematics Assessment Resource Service, University of Nottingham reuse under Creative Commons License)

Table 2.7 Design principles for Formative Assessment for Problem-Solving

Design principle	Illustrated by the case
Tasks for formative assessment of problem-solving strategies need to be unfamiliar for students but at the same time offer opportunities to start the solving process in order to elicit students' different intuitive approaches.	The counting trees task is unfamiliar to students, but students can start reasoning using mathematical concepts related to area and proportion.
Follow-up activities are intended to support reflection on intuitions and to help all students to move their thinking forward.	The well-chosen sample work handed over to students encourages them to reflect on possible mistakes (Laura's work) and to consider more sophisticated methods (Amber's work).
Formative assessment includes offering students opportunities to revise and improve their initial responses (e.g., based upon individual feedback or feedback through sample work).	After evaluating and discussing sample work in small groups, the students get the opportunity to revise their initial responses to the Counting Trees task.
A sequence of formative assessment activities asks for an explicit reflection and conclusion on the content as well as on the problem-solving strategies.	The example lesson finishes with a whole-class discussion of students' revised responses so as to draw out lessons learned from the approaches used and about the power of sampling.

After critiquing the work, students are offered the opportunity to refine their own approaches. This process of successive refinement in which methods are tried, critiqued, and adapted has been found to be extremely profitable for developing problem-solving strategies. The lesson concludes with a whole-class discussion that is intended to draw out some comparisons of the approaches used and the power of sampling.

Principles for task design underlying this example (Table 2.7) relate to both the design of the actual task and the supporting materials, including the student work and the lesson plan.

This example illustrates how a series of lesson activities (tackling the problem alone and then in groups, evaluating sample work, refining solutions, and whole-class discussion) may be designed to foster reflective, metacognitive behavior in which students step back from their own approaches and compare them with alternatives. The carefully designed nature of these lessons allows teachers to respond to student learning needs more sensitively and in a planned manner. The principles that are described for this example offer vital theoretical tools for designing formative assessment that can enhance the development of students' approaches to nonroutine problem-solving in mathematics.

2.3.2.7 Case 7: Japanese Lesson Study

This case is drawn from a videotaped Lesson Study (Tejima, 1987) at an elementary school affiliated to the University of Tsukuba, the oldest normal experimental school in Japan. The case illustrates principles related to task design within Lesson Study, a "craft-based" intermediate-level frame, and shows how these principles

Fig. 2.7 A textbook task (adapted from Sawada & Sakai, 2013)

encompass the various phases related to task design. These phases include an analysis of existing practices and instructional materials, a consideration of alternatives for reaching a new goal or for solving an educational problem, the actual task design, the teaching of the lesson, and the evaluation of the lesson and, in particular, the task (see also Sect. 2.2.2.3). As will be seen from this case, the task that is the basis for a Lesson Study can originate with a textbook task that is adapted for the research lesson.

This case focuses on how an expert mathematics teacher organized his research lesson. The task design challenge for this teacher was to support second-grade students in developing their understanding of addition and subtraction for ordinal numbers. So far, these children have learned to add and subtract in many contexts dealing with cardinal numbers. The first task in a textbook that uses a situation with ordinal numbers deals with a row of children (see Fig. 2.7).

The textbook task illustrated in Fig. 2.7 was originally preceded by the question: *There are twelve children standing in line. Hiroshi is the fifth from the front. How many children are standing after Hiroshi?* From a mathematical point of view, this question is easier than the task shown in Fig. 2.7. However, the teacher (Mr. Tejima) thought it important to start immediately with Fig. 2.7 task so as to challenge the children, to induce their naïve ideas (e.g., adding the numbers in the text), and to create opportunities for learning. The teacher also decided not to use the box representation of the textbook (i.e., the row of cubes with the double arches overlapping at the Hiroshi cube, in the bottom-right corner of Fig. 2.7). He wanted the students to think about this critical aspect of the problem for themselves.

Fig. 2.8 An intuitive (wrong) strategy and the emergence of a representation

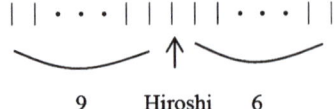

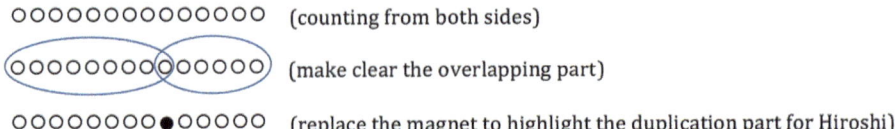

(counting from both sides)

(make clear the overlapping part)

(replace the magnet to highlight the duplication part for Hiroshi)

Fig. 2.9 Several strategies resulting in 14 with variations on a row representation

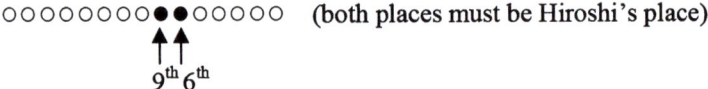

(both places must be Hiroshi's place)

Fig. 2.10 Explaining that 15 is wrong

Next, the teacher taught the lesson with the adapted task. He started by presenting the situation: *Children are standing in a line. Hiroshi is 9th from the front and 6th from the back*. Then the teacher asked the students to formulate a mathematical question for this situation. Next, the teacher took the question of the number of children in the row (the question emerged in the class) and asked all students to solve it. He assessed the students' answers as he circulated around the class, observing them working on the problem, and made a provisional plan for the following classroom discussion. The teacher intentionally picked up the dominant and wrong idea of $9+6=15$. He asked volunteers to explain their answer. One student explained the reason why 9 plus 6 equals 15. Then the teacher asked for an explanation from a student who thought the answer should be 16. The student illustrated his explanation on the blackboard (see Fig. 2.8).

In reaction to this idea, the proponents of 14 displayed their ideas in different representations with progressive sophistication (see Fig. 2.9).

Basically, the order in which the teacher nominated students to expose their reasoning was based on his provisional plan for the classroom discussion. In the course of this discussion, a student who usually struggled with mathematics said loudly: "I got it, I got it." He came to the blackboard by himself and explained the reason why the answer should be 14 (see Fig. 2.10): "Assume the answer is 15, there must be two Hiroshis and it is impossible."

This case illustrates how the alternate design generated by the teacher for the research lesson helped the students to come to understand the problem and to use a row representation for solving addition and subtraction tasks with ordinal numbers. The case also illustrates various design principles related to Japanese Lesson Study (Table 2.8).

Table 2.8 Design principles related to Japanese Lesson Study

Design principle	Illustrated by the case
An examination of existing practices and instructional materials. The identification of an issue worth studying and the design of an alternative task and a structured lesson plan (*kyozaikenkyu*).	An analysis of the textbook task and a consideration of possible alternatives. Rearrange the task to evoke multiple solutions and to support the students in developing correct conceptions of ordinal situations.
Teaching the lesson. Evoke students' naïve ideas and create opportunities for learning. Guide the students to critically analyze, compare, and contrast emerging ideas (*neriage*).	The teacher presents the task. After seeing how the students are solving the problem, he makes a provisional plan for them to share their work, starting with the more common, erroneous approach. Students present their reasoning at the blackboard and are encouraged to compare and discuss their ideas.
Lesson evaluation.	The lesson was videotaped and afterwards discussed with colleagues. During the discussion, the task and the effects of the initial task design were evaluated.

The reorganization of the task sequence actually challenged the students and created opportunities for learning. The teacher's assessment of the students' individual naïve ideas allowed him to make a provisional plan for a whole-class discussion of possible strategies for tackling the task. His order of nominating students to expose and discuss their ideas supported all students in developing an understanding of the structure of the task situation and the emerging solution procedure.

2.3.3 Discussion

The cases in this section illustrate a variety of design principles and frameworks for task design in relation to specific starting points or learning aims. Each case shows how task design can start from learners with particular characteristics and needs or from (new) knowledge, skills, attitudes, and competences that are aimed at. The application of the principles to a particular content area or mathematical topic renders specific the rather general principles and frameworks for task design. The resulting implementation of the starting frame is domain specific, and the design process tells the story of converting the general to the particular. As already remarked in the introduction of this section, it must be said that the cases are discussed too briefly to do justice to the richness of the underlying theory, to the whole design and design process, and to the context in which the task was designed and used. However, the cases reflect aspects of the current state of the art in task design and offer possibilities for reflecting on the two starting questions of this section:

(a) What do tasks look like when designed within a given theoretical frame or according to given design principles?
(b) Why do they look the way they do?

The cases describe a design challenge and how a framework for task design guides the design process and results in specific task characteristics. As such, these cases illuminate what is asked for with the first question. The second question is used to reflect on task characteristics and similarities and differences among the various cases. In some cases, the domain-specific implementation of a general framework is explicitly enriched by prior domain-specific findings. For example, the task design in the Conceptual Change case (case 3) is informed by previous research on students' conceptions of rational numbers.

All cases have in common a view of mathematics learning as being driven by doing mathematics. Especially, the ATD and RME cases (cases 1 and 5) emphasize the importance of interpreting mathematics as a human activity. Learning mathematics involves starting with students' current understandings and aiming at extending (e.g., by using rich realistic contextual problems in ATD and RME) their mathematical knowledge and skills in connection with their common sense understanding of everyday phenomena (e.g., questioning the world in ATD).

The size of the design problem being addressed is different in the presented cases. Some cases offer principles to solve a local problem in an existing task sequence. For example, the VT case (case 2) is oriented toward improving the anticipation and organization of learners' experiences in an existing task setting. The Japanese Lesson Study case (case 7) focuses on improving an existing textbook task. Still other cases describe the design of tasks as part of a task sequence that covers an entire topic. For instance, the ATD describes a task as part of a design for cardinal numbers (case 1) and the RME case describes one task in a task sequence on percentages (case 5).

With respect to characteristics of the resulting tasks, we can distinguish between the context of the task and the learning opportunities offered by the task. Two cases describe explicitly characteristics of the context for the task. The ATD case (case 1) stresses the importance of looking for a context that offers generating questions that go beyond school mathematics (e.g., how to take care of silkworms?). The context of the task in the RME case (case 5) has a slightly different focus. It stresses the notion that contexts should "beg to be organized" from a mathematical perspective and thus evoke solution strategies that have the potential to be mathematized (from "model of" a situation to "model for" mathematical reasoning). The case dealing with Formative Assessment for Problem-Solving (case 6) asks for unfamiliar contexts, contexts that do not immediately refer to well-known solution procedures but that also provide opportunities to start solving the problem and require processes like planning, representing, and collaborating.

All cases have similarities with respect to the learning opportunities offered by the task. They all stress the importance of tasks that create opportunities to build upon students' current understandings. For instance, the Conceptual Change example (case 3) and the Conceptual Learning example (case 4) both take students' current ways of reasoning into account. The Conceptual Change case takes an inevitable misconception as starting point, while the latter case starts from a potential activity that is already available for abstraction. In addition, most cases reflect the principle that tasks offer opportunities to share initial ideas and strategies. For instance, in the

Formative Assessment example (case 6), students are offered sample work done by other students; in the Lesson Study example (case 7), students are asked to share their ideas at the blackboard in an order that the teacher thinks will be most conducive to supporting all students in developing the intended understandings.

The cases show differences in the balance between learning opportunities aiming at mathematical content or aiming at more general regulative or motivational learning aims. For instance, the silkworm context in the ATD case (case 1) seems to be rather sophisticated in comparison with the mathematics involved, whereas the VT case (case 2) and the RME case (case 5) are explicitly oriented toward a specific mathematical object of learning. Design theories that focus on how an object of learning can be handled are important for helping teachers in their classrooms. In this respect, Variation Theory provides an effective instrument for design studies that also aim at promoting teachers' professional development (Cheng & Lo, 2013).

In several cases, tasks were designed so as to foster the development of representations and models that support the learning of mathematical concepts and skills, for instance, the rubber band for a number line in case 3, the diagram for reasoning with fractions in case 4, and the row representation for reasoning with ordinal numbers in case 7. This feature of tasks is further utilized in task sequences that foster the constitution of mathematics by exploiting didactical models that emerge from the activity of students, as in the development of the percentages concept in case 5.

Another aspect that arises when considering all cases is the degree of "challenge" offered to the students. Not all tasks have to be very challenging, but to foster students' learning, we need to provide them at some point with tasks that have (for them) some degree of challenge. This notion of challenge is not explicitly discussed in the cases presented in this section. It shows up in a somewhat incidental way in the discussion of VT (case 2) and of Japanese Lesson Study (case 7). Perhaps it is evident that tasks for students need to be challenging, but what "challenge" means for diverse classrooms with students having mixed abilities is not trivial at all. How to address this aspect in task design is an issue for future development related to design frameworks.

A theoretically important aspect that we can draw from the cases in this section is that the distinction between task design and lesson design is indeed blurred. In fact, the boundaries between them have been found to be extremely fluid. Almost all the cases presented in this section illustrated principles for task design that extended considerably beyond that of a task narrowly conceived as a question or sequence of questions proposed by a teacher or alternatively by a student. The task or task sequence, while treated as the main focus, was clearly conceived of within an orchestrated classroom activity — one where principles related to the actual classroom processes and instructional support that would make it possible to experience the potential of the task(s) were explicitly included as part of the task design. Design as intention was inherent to these cases, even if some of them could also be characterized by design as implementation. The current state of the art of task design in mathematics education would appear to suggest that designing a task or task sequence in isolation from consideration of the design of the instructional culture in which the task is to be integrated may be of quite limited value — somewhat analogous to expecting a bird to fly with just one wing.

Finally, in the process of moving from frameworks and principles to their actual application in the design of tasks, a great many decisions need to be made by the designer. How tasks look is largely determined by the hand of the artist! Nevertheless, these cases have shed some light upon the relation between these rather general starting points and the resulting tasks with their aimed-at learning processes.

2.4 Frameworks and Principles Do Not Tell the Whole Story in Task Design

2.4.1 Introduction

The two previous sections have examined the design process, and task design in particular, from the perspective of the frameworks and principles that have underpinned much of the design-oriented research in mathematics education. The particular perspective that was used was that of grand, intermediate, and domain-specific levels of frames—a perspective that aimed at elaborating the ways in which frames and task design are related. However, such frameworks and principles do not tell the whole story of task design. Some of the scholars and reflective practitioners who engage in task design see it as being a much more eclectic activity than has been suggested thus far. In addition, several factors have not yet been accounted for in the design process, such as its artistic and value-laden aspects. With the aim of providing a more balanced picture of the state of the art, this section of the chapter addresses task design from a variety of other perspectives, including the tension between *design as science* and *design as art* and the lens of "basic" versus "applied" research. Elaborating on these various perspectives opens up the idea of the value of collaborative work across different groups and leads to a discussion of some of the recent collaborative efforts in task design across professional communities.

2.4.2 Additional Factors Related to Task Design

2.4.2.1 The Tension Between *Design as Science* and *Design as Art*

In the educational design community at large, there is a tension between centripetal and centrifugal forces. Centripetal forces urge stasis and system and desire consistency. Centrifugal forces urge change and feed the need for diversity (Clark & Holquist, 1986). The yin and yang of such forces are at work in the world of the educational design community. Schein (1972) characterized science and practice as "convergent" and "divergent", respectively, and remarked that there is a gap between the two. Schoenfeld (2009) saw a similar gap between the theoretical aims of educational research and the practical aims of designers and consequently recommended the unpacking of designers' productive practices and a sharpening of the notion of *professional vision*—an elaboration that would be of value not only to the design community itself but also to educational design researchers. In addition, he pointed to the mutual benefits to be derived from the collaboration of the educational researcher and the designer-practitioner.

In contrast to proponents for delineating the explicit and rational frameworks and principles for task design, some colleagues insist that educational design research cannot help to sharpen the notion of professional vision. According to de Lange (2012), educational design research hardly offers usable knowledge for designers and practical suggestions for design nor does it offer a theoretical underpinning for educational design at the microlevel. De Lange (2013; also Chap. 10) argues that this limitation of theory is due to the artistic aspects of task creation. He underlines his reasoning by quoting Hilton (1976):

> Since mathematics [analogous to educational design theory/science] incorporates a systematic body of knowledge and involves cumulative reasoning and understanding, it is to that extent a science. And since applied mathematics (analogous to the actual practice of designers) involves *choices which must be made on the basis of experience, intuition, and even inspiration, it partakes the quality of art*. (p. 95, emphasis added)

It thus comes as no surprise that some educational designers prefer the powers that inhere in centrifugal forces: design activity in flux, simultaneity, diversity, and heterogeneity. Moreover, according to Schunn (2008), the educational design community has no communal mechanisms for codifying craft knowledge. Codifying design thinking is said to threaten its central value of flexibility (Collopy, 2009). Schön (1983) explicitly challenged the positivist doctrine underlying much of the *design science* movement and offered instead a constructivist paradigm. He criticized H.A. Simon's view of a *science of design* for being based on approaches to solving well-formed problems, whereas professional practice throughout design and technology and elsewhere has to face and deal with "messy, problematic situations". As pointed out by Cross (2001), Schön proposed instead to search for "an epistemology of practice implicit in the artistic, intuitive processes which some practitioners do bring to situations of uncertainty, instability, uniqueness, and value conflict" and which he characterized as "reflective practice" (p. 54).

Based on his personal reflection, including experiences with the HEWET project (1981–1985), which involved designing a quite new Dutch secondary mathematics course (*Wiskunde A—Mathematics A*) for humanity and social majors, de Lange (2013) describes what he refers to as a "slow design process." A *slow design process* involves several cyclic stages with rich partnership among researchers, designers, teachers, and students. It includes: selecting the subject, duration, and level; designing a mental sketch of flow while using intuitions; choosing a context for mathematization with the help of inspirations gained from random search; refining the design for a classroom experiment; discussing with experienced teachers; observing classroom activities and checking students' reactions while walking around; taking discrepancies between "intended" and "achieved" seriously; and concentrating on essential conceptual development. According to de Lange, slow design is possible under the following conditions: freedom of choice of what to design, freedom in time, freedom of thought, and freedom to explore with certain restrictions according to contextual and theoretical conditions.

To illustrate the slow design process, de Lange (1979) has described an example related to the topic of "Exponentials and Logarithms" for humanity and social majors, the design of which was guided by the philosophy of Realistic Mathematics Education. He progressively designed a task situation that functions as a model for a mathematical concept (de Lange, 1987). In this situation, "propagation of water plants" is chosen as the introductory task situation and the concept of logarithm is defined by growth factor and time: $\log_3 10$ is defined as the time needed to get 10 times the spread of water plants when the growth factor per month is 3 (i.e., a bit more than 2 months). With this situation and language in mind, students can interpret basic logarithmic relations such as $\log_3 10 + 1 = \log_3 30$ as follows: with this 1 extra month, you get 3 times more than 10, which equals 30. Experimental textbooks were developed to elaborate, try out, and optimize this approach (de Lange & Kindt, 1984); eventually, the approach entered Dutch curriculum descriptions and was adopted by commercial textbooks (e.g., Boer et al., 2004, p. 30).

De Lange (2013) argues that such a design and implementation process asks for slow design. It illustrates the need for extensive design processes that can do justice to both scientific and artistic aspects of task design. The tension between *design as science* and *design as art* is not easily solved, if it can be solved at all, and emphasizes a reconsideration of the time allocated for task design in educational research and in curriculum innovation projects.

2.4.2.2 Values in Task Design

Frameworks and principles for task design will vary relative to philosophies of mathematics education. Different philosophies of mathematics education mediate different values with respect to task design. Ernest (1991) distinguishes four sets of issues related to one's philosophy of mathematics education: the philosophy of mathematics, the nature of learning, the aims of education, and the nature of teaching. In this regard, Burkhardt (2014) points out that different groups of people have different priorities with respect to curricular aims or goals in mathematics: "basic skills people", "mathematical literacy people", "technology people", and "investigation people". Likewise, Treffers (1987) distinguished four trends in instructional approaches to mathematics in terms of "horizontal" and "vertical" mathematization: mechanistic, empiricist, structuralist, and realistic, with each instructional approach drawing upon different psychological backgrounds—Gagné's cumulative learning for the mechanistic, Piaget's constructivism for the empiricist, Bruner's modes of representation for the structuralist, and Gestalt psychology for the realistic.

The role of values in task design is illuminated by contrasting the approaches described in two recent studies that were presented at the ICMI Study-22 Conference on Task Design: *Concept-development task design* by Koichu et al. (2013) and *Competence-based task design* by Aizikovitsh-Udi, Clarke, and Kuntze (2013).

In the first study, based on Variation Theory, the task designer values delineating a variation space for the intended object of learning by eliminating or excluding hindering experience factors. This is done to direct the learner's attention to certain aspects that constitute the defining characteristic of the concept. In the second study, the competence-based task design proposes the idea of a "hybrid task" that stimulates different forms of thinking through a single task: the discipline-specific thinking of statistics and more generic forms of higher-order thinking, such as critical thinking. A hybrid task is characterized as having a structure that can offer to the learner superfluous and sometimes contradictory information. These two examples serve to illustrate that the frames and principles used in task design are intimately related to aims of mathematics education, which can in turn prioritize either the acquisition of conceptual and procedural knowledge or competence in dealing critically with information. A somewhat different perspective on the role of values in task design is exemplified by the explicit integration of "educating for values" into the teaching of mathematics. For instance, Movshovitz-Hadar and Edri (2013) conducted a multifaceted study to investigate the possibilities of combining social and personal values like equity, tolerance, social justice, rationality, and achievement and reaching one's intellectual potential—all within a designed approach to learning mathematics (see Chap. 5 for an elaborated example).

2.4.3 Diversity of Design Approaches Through the Lens of Basic Versus Applied Research

2.4.3.1 A Two-Dimensional Scheme for Classifying Research

In describing diverse approaches to task design, a useful perspective is offered by Schoenfeld's (1999) text on the synergy between theory and applications. Here he discusses the productive dialectical relationship between pure and applied work in education and makes use of Stokes's (1997) two-dimensional scheme of research in science and technology (see Fig. 2.11). Despite its formulation for the field of science, its applicability to the area of task design in mathematics education makes it of interest, especially with respect to situating the purely artistic position on task design as well as informing a potential bridging between the design-as-science and design-as-art tensions that were previously discussed.

In the two-dimensional representation, Niels Bohr and Thomas Edison are located as paradigmatic figures of pure basic and pure applied research. Louis Pasteur is located differently, as the paradigmatic figure of "use-inspired basic research": he not only engaged in germ theory for solely basic biological interests but was also motivated by problems of spoilage of drinks and curing diseases.

By its nature, educational research generally and educational design research especially aim at conducting "use-inspired basic research." According to McKenney and Reeves (2012): "Educational design research describes a family of approaches that strive toward the dual goal of developing theoretical understanding that can be of

Research is inspired by:

		Considerations of Use?	
		No	Yes
Quest for Fundamental Understanding?	Yes	Pure Basic Research (Bohr)	Use-Inspired Basic Research (Pasteur)
	No		Pure Applied Research (Edison)

Fig. 2.11 Stokes's (1997, p. 73) quadratic scheme for categorizing science research (Copyright 1997 by the Brookings Institution)

use to others while also designing and implementing interventions to address problems in practice" (p. 17). However, in the development of educational research, it may be difficult for any work to contribute simultaneously to both theory and practice: "Sometimes the state of theory is such that it may best be nurtured, temporarily, aside from significant considerations of use; sometimes the need to solve practical problems seems so urgent that theoretical considerations may be given secondary status" (Schoenfeld, 1999, p. 9). In his seminal work on problem-solving, Schoenfeld (1985) describes the dialectical relationship of give and take between theory and practice. Purely theoretical research in a laboratory setting can suggest some substantial ideas for designing practical courses in problem-solving and vice versa; actually teaching a course can raise theoretical issues to be pursued in an experimental setting. In order to have a relatively comprehensive "use-inspired basic research", it is necessary to move between such carefully designed laboratory settings and those settings that represent daily teaching practice.

2.4.3.2 Design Frames in the Light of Distinctions Between Basic and Applied Research

The principles and frameworks for task design that were described in earlier sections of this chapter in terms of levels of, or rootedness in, theory could alternatively be characterized according to their situatedness with respect to basic and applied research. In this spirit, we explore the use of the Stokes two-dimensional scheme as a lens for reflecting upon some of the various task design frames that were presented in Sect. 2.3 but in an alternative, albeit complementary, light.

Adherents of Variation Theory (VT) can be associated with the "basic research" cell of Stokes's two-dimensional scheme. A paradigmatic example is the research aimed at identifying the critical features of designated objects of learning and at ensuring that the designed task situations impart these critical features (Koichu, 2013). As pointed out by Cheng and Lo (2013), designers "must first identify a worthwhile object of learning and the critical features that the students must discern in order to see the object of learning in the intended way; they would then design patterns of variation (what to vary and what to keep invariant) to help the students to discern the critical features/aspects" (p. 10). Clinical observations in a laboratory setting are often used for identifying the affordances of task variations (see the VT case in Sect. 2.3). Such studies illustrate the complex relationships among the intended, enacted, and lived objects of learning and the need for clinical research settings to investigate these relationships. Task design for Conceptual Change (Van Dooren et al., 2013) and Conceptual Learning through Reflective Abstraction (M. Simon, 2013) are also associated with "basic research." These two frames rely on laboratory and clinical settings for studying the kind of experiences a task affords and the extent to which those experiences are beneficial for conceptual change and salient abstraction, respectively.

The Japanese Lesson Study (LS) approach to task design, which is based on deep craft knowledge and expertise, can be located in the "pure applied research" cell. It does not aim at developing substantial theoretical understanding. Rather, LS aims at teachers' professional development through *kyozaikenkyu* (Fujii, 2013): building up insights into children's learning trajectories, decision-making competency with respect to carrying out tactical interventions during classroom interactions, organizing provocative discourse, establishing a productive classroom microculture, and so forth. The setting of LS in daily teaching practice mediates the activity of the various participants involved in LS and leads to the co-construction of deep craft knowledge.

Research frames such as the ATD and Realistic Mathematics Education (RME) can be viewed as paradigmatic examples of the "use-inspired basic research". Both paradigms can contribute simultaneously to theory and practice—the contributions to theory occurring especially during the early period of development of these intermediate-level research frames, as well as during their later application phases when the intermediate-level frame is particularized to the learning of domain-specific concepts and processes. ATD-based task sequences, or study research paths, are developed and implemented across many years of schooling from pre-primary to university. Many of the ATD-based tasks are characterized as open-ended mathematical modeling activities that address social issues. Likewise, RME has served in the development of different types of curriculum projects at all school levels varying from kindergarten (e.g., van Nes & Doorman, 2011) to upper secondary education (e.g., Doorman & Gravemeijer, 2009). As well, rich RME-based, problem-solving, assessment tasks (A-lympiade and Math B-day) are elaborated and implemented annually by teachers in their daily practice (Goddijn, 2008; Goris, 2006).

As a means of fostering further development of design principles that might contribute simultaneously to theory and practice, as well as exploring whether in fact some of the more theory-oriented and more practice-oriented frames for task design are in fact amenable to joint articulation, some researchers (e.g., Artigue,

Cerulli, Haspekian, & Maracci, 2009; Artigue & Mariotti, 2014; Kieran, Krainer, & Shaughnessy, 2013) have already begun to study the ways in which different groups, cultures, and communities work together productively on advancing such issues. As has been suggested by the examples in the previous paragraphs, basic laboratory research in VT, Conceptual Change, and Concept Learning can reveal some substantial ideas of potential interest to designers with a practical research orientation. Conversely, difficulties and dilemmas that have emerged in actual Lesson Studies raise salient theoretical issues that could be pursued in less complex clinical settings. Immediately below, we address collaborative work across professional communities, work that has begun to grapple with the theoretical-practical interface of design activity.

2.4.4 *Design Activity Across Professional Communities*

2.4.4.1 The Stakeholder Approach as a Foundation for Thinking About Collaborative Design Activity

In order to realize collaboration across the research and teaching communities (as well as collaboration involving the educational researcher and the educational designer-practitioner), a wise strategy is to establish a transparent context between researchers and practitioners, not forgetting that some practitioners are researchers themselves. A provisional theoretical perspective would be Krainer's (2011) notion of the *stakeholder approach*, which avoids privileging theory over practice in the design process. According to Krainer, the term *stakeholder approach* is intended to capture the idea that teachers are key stakeholders in the design research enterprise, not mere users of research. It is teachers who are in a position to achieve one of the main purposes of that enterprise, which is the improvement of students' learning of mathematics. Developing a stakeholder approach is central to establishing the kind of collaboration between these two communities that will facilitate mathematical learning with rich task design. Krainer asserts that researchers should highlight teachers' reflective and creative practice and offer viable opportunities that encourage them to get interested in being involved in such research. The stakeholder approach asks of task design not *what* but *where* it is. With this approach, task design is situated in the interaction between practitioners and researchers.

2.4.4.2 Task Design Involving Practitioners and Researchers

In Sects. 2.2 and 2.3, our attention was fixed on the nature of the frameworks and principles used in the activity of task design research, without focusing on the nature of collaborative work in this area. We now take a closer look at this aspect and discuss some recent design efforts involving cross-community collaboration. A few of the research papers presented at the ICMI Study-22 Conference on Task Design reflected a rethinking of the boundaries between theory and practice and the relative roles of researchers and practitioners (e.g., Ding et al., 2013; Morselli, 2013; Ponte et al., 2013; Stephan & Akyuz, 2013). Ding and her colleagues report on the process of design and implementation of tasks within a team consisting of academic researchers, teachers, and a teacher educator who was also an expert teacher, in a school-based teacher professional development program in Shanghai, China. In their report (Ding et al., 2013), they highlight, in particular, the role played by the expert teacher, who contributed to the development of a "hypothetical learning structure" for a particular topic (decimal value) and to creating tasks within a web-like structure of knowledge constructions.

Morselli (2013) describes a collaborative project aiming at designing, experimenting, and refining task sequences for a smooth and meaningful approach to proof in lower secondary school in Genoa, Italy. The project was supported by an initiative of the Italian Ministry of Education aimed at fostering and stimulating young students' interest in studying science. Within this context, collaborative work between university researchers and school teachers was set up. Teams were created for each school level and these teams met regularly in order to share theoretical references on argumentation and to discuss theoretical tools and their didactical and methodological potential. Productive cycles of task design, experimentation, analysis, refinement, and modification emerged.

Ponte et al. (2013) address the design of exploratory tasks that were developed and implemented in collaboration between researchers and a group of teachers in Lisbon, Portugal. A new mathematics curriculum for basic education required teachers to develop and use exploratory tasks designed to support students' mathematical reasoning and the growth of their problem-solving abilities. With such an institutional context, developmental work on task design was conducted by using a combination of research expertise and classroom teaching expertise. The team started with an overall plan for a teaching unit, which included the formulation of the learning objectives, assumed previous knowledge of students, time available, and organization of a schedule. Tasks were later selected to fit the overall planning of the teaching unit, followed by a dialectical movement of adjustments at the macrolevel of the unit and at the specific level of the tasks. Usually, the first idea for an exploratory task was provided by a classroom teacher, and the subsequent refinement was carried out in interaction with the other teachers and researchers.

Stephan and Akyuz (2013) describe a design study involving a collaborative community in one middle school, consisting of two mathematics teachers, a special education teacher, a researcher, and a graduate student. After the researcher introduced the main idea of the hypothetical learning trajectory (HLT), the members of the collaborative group worked together to create a six-phase instructional sequence, based on RME heuristics, for the learning of integers and integer operations (see also Sect. 2.2.2.4). The community met on several occasions before instruction began and then almost daily throughout the implementation. The pivotal contributions of various members of the group included anticipating supportive mathematical imagery, creating challenging formative assessment, using their mathematical knowledge to alter the instructional sequence, and working and revising already created tasks or the sequence of the instruction. This collaborative research showed that, "during the implementation of the instruction, the practitioners began to discuss more theoretical issues while the researchers began to think more about teaching practices" (Stephan & Akyuz, 2013, p. 515).

In introducing this discussion on cross-community design activity, we were reminded of Krainer's (2011) elaboration of the stakeholder approach and the related query that it is "not *what* task design is, but *where* it is situated" that needs to be considered. The design projects that have just been exemplified and which integrated the cross-professional communities of practitioners, researchers, and, in some cases, teacher educators allow us to respond more fully to this query. Sections 2.2 and 2.3 situated task design in the nexus between principles and frameworks and their application to particular content areas or mathematical topics. The above collaborative projects have succeeded in showing that task design is also situated both in the interactions among its cross-community participants and in the interface between theory and practice.

2.4.5 Toward the Resolving of Perceived Tensions

By examining a variety of alternative perspectives, this section of the chapter has touched upon a range of additional factors affecting task design and its diversity—factors that might suggest a certain inherent tension between opposing forces—but at the same time has offered avenues for resolving the perceived tensions. These alternative perspectives have allowed us to see that the structured frameworks and principles that characterize much of the design research in the mathematics education research community do not capture the eclectic nature of design activity as engaged in by some of its scholars. For example, some of the proponents of *design as art* espouse a quite different set of starting points from the proponents of *design as science*. At the heart of this tension, as we have noted, is Schön's (1983) criticism of H.A. Simon's view of a *science of design* as being based on approaches to solving well-formed problems, whereas professional design practice has to deal with "messy" problems. With positivist approaches to design practice being found to be of limited utility during the 1980s (Lincoln & Guba, 1985), design had to be reconsidered as a process in which uncertainty must be grappled with. Artigue (2009) reminded us of this when she remarked that theory-based intervention programs have faced difficulty because of the many design factors that are not under theoretical control.

As was seen, one of the ways of approaching this dilemma is to consider the practitioner as a full actor in the design process. The stakeholder approach (Krainer, 2011) embodies this perspective but also acknowledges that there are significant differences in the guiding principles specific to different communities. Stokes's (1997) two-dimensional scheme enabled us to situate various orientations in design research, including design activity that is motivated as much by artistic as by theoretical concerns. As was also seen in this section, frameworks and principles constitute a communal practice of task design, with innovation and use of specific frameworks and principles for task design being a reflexive activity. By means of interaction among diverse cultures and communities, frameworks and principles are progressively developed in the light of task implementation. Therefore, the interactions between diverse communities and the concomitant grappling with diverse design principles would seem crucial to moving forward in the area of design. However, such interactions may not be straightforward or easy to orchestrate. As pointed out by Artigue and Mariotti (2014) in their discussion of the networking efforts engaged in by researchers from different cultures in the ReMath project: "[When] the possibilities of networking are examined in terms of potential for guiding design, ... the activity is much more demanding ... [but can be] especially insightful; ... such advances are especially important considering that design in mathematics education lies at the interface between theory and practice" (pp. 350–351). Although collaboration involving the diverse actors engaged in the enterprise of task design in mathematics education may be challenging, the process can yield not only an enhancement of the quality of the designed tasks and task sequences but also a narrowing of the perceived divide between design as artful practice and design as theory building. Recall Stephan and Akyuz's (2013) earlier remark: "It is interesting to note that during the implementation of the instruction, the practitioners began to discuss more theoretical issues while the researchers began to think more about teaching practices" (p. 515).

2.5 Concluding Discussion: Progress Thus Far and Progress Still Needed

The objective of this chapter was to give an overview of the current state of the art related to frameworks and principles for task design so as to provide a better understanding of the design process and the various interfaces between teaching, researching, and designing. The chapter started with a description of the history of task design in mathematics education. The 1970s reflected the beginnings of the new community of mathematics education researchers' efforts to grapple with the interaction between curriculum materials and the quality of mathematical teaching and learning. We noted, for example, that Alan Bell was one of the first colleagues who explicitly referred to the importance of design principles for the transition from a situation that embodies the concepts and relations of the conceptual field to the design of tasks that bring into play these concepts and relations.

What progress have we—as a community—made over the past four decades? This chapter has described in which directions we have made some progress in understanding and articulating aspects related to task design. These aspects have included aims, levels, communities, and values that influence and are influenced by frameworks and principles for task design in educational practice and in educational research. Topics that were addressed related to levels of frameworks for task design, the distinction between theories as resource for and as product of design research, the tension between design as science and design as art, and the relations among the professional communities that develop and use specific frameworks for task design. So what have we learned about this field, about the topics, and about ourselves, as a result of coming together at the ICMI Study-22 Conference and developing this chapter on frameworks and principles?

One aspect that became more clear concerns the nature and levels of the frames that guide the process of task design. Using the lens of grand, intermediate, and domain-specific levels allowed us to see that our frames tend to be either holistic or multidimensional in nature. That is, the inspiration for our designs can come primarily from one quite global, intermediate-level framework (e.g., TDS, ATD, VT) or from a constellation of theories of different levels and different types (e.g., the various examples of domain-specific frames for the learning of particular concepts or processes). We saw that drawing from a combination of theoretical foundations can present advantages that may not be available when we rely on just one overall frame and its design tools—advantages such as being able to delineate not only a broad set of principles for the design of tasks or task sequences but also a related set of principles for the design of the instructional culture in which the task is to be integrated. In fact, a significant number of the task design studies presented within the conference theme group on principles and frameworks relied upon principles for task design that extended considerably beyond that of a task narrowly conceived. While the task or task sequence was seen as being central, it was clearly viewed as taking place within an orchestrated classroom activity—one where principles related to the actual classroom processes and instructional support that would make it possible to experience the potential of the task were explicitly included as part of the task design. Thus, the distinction between task design and lesson design was found, indeed, to be quite blurred.

Another aspect that has emerged is that theories are both a resource for and a product of the design process. As a resource, they provide theoretical tools and principles to support the design of a teaching sequence. As a product of design research, theories inform us about both the processes of learning and the means that have proven to be effective for supporting that learning. Related to this dual role of theory is the distinction between *design as intention* and *design as implementation*—*design as intention* addressing specifically the initial formulation of the design and *design as implementation* focusing attention on the process by which a designed sequence is integrated into the classroom environment, subsequently refined, and then theorized about. This distinction highlights the relative nature of the significance given to the design of the task sequence or task itself within the design process.

Although the major part of this chapter has been devoted to the theoretical frames that underlie task design, not all design is based on theory. The Lesson Study frame is a classic example of craft-based task design based on teaching practice, one where teachers with their deep, experiential knowledge are central to the process. Fundamental to teachers' ability to design, implement, and study high-quality mathematics lessons is a detailed, widely shared conception of what constitutes effective mathematics pedagogy and professional development. The planning of the research lesson, which is the main component of Lesson Study, includes not only the task and its materials but also anticipated student thinking, the teacher's planned questioning and intervention activities, and the points to be noticed and evaluated.

At the same time, we have become aware that the grain size for describing principles for task design is an area for further reflection and development. While the cases presented in this chapter took account of principles related to grand, intermediate, and domain-specific levels of theories, as well as instructional and tool-related principles, the work of the educational designer Kali (2008) suggests the feasibility of considering, and possibly integrating, a much finer grain size of levels of principles into our design work. We are reminded of the critical question that, according to Cobb et al. (2003), must be asked of our frames, that is, whether their principles inform prospective design and, if so, in precisely what way. As seen in Sect. 2.3, current work in task design indicates that there is a great deal of variation in the nature of the principles and heuristics being adopted for task design, with only a few points of convergence across the broad set of principles informing task design. Many of the principles tend to be phrased in rather general terms that are subject to broad interpretation and thus cannot be said to inform prospective design in highly specific ways. Clearly, further theoretical work on grain size of principles for task design is needed. For example, in applying general task design principles to the learning of particular mathematical content areas or reasoning processes, being more explicit with respect to the way in which past research in that area is being woven into the design of the task or task sequence would surely be useful.

Tasks play a crucial role in forwarding the process of improving the educational system. While, for instance, in the mathematics education community, competencies like creativity, critical thinking, and problem-solving are highly valued, tasks presented by high-stakes examinations tend to address basic skills. Such examination tasks largely determine the types of tasks that are used in classrooms. Curriculum innovation can be moved forward with illustrative alternative tasks and explicit attention to the underlying principles and frameworks used to design them, without losing consideration of skill development, fluency, and flexibility. A vital component often missing in curriculum innovation documents is the vivid exemplification that is necessary to show exactly what tasks might look like and how they relate to improving teaching and learning.

In addition, current changes in educational systems and trends in mathematics education ask for a reconsideration of design principles. Trends in education that are related to task design are, for instance, beginning to show an increasing focus on interdisciplinarity and authentic practices. Trying to better connect mathematics education to other subjects like physics, biology, and economics requires a reconsideration of the role of contexts and bridging concepts. A serious consideration of the use of authentic practices and the world of work in mathematics education calls for tasks with a purpose and utility, shifting from solving a school mathematics problem to asking for a product as a final result. New task characteristics emerge and others might become less relevant in the near future.

From this Study conference and its follow-up exchanges and research for the preparation of this chapter, we have also learned that knowledge about design grows in the community as design principles are explicitly described, discussed, and refined. Although the papers presented within the Principles and Frameworks theme group of this conference all specified the frames and principles underlying their designs and illustrated how these were being implemented in the resulting tasks, such is not common in the majority of papers presented at mathematics education research conferences (Sierpinska, 2003). Despite the recent growth spurt of design studies within mathematics education, the specificity of the principles that inform task design in a precise way remains both underdeveloped and, even when somewhat developed, underreported. A possible obstacle that stands in the way of specificity can be traced to length constraints on published papers and the extended amount of space that the provision of specific details requires. Were it not for websites such as *Educational Designer* (http://www.educationaldesigner.org), there are few avenues for presenting the explicit and detailed thinking that lies behind the final versions of designed tasks. Nevertheless, it seems reasonable to expect that mathematics education researchers could be more explicit in their published research papers about the principles that underlie the tasks they design for their research studies. Clearly, more work remains to be done in encouraging such practice. This chapter provides a starting point for future efforts that aim at a further and deeper investigation of task design, its frameworks, and its principles, so that design might become a mature element in mathematics education research and practice.

References

Ainley, J., & Pratt, D. (2005). The significance of task design in mathematics education: Examples from proportional reasoning. In H. L. Chick & J. L. Vincent (Eds.), *Proceedings of the 29th Conference of the International Group for the Psychology of Mathematics Education* (Vol. 1, pp. 103–108). Melbourne: PME.

Aizikovitsh-Udi, E., Clarke, D., & Kuntze, S. (2013). Hybrid tasks: Promoting statistical thinking and critical thinking through the same mathematical activities. In C. Margolinas (Ed.), *Task design in mathematics education* (Proceedings of ICMI Study 22, pp. 451–460). Available from hal.archives-ouvertes.fr/hal-00834054

Anderson, J. R. (1995/2000). *Learning and memory*. New York: Wiley.

Arcavi, A., Kessel, C., Meira, L., & Smith, J. P. (1998). Teaching mathematical problem solving: An analysis of an emergent classroom community. *Research in Collegiate Mathematics Education III, 7*, 1–70.

Artigue, M. (1992). Didactical engineering. In R. Douady & A. Mercier (Eds.), *Recherches en Didactique des Mathématiques, Selected papers* (pp. 41–70). Grenoble: La Pensée Sauvage.

Artigue, M. (2009). Didactical design in mathematics education. In C. Winslow (Ed.), *Nordic research in mathematics education: Proceedings from NORMA08 in Copenhagen* (pp. 7–16). Rotterdam: Sense Publishers.

Artigue, M., Cerulli, M., Haspekian, M., & Maracci, M. (2009). Connecting and integrating theoretical frames: The TELMA contribution. *International Journal of Computers for Mathematical Learning, 14*, 217–240.

Artigue, M., & Mariotti, M. A. (2014). Networking theoretical frames: The ReMath enterprise. *Educational Studies in Mathematics, 85*, 329–355.

Arzarello, F., & Sabena, C. (2011). Semiotic and theoretic control in argumentation and proof activities. *Educational Studies in Mathematics, 77*, 189–206. doi:10.1007/s10649-010-9280-3.

Barquero, B., & Bosch, M. (2015). Didactic engineering as a research methodology: From fundamental situations to study and research paths. In A. Watson & M. Ohtani (Eds.), *Task design in mathematics education: An ICMI Study 22*. New York: Springer.

Bartolini Bussi, M. (1991). Social interaction and mathematical knowledge. In F. Furinghetti (Ed.), *Proceedings of the 15th Conference of the International Group for the Psychology of Mathematics Education* (Vol. I, pp. 1–16). Assisi: PME.

Bell, A. W. (1979). Research on teaching methods in secondary mathematics. In D. Tall (Ed.), *Proceedings of the Third Conference of the International Group for the Psychology of Mathematics Education* (pp. 4–12). Warwick: PME.

Bell, A. (1993a). Guest editorial. *Educational Studies in Mathematics, 24*, 1–4.

Bell, A. (1993b). Principles for the design of teaching. *Educational Studies in Mathematics, 24*, 5–34.

Bishop, A. J. (1979). Visual abilities and mathematics learning. In D. Tall (Ed.), *Proceedings of the Third Conference of the International Group for the Psychology of Mathematics Education* (pp. 17–28). Warwick: PME.

Black, P., Harrison, C., Lee, C., Marshall, B., & Wiliam, D. (2003). *Assessment for learning: Putting it into practice*. Buckingham: Open University Press.

Black, P., & Wiliam, D. (1998). Inside the black box: Raising standards through classroom assessment. *Phi Delta Kappan, 80*(2), 139–148. Also available from: http://weaeducation.typepad.co.uk/files/blackbox-1.pdf

Boer, W., et al. (Eds.). (2004). *Moderne wiskunde (edite 8), vwo B1 deel 2*. Groningen: Wolters-Noordhoff.

Bransford, J. D., Brown, A. L., & Cocking, R. R. (Eds.). (1999). *How people learn*. Washington, DC: National Academy Press.

Brousseau, G. (1997). *Theory of didactical situations in mathematics* (N. Balacheff, M. Cooper, R. Sutherland, & V. Warfield, Eds. & Trans.). Dordrecht, The Netherlands: Kluwer.

Brown, A. L. (1992). Design experiments: Theoretical and methodological challenges in creating complex interventions in classroom settings. *Journal of the Learning Sciences, 2*(2), 141–178.

Burkhardt, H. (2014). Curriculum design and systemic change. In Y. Li & G. Lappen (Eds.), *Mathematics curriculum in school education* (pp. 13–34). New York: Springer.

Burkhardt, H., & Swan, M. (2013). Task design for systemic improvement: principles and frameworks. In C. Margolinas (Ed.), *Task design in mathematics education* (Proceedings of ICMI Study 22, pp. 431–440). Available from hal.archives-ouvertes.fr/hal-00834054

Cheng, E. C., & Lo, M. L. (2013). *Learning study: Its origins, operationalisation, and implications* (OECD Education Working Papers, No. 94). Paris: OECD Publishing. Available from doi:10.1787/5k3wjp0s959p-en

Chevallard, Y. (1999). L'analyse des pratiques enseignantes en théorie anthropologique du didactique. *Recherches en Didactique des Mathématiques, 19*, 221–266.

Chevallard, Y. (2012). *Teaching mathematics in tomorrow's society: A case for an oncoming counterparadigm*. Plenary at ICME 12. Retrieved from http://www.icme12.org/upload/submission/1985_F.pdf

Clark, K., & Holquist, M. (1986). *Mikhail Bakhtin*. Cambridge, MA: Harvard University Press.

Clements, D. H., & Battista, M. T. (1992). Geometry and spatial reasoning. In D. A. Grouws (Ed.), *Handbook of research on mathematics teaching and learning* (pp. 420–464). New York: Macmillan.

Cobb, P. (2007). Putting philosophy to work: Coping with multiple theoretical perspectives. In F. K. Lester Jr. (Ed.), *Second handbook of research on mathematics teaching and learning* (pp. 3–67). Charlotte, NC: Information Age.

Cobb, P., Confrey, J., diSessa, A., Lehrer, R., & Schauble, L. (2003). Design experiments in educational research. *Educational Researcher, 32*(1), 9–13.

Cobb, P., & Gravemeijer, K. (2008). Experimenting to support and understand learning processes. In A. E. Kelly, R. A. Lesh, & J. Y. Baek (Eds.), *Handbook of design research methods in education* (pp. 68–95). London: Routledge.

Cobb, P., & Steffe, L. P. (1983). The constructivist researcher as teacher and model builder. *Journal for Research in Mathematics Education, 14*, 83–95.

Cobb, P., & Yackel, E. (1996). Constructivist, emergent, and sociocultural perspectives in the context of developmental research. *Educational Psychologist, 31*, 175–190.

Collins, A. (1992). Toward a design science of education. In E. Scanlon & T. O'Shea (Eds.), *New directions in educational technology* (pp. 15–22). Berlin: Springer.

Collins, A., Brown, J. S., & Newman, S. (1989). Cognitive apprenticeship: Teaching the craft of reading, writing, and mathematics. In L. B. Resnick (Ed.), *Knowing, learning, and instruction: Essays in honor of Robert Glaser*. Hillsdale, NJ: Lawrence Erlbaum Associates.

Collins, A., Joseph, D., & Bielaczyc, K. (2004). Design research: Theoretical and methodological issues. *Journal of the Learning Sciences, 13*, 15–42.

Collopy, F. (2009). Lessons learned – Why the failure of systems thinking should inform the future of design thinking. *Fast Company blog*. http://www.fastcompany.com/1291598/lessons-learned-why-failure-systems-thinking-should-inform-future-design-thinking

Corey, D. L., Peterson, B. E., Lewis, B. M., & Bukarau, J. (2010). Are there any places that students use their heads? Principles of high-quality Japanese mathematics instruction. *Journal for Research in Mathematics Education, 41*, 438–478.

Cross, N. (2001). Designerly ways of knowing: Design discipline versus design science. *Design Issues, 17*(3), 49–55.

de Lange, J. (1979). *Exponenten en logaritmen*. Utrecht: I.O.W.O.

de Lange, J. (1987). *Mathematics, insight, and meaning: Teaching, learning and testing of mathematics for the life and social sciences*. Utrecht: OW & OC.

de Lange, J. (2012). *Dichotomy in design: And other problems from the swamp*. Plenary address at ISDDE. Utrecht: Freudenthal Institute.

de Lange, J. (2013). *There is, probably, no need for this presentation*. Plenary presentation at the ICMI Study-22 Conference, The University of Oxford. http://www.mathunion.org/icmi/digital-library/icmi-study-conferences/icmi-study-22-conference/

de Lange, J., & Kindt, M. (1984). *Groei*, (Een produktie ten behoove van de experiment in het kader van de Herverkaveling Eindexamenprogramma's Wiskunde I en II VWO, 1e herziene versie). Utrecht: OW & OC.

Ding, L., Jones, K., & Pepin, B. (2013). Task design in a school-based professional development programme. In C. Margolinas (Ed.), *Task design in mathematics education* (Proceedings of ICMI Study 22, pp. 441–450). Available from hal.archives-ouvertes.fr/hal-00834054

Ding, L., Jones, K., Pepin, B., & Sikko, S. A. (2014). How a primary mathematics teacher in Shanghai improved her lessons: A case study of 'angle measurement'. In S. Pope (Ed.), *Proceedings of the 8th British Congress of Mathematics Education* (pp. 113–120). Nottingham: BCME.

Doig, B., Groves, S., & Fujii, T. (2011). The critical role of task development in lesson study. In L. C. Hart, A. Alston, & A. Murata (Eds.), *Lesson study research and practice in mathematics education. Learning together* (pp. 181–199). New York: Springer.

Doorman, L. M., & Gravemeijer, K. P. E. (2009). Emergent modeling: Discrete graphs to support the understanding of change and velocity. *ZDM: The International Journal on Mathematics Education, 41*(1), 199–211.

Duval, R. (2006). Les conditions cognitives de l'apprentissage de la géométrie: développement de la visualisation, différenciation des raisonnements et coordination de leur fonctionnement [Cognitive conditions of learning geometry: development of visualization, differentiation of reasoning and coordination of its functioning]. *Annales de Didactique et de Sciences Cognitives, 10*, 5–53.

Ernest, P. (1991). *Philosophy of mathematics education*. London: Falmer Press.

Fernandez, C., & Yoshida, M. (2004). *Lesson study: A Japanese approach to improving mathematics teaching and learning*. Mahwah, NJ: Lawrence Erlbaum Associates.

Filloy, E., & Rojano, T. (1989). Solving equations: The transition from arithmetic to algebra. *For the Learning of Mathematics, 9*(2), 19–25.

Fisk, A. D., & Gallini, J. K. (1989). Training consistent components of tasks: Developing an instructional system based on automatic-controlled processing principles. *Human Factors, 31*, 453–463.

Freudenthal, H. (1973). *Mathematics as an educational task*. Dordrecht: Reidel.

Freudenthal, H. (1978). *Weeding and sowing*. Dordrecht: Reidel.

Freudenthal, H. (1979). How does reflective thinking develop? In D. Tall (Ed.), *Proceedings of the Third Conference of the International Group for the Psychology of Mathematics Education* (pp. 92–107). Warwick: PME.

Freudenthal, H. (1983). *Didactical phenomenology of mathematical structures*. Dordrecht: Reidel.
Freudenthal, H. (1991). *Revisiting mathematics education. China lectures*. Dordrecht: Kluwer.
Fujii, T. (2013, July). *The critical role of task design in lesson study*. Plenary paper presented at the ICMI Study 22 Conference on Task Design in Mathematics Education, Oxford. http://www.mathunion.org/icmi/digital-library/icmi-study-conferences/icmi-study-22-conference/
Gagné, R. M. (1965). *The conditions of learning*. New York: Holt, Rinehart & Winston.
García, F. J., & Ruiz-Higueras, L. (2013). Task design within the Anthropological Theory of the Didactics: Study and research courses for pre-school. In C. Margolinas (Ed.), *Task design in mathematics education* (Proceedings of ICMI Study 22, pp. 421–430). Available from hal.archives-ouvertes.fr/hal-00834054
Glaser, R. (1976). Components of a psychology of instruction: Toward a science of design. *Review of Educational Research, 46*(1), 1–24.
Goddijn, A. (2008). Polygons, triangles and capes: Designing a one-day team task for senior high school. In *ICME-11 – Topic Study Group 34: Research and development in task design and analysis*. Available from http://tsg.icme11.org/tsg/show/35
Goldenberg, E. P. (2008). Task Design: How? In *ICME-11 – Topic Study Group 34: Research and development in task design and analysis*. Available from http://tsg.icme11.org/tsg/show/35
Goris, T. (2006). Math B day, Olympiad and a few words of Japanese. *Nieuwe Wiskurant, 26*(2), 4–5.
Gravemeijer, K. (1994). Educational development and developmental research. *Journal for Research in Mathematics Education, 25*, 443–471.
Gravemeijer, K. (1998). Developmental research as a research method. In J. Kilpatrick & A. Sierpinska (Eds.), *What is research in mathematics education and what are its results?* (Vol. 2, pp. 277–295). Dordrecht: Kluwer.
Gravemeijer, K., & Cobb, P. (2006). Design research from a learning design perspective. In J. van den Akker, K. Gravemeijer, S. McKenney, & N. Nieveen (Eds.), *Educational design research* (pp. 45–85). Available from http://www.fisme.science.uu.nl/publicaties/literatuur/EducationalDesignResearch.pdf
Gravemeijer, K., & Cobb, P. (2013). Design research from the learning design perspective. In T. Plomp & N. Nieveen (Eds.), *Educational design research* (pp. 72–113). London: Routledge.
Gravemeijer, K., & Stephan, M. (2002). Emergent models as an instructional design heuristic. In K. P. E. Gravemeijer, R. Lehrer, B. V. Oers, & L. Verschaffel (Eds.), *Symbolizing, modeling and tool use in mathematics education* (pp. 145–169). Dordrecht: Kluwer.
Gravemeijer, K., van Galen, F., & Keijzer, R. (2005). Designing instruction on proportional reasoning with average speed. In H. L. Chick & J. L. Vincent (Eds.), *Proceedings of the 29th Conference of the International Group for the Psychology of Mathematics Education* (Vol. 1, pp. 103–108). Melbourne: PME.
Hadas, N., Hershkowitz, R., & Schwarz, B. B. (2001). The role of surprise and uncertainty in promoting the need to prove in computerized environment. *Educational Studies in Mathematics, 44*, 127–150.
Hart, L., Alston, A., & Murata, A. (Eds.). (2011). *Lesson study research and practice in mathematics education: Learning together*. New York: Springer.
Hershkowitz, R. (1990). Psychological aspects of geometry learning – Research and practice. In P. Nesher & J. Kilpatrick (Eds.), *Mathematics and cognition* (pp. 70–95). Cambridge: Cambridge University Press.
Hilton, P. (1976). Education in mathematics and science today: The spread of false dichotomies. In H. Athen & H. Kunle (Eds.), *Proceedings of the Third International Congress on Mathematical Education* (pp. 75–97). Karlsruhe, FRG: University of Karlsruhe.
Huang, R., & Bao, J. (2006). Towards a model for teacher professional development in China: Introducing *Keli*. *Journal of Mathematics Teacher Education, 9*, 279–298.
Jacobs, J. K., & Morita, E. (2002). Japanese and American teachers' evaluations of videotaped mathematics lessons. *Journal for Research in Mathematics Education, 33*, 154–175.
Janvier, C. (1979). The use of situations for the development of mathematical concepts. In D. Tall (Ed.), *Proceedings of the Third Conference of the International Group for the Psychology of Mathematics Education* (pp. 135–143). Warwick: PME.

Johnson, D. C. (1980). The research process. In R. J. Shumway (Ed.), *Research in mathematics education* (pp. 29–46). Reston, VA: National Council of Teachers of Mathematics.

Kali, Y. (2008). The design principles database as a means for promoting design-based research. In A. E. Kelly, R. A. Lesh, & J. Y. Baek (Eds.), *Handbook of design research methods in education* (pp. 423–438). London: Routledge.

Kalmykova, Z. I. (1966). Methods of scientific research in the psychology of instruction. *Soviet Education, 8*(6), 13–23.

Kelly, A. E., Lesh, R. A., & Baek, J. Y. (Eds.). (2008). *Handbook of design research methods in education*. London: Routledge.

Kieran, C., & Drijvers, P. (2006). The co-emergence of machine techniques, paper-and-pencil techniques, and theoretical reflection: A study of CAS use in secondary school algebra. *International Journal of Computers for Mathematical Learning, 11*, 205–263.

Kieran, C., Krainer, K., & Shaughnessy, J. M. (2013). Linking research to practice: Teachers as key stakeholders in mathematics education research. In M. A. Clements, A. Bishop, C. Keitel, J. Kilpatrick, & F. Leung (Eds.), *Third international handbook of mathematics education* (pp. 361–392). New York: Springer.

Kilpatrick, J. (1992). A history of research in mathematics education. In D. A. Grouws (Ed.), *Handbook of research on mathematics teaching and learning* (pp. 3–38). New York: Macmillan.

Koedinger, K. R. (2002). Toward evidence for instructional design principles: Examples from Cognitive Tutor Math 6. In D. S. Mewborn, et al. (Eds.), *Proceedings of the 24th Annual Meeting of the North American Chapter of the International Group for the Psychology of Mathematics Education* (Vol. 1, pp. 1–20). Columbus, OH: ERIC Clearinghouse for Science, Mathematics, and Environmental Education.

Koichu, B. (2013). *Variation theory as a research tool for identifying learning in the design of tasks*. Plenary panel at the ICMI Study-22 Conference, The University of Oxford. http://www.mathunion.org/icmi/digital-library/icmi-study-conferences/icmi-study-22-conference/

Koichu, B., Zaslavsky, O., & Dolev, L. (2013). Effects of variations in task design using different representations of mathematical objects on learning: A case of a sorting task. In C. Margolinas (Ed.), *Task design in mathematics education* (Proceedings of ICMI Study 22, pp. 461–470). Available from: hal.archives-ouvertes.fr/hal-00834054

Komatsu, K., & Tsujiyama, Y. (2013). Principles of task design to foster proofs and refutations in mathematical learning: Proof problem with diagram. In C. Margolinas (Ed.), *Task design in mathematics education* (Proceedings of ICMI Study 22, pp. 471–480). Available from hal.archives-ouvertes.fr/hal-00834054

Komatsu, K., Tsujiyama, Y., Sakamaki, A., & Koike, N. (2014). Proof problems with diagrams: An opportunity for experiencing proofs and refutations. *For the Learning of Mathematics, 34*(1), 36–42.

Krainer, K. (2011). Teachers as stakeholders in mathematics education research. In B. Ubuz (Ed.), *Proceedings of the 35th Conference of the International Group for the Psychology of Mathematics Education* (Vol. 1, pp. 47–62). Ankara: PME.

Lakatos, I. (1976). *Proofs and refutations: The logic of mathematical discovery*. Cambridge: Cambridge University Press.

Leikin, R. (2013). On the relationships between mathematical creativity, excellence and giftedness. In S. Oesterle & D. Allen (Eds.), *Proceedings of 2013 Annual Meeting of the Canadian Mathematics Education Study Group/Groupe Canadien d'Étude en Didactique des Mathématiques* (pp. 3–17). Burnaby, BC: CMESG/GCEDM.

Lerman, S. (1996). Intersubjectivity in mathematics learning: A challenge to the radical constructivist paradigm? *Journal for Research in Mathematics Education, 27*, 133–150.

Lerman, S., Xu, G., & Tsatsaroni, A. (2002). Developing theories of mathematics education research: The ESM story. *Educational Studies in Mathematics, 51*, 23–40.

Lesh, R. A. (2002). Research design in mathematics education: Focusing on design experiments. In L. English (Ed.), *Handbook of international research in mathematics education* (pp. 27–50). Hillsdale, NJ: Lawrence Erlbaum Associates.

Lesh, R., Hoover, M., Hole, B., Kelly, A., & Post, T. (2000). Principles for developing thought-revealing activities for students and teachers. In A. E. Kelly & R. A. Lesh (Eds.), *Handbook of research design in mathematics and science education* (pp. 591–645). Mahwah, NJ: Lawrence Erlbaum Associates.

Levav-Waynberg, A., & Leikin, R. (2009). Multiple solutions for a problem: A tool for evaluation of mathematical thinking in geometry. In V. Durand-Guerrier, S. Soury-Lavergne, & F. Arzarello (Eds.), *Proceedings of the Sixth Congress of the European Society for Research in Mathematics Education* (pp. 776–785). Lyon, FR: CERME6.

Lewis, C. (2002). *Lesson study: A handbook of teacher-led instructional change.* Philadelphia, PA: Research for Better Schools.

Limón, M. (2001). On the cognitive conflict as an instructional strategy for conceptual change. *Learning & Instruction, 11,* 357–380. doi:10.1016/S0959-4752(00)00037-2.

Lin, F.-L., Yang, K.-L., Lee, K.-H., Tabach, M., & Stylianides, G. (2012). Principles of task design for conjecturing and proving. In G. Hanna & M. de Villiers (Eds.), *Proof and proving in mathematics education* (pp. 305–325). New York: Springer.

Lincoln, Y. S., & Guba, E. G. (1985). *Naturalistic inquiry.* Newbury Park, CA: Sage.

Margolinas, C. (Ed.). (2013). *Task design in mathematics education* (Proceedings of ICMI Study 22). Available from hal.archives-ouvertes.fr/hal-00834054

Martinez, M. V., & Castro Superfine, A. (2012). Integrating algebra and proof in high school: Students' work with multiple variables and a single parameter in a proof context. *Mathematical Thinking and Learning, 14,* 120–148.

Marton, F., Runesson, U., & Tsui, B. M. (2004). The space of learning. In F. Marton & A. B. Tsui (Eds.), *Classroom discourse and the space of learning* (pp. 3–40). Mahwah, NJ: Lawrence Erlbaum.

Mathematics Assessment Resource Service (MARS). (2012). *Estimating: Counting trees* (p. T-2). Nottingham: Shell Centre. Available from http://map.mathshell.org

McKenney, S., & Reeves, T. (2012). *Conducting educational design research.* London: Routledge.

Menchinskaya, N. A. (1969). Fifty years of Soviet instructional psychology. In J. Kilpatrick & I. Wirszup (Eds.), *Soviet studies in the psychology of learning and teaching mathematics* (Vol. 1, pp. 5–27). Stanford, CA: School Mathematics Study Group.

Morselli, F. (2013). The "Language and argumentation" project: researchers and teachers collaborating in task design. In C. Margolinas (Ed.), *Task design in mathematics education* (Proceedings of ICMI Study 22, pp. 481–490). Available from hal.archives-ouvertes.fr/hal-00834054

Movshovitz-Hadar, N., & Edri, Y. (2013). Enabling education for values with mathematics teaching. In C. Margolinas (Ed.), *Task design in mathematics education* (Proceedings of ICMI Study 22, pp. 377–388). Available from hal.archives-ouvertes.fr/hal-00834054

Newell, A., & Simon, H. A. (1972). *Human problem solving.* Englewood Cliffs, NJ: Prentice Hall.

Ni, Y., & Zhou, Y.-D. (2005). Teaching and learning fraction and rational numbers: The origins and implications of whole number bias. *Educational Psychologist, 40*(1), 27–52.

Ohtani, M. (2011). Teachers' learning and lesson study: Content, community, and context. In B. Ubuz (Ed.), *Proceedings of the 35th Conference of the International Group for the Psychology of Mathematics Education* (Vol. 1, pp. 63–66). Ankara: PME.

Okamoto, K., Koseki, K., Morisugi, K., Sasaki, T., et al. (2012). *Mathematics for the future.* Osaka: Keirinkan (in Japanese).

Piaget, J. (1971). *Genetic epistemology.* New York: W.W. Norton.

Pirie, S., & Kieren, T. E. (1994). Growth in mathematical understanding: How can we characterize it and how can we represent it? *Educational Studies in Mathematics, 26,* 61–86.

Pólya, G. (1945/1957). *How to solve it: A new aspect of mathematical method.* Princeton, NJ: Princeton University Press.

Ponte, J. P., Mata-Pereira, J., Henriques, A. C., & Quaresma, M. (2013). Designing and using exploratory tasks. In C. Margolinas (Ed.), *Task design in mathematics education* (Proceedings of ICMI Study 22, pp. 491–500). Available from hal.archives-ouvertes.fr/hal-00834054

Prediger, S., Bikner-Ahsbahs, A., & Arzarello, F. (2008). Networking strategies and methods for connecting theoretical approaches: First steps towards a conceptual framework. *ZDM: The International Journal on Mathematics Education, 40,* 165–178.

Prusak, N., Hershkowitz, R., & Schwarz, B. B. (2013). Conceptual learning in a principled design problem solving environment. *Research in Mathematics Education*. doi:10.1080/14794802.20 13.836379

Radford, L. (2003). Gestures, speech, and the sprouting of signs: A semiotic-cultural approach to students' types of generalization. *Mathematical Thinking and Learning, 5*(1), 37–70.

Runesson, U. (2005). Beyond discourse and interaction. Variation: A critical aspect for teaching and learning mathematics. *The Cambridge Journal of Education, 35*(1), 69–87.

Ruthven, K., Laborde, C., Leach, J., & Tiberghien, A. (2009). Design tools in didactical research: Instrumenting the epistemological and the cognitive aspects of the design of teaching sequences. *Educational Researcher, 38*, 329–342.

Sawada, T., & Sakai, Y. (Eds.). (2013). *Elementary mathematics 2 (Part 1)*. Tokyo: Kyoiku Shuppan (in Japanese).

Schein, E. (1972). *Professional education: Some new directions*. New York: McGraw-Hill.

Schoenfeld, A. H. (1985). *Mathematical problem solving*. New York: Academic.

Schoenfeld, A. H. (1994). Reflections on doing and teaching mathematics. In A. H. Schoenfeld (Ed.), *Mathematical thinking and problem solving* (pp. 53–69). Hillsdale, NJ: Lawrence Erlbaum Associates.

Schoenfeld, A. H. (1999). Looking toward the 21st century: Challenge of educational theory and practice. *Educational Researcher, 28*(7), 4–14.

Schoenfeld, A. H. (2009). Bridging the cultures of educational research and design. *Educational Designer, 1*(2). http://www.educationaldesigner.org/ed/volume1/issue2/article5/pdf/ed_1_2_schoenfeld_09.pdf

Schön, D. (1983). *The reflective practitioner: How professionals think in action*. London: Basic Books.

Schunn, C. (2008). Engineering educational design. *Educational Designer, 1*(1). http://www.educationaldesigner.org/ed/volume1/issue1/article2/index.htm

Sfard, A. (1991). On the dual nature of mathematical conceptions: Reflections on processes and objects as different sides of the same coin. *Educational Studies in Mathematics, 22*, 1–36.

Sfard, A. (2008). *Thinking as communicating*. New York: Cambridge University Press.

Shimizu, S. (1981). Characteristics of "problem" in mathematics education (II). *Epsilon: Bulletin of Department of Mathematics Education, Aichi University of Education, 23*, 29–43 (in Japanese).

Sierpinska, A. (2003). Research in mathematics education: Through a keyhole. In E. Simmt & B. Davis (Eds.), *Proceedings of the 2003 Annual Meeting of the Canadian Mathematics Education Study Group/Groupe Canadien d'Étude en Didactique des Mathématiques* (pp. 11–35). Edmonton, AB: CMESG/GCEDM.

Simon, H. A. (1969). *The sciences of the artificial*. Cambridge, MA: MIT Press.

Simon, M. (2013). Developing theory for design of mathematical task sequences: Conceptual learning as abstraction. In C. Margolinas (Ed.), *Task design in mathematics education* (Proceedings of ICMI Study 22, pp. 501–508). Available from hal.archives-ouvertes.fr/hal-00834054

Simon, M. A., Saldanha, L., McClintock, E., Karagoz Akar, G., Watanabe, T., & Ozgur Zembat, I. (2010). A developing approach to studying students' learning through their mathematical activity. *Cognition and Instruction, 28*, 70–112.

Skemp, R. R. (1979). Goals of learning and qualities of understanding. In D. Tall (Ed.), *Proceedings of the Third Conference of the International Group for the Psychology of Mathematics Education* (pp. 250–261). Warwick: PME.

Steffe, L. P., & Kieren, T. E. (1994). Radical constructivism and mathematics education. *Journal for Research in Mathematics Education, 25*, 711–733.

Stephan, M., & Akyuz, D. (2012). A proposed instructional theory for integer addition and subtraction. *Journal for Research in Mathematics Education, 43*, 428–464.

Stephan, M., & Akyuz, D. (2013). An instructional design collaborative in one middle school. In C. Margolinas (Ed.), *Task design in mathematics education* (Proceedings of ICMI Study 22, pp. 509–518). Available from hal.archives-ouvertes.fr/hal-00834054

Stigler, J. W., & Hiebert, J. (1999). *The teaching gap*. New York: Free Press.
Stokes, D. E. (1997). *Pasteur's quadrant: Basic science and technical innovation*. Washington, DC: Brookings.
Streefland, L. (1990). *Fractions in realistic mathematics education, a paradigm of developmental research*. Dordrecht: Kluwer.
Streefland, L. (1993). The design of a mathematics course. A theoretical reflection. *Educational Studies in Mathematics, 25*(1–2), 109–135.
Swan, M. (2008). The design of multiple representation tasks to foster conceptual development. In *ICME-11 – Topic Study Group 34: Research and development in task design and analysis*. Available from http://tsg.icme11.org/tsg/show/35
Swan, M., & Burkhardt, H. (2012). Designing assessment of performance in mathematics. *Educational Designer, 2*(5). Available from http://www.educationaldesigner.org/ed/volume2/issue5/article19/
Sweller, J., van Merriënboer, J. J. G., & Paas, F. G. W. C. (1998). Cognitive architecture and instructional design. *Educational Psychology Review, 10*, 251–296.
Tejima, K. (1987). *How many children in a line?: Task on ordinal numbers (video)*. Tokyo: Tosho Bunka Shya (in Japanese).
Treffers, A. (1987). *Three dimensions: A model of goal and theory description in mathematics instruction – The Wiskobas project*. Dordrecht: Reidel.
Vamvakoussi, X., & Vosniadou, S. (2012). Bridging the gap between the dense and the discrete. The number line and the "rubber line" bridging analogy. *Mathematical Thinking & Learning, 14*(4), 265–284.
Van den Heuvel-Panhuizen, M. (2003). The didactical use of models in realistic mathematics education: An example from a longitudinal trajectory on percentage. *Educational Studies in Mathematics, 54*(1), 9–35.
Van den Heuvel-Panhuizen, M., & Drijvers, P. (2013). Realistic mathematics education. In S. Lerman (Ed.), *Encyclopedia of mathematics education* (pp. 521–525). New York: Springer.
Van Dooren, W., Vamvakoussi, X., & Verschaffel, L. (2013). Mind the gap – Task design principles to achieve conceptual change in rational number understanding. In C. Margolinas (Ed.), *Task design in mathematics education* (Proceedings of ICMI Study 22, pp. 519–527). Available from: hal.archives-ouvertes.fr/hal-00834054
van Merriënboer, J. J. G., Clark, R. E., & de Croock, M. B. M. (2002). Blueprints for complex learning: The 4C/ID-model. *Educational Technology Research and Development, 50*(2), 39–64.
van Nes, F. T., & Doorman, L. M. (2011). Fostering young children's spatial structuring ability. *International Electronic Journal of Mathematics Education, 6*(1), 27–39.
von Glasersfeld, E. (1987). Learning as a constructive activity. In C. Janvier (Ed.), *Problems of representation in the teaching and learning of mathematics* (pp. 3–17). Hillsdale, NJ: Lawrence Erlbaum Associates.
Vosniadou, S., Vamvakoussi, X., & Skopeliti, I. (2008). The framework theory approach to conceptual change. In S. Vosniadou (Ed.), *International handbook of research on conceptual change* (pp. 3–34). Mahwah, NJ: Lawrence Erlbaum Associates.
Watson, A., & Mason, J. (2006). Seeing an exercise as a single mathematical object: Using variation to structure sense-making. *Mathematical Thinking and Learning, 8*(2), 91–111.
Watson, A., et al. (2013). Introduction. In C. Margolinas (Ed.), *Task design in mathematics education* (Proceedings of ICMI Study 22, pp. 7–14). Available from: hal.archives-ouvertes.fr/hal-00834054
Wittmann, E. (1984). Teaching units as the integrating core of mathematics education. *Educational Studies in Mathematics, 15*, 25–36.
Wittmann, E. C. (1995). Mathematics education as a 'design science'. *Educational Studies in Mathematics, 29*, 355–374.
Yang, Y., & Ricks, T. E. (2013). Chinese lesson study: Developing classroom instruction through collaborations in school-based teaching research group activities. In Y. Li & R. Huang (Eds.), *How Chinese teach mathematics and improve teaching* (pp. 51–65). London: Routledge.

Chapter 3
The Relationships Between Task Design, Anticipated Pedagogies, and Student Learning

Peter Sullivan, Libby Knott, and Yudong Yang
with *Mike Askew, Laurinda Brown, Maria G. Bartolini Bussi, Haiwen Chu, María J. González, Merrilyn Goos, Angelika Kullberg, Yuly Marsela Vanegas, Irit Peled, Gila Ron, Orit Zaslavsky* and additional contributions from *Anne Adams, Jonei Barbosa, Ruthi Barkai, Lisa Canty, Alf Coles, Shelley Dole, Rob Ely, Vicenç Font, Vince Geiger, Joaquin Giménez, The GEMAD Group, Esther Levenson, Pernilla Mårtensson, Andreia Maria Pereira de Oliveira, Jo Olson, Kathryn Omoregie, Alessandro Ramploud, Ulla Runesson, Xuhua Sun, Anat Suzan, Michal Tabach, Dina Tirosh, Pessia Tsamir, Iris Zodik*

3.1 Introduction

This chapter seeks to synthesize research and scholarship about the relationship between the design of classroom tasks, the pedagogies associated with the effective implementation of tasks, and the learning of mathematics. We use the term *classroom tasks* similarly to Watson and Sullivan (2008) who describe tasks as the questions, situations, and instructions that might be used when teaching students. Tasks prompt activity which offers students opportunities to encounter mathematical concepts, ideas, and strategies. The role of the teacher is to select, modify, design, redesign, sequence, implement, and evaluate the tasks.

The intended task and the enacted task may differ considerably. Even though, as argued by Hiebert and Wearne (1997), "what students learn is largely defined by the tasks they are given" (p. 395), Christiansen and Walther (1986) note that "even when students work on assigned tasks supported by carefully established educational contexts and by corresponding teacher-actions, learning as intended does not follow automatically from their activity on the tasks" (p. 262). Christiansen and Walther differentiate between the task as set and the activity that follows, including

---

P. Sullivan (✉)
Monash University, Melbourne, VIC, Australia
e-mail: peter.sullivan@monash.edu

L. Knott
Washington State University, Pullman, WA, USA

Y. Yang
Shanghai Academy of Educational Sciences, Shanghai, China

This chapter has been made open access under a CC BY-NC-ND 4.0 license. For details on rights and licenses please read the Correction https://doi.org/10.1007/978-3-319-09629-2_13

students' interpretations of the purpose of the task, ways of working, teacher interventions, how language and symbols are used, and what are seen as valuable mathematical actions. The relationship between task and activity can develop in a variety of ways: in some cultures it is the norm for teachers to take a given task and develop it into a whole lesson plan with challenging goals, whereas Stein, Grover, and Henningsen (1996) observe that teachers and students can also act together to reduce a classroom task to a mere sequence of actions.

This chapter first addresses factors that influence the task design process and accompanying pedagogical considerations. It then presents three tasks chosen to provide context for various discussions within the chapter, along with a consideration of the age range of students at which tasks are appropriate. Subsequent sections describe:

1. The *interactions among aspects of task design*: design elements of tasks, the nature of the mathematics that is the focus of the tasks, and the task design processes
2. *Pedagogies*: the nature of the authority and autonomy of the teacher in creating and implementing tasks and problematic aspects of converting tasks from one culture to another
3. *Student learning*: consideration of students' responses in anticipating the pedagogies

A crosscutting theme is that tensions occur in making decisions on culture, mathematics, language, context, and pedagogy. Designers and teachers make decisions among competing options at both the design and implementation stages. In many cases, the decisions on whether, for example, to foster challenge or success, to focus on abstract mathematical ideas or their applications, on whether to exemplify the dominant culture or to introduce perspectives of marginalized groups, may be secondary to the teacher's awareness of those decisions and his/her capacity to interact with the students to explore all aspects of the task potential.

3.2 Factors Influencing Task Design and Pedagogies

Of course, tasks do not exist separately from the pedagogies associated with their use nor are the pedagogies independent of the task. Knowledge for teaching mathematics and anticipatory pedagogical decision-making are two key and complementary elements that are central issues in task design. Jaworski (2014), for example, in elaborating an inquiry stance by teachers in her projects, described a difference between didactics and pedagogy as often used in Europe. She described didactics as being about "the transformation of the subject (mathematics) into activity and tasks through which learners can gain access to the mathematics, engage with mathematics, and come to know mathematical concepts" (p. 2). She described pedagogy as about "creating the learning environment through which learners' engagement with mathematics can take place effectively" (p. 2). Clearly the process of connecting task design with pedagogy involves consideration of both aspects. These are elaborated further in the following sections.

3.2.1 Knowledge for Teaching Mathematics that Informs Task Design

In describing teacher knowledge, we present the categorization proposed by Hill, Ball, and Schilling (2008) who described two aspects of knowledge associated with converting tasks for the use in one's classroom: subject matter knowledge and pedagogical content knowledge. Included within the former are common content knowledge, specialized content knowledge, and knowledge at the mathematical horizon. Included within the latter are knowledge of content and teaching, knowledge of content and students, and knowledge of curriculum.

Perhaps the most critical for task design is *specialized content knowledge* or "the knowledge that allows teachers to engage in particularly *teaching* tasks, including how to accurately represent mathematical ideas, provide mathematical explanations for common rules and procedures, and examine and understand unusual solution methods to problems" (Hill et al., 2008, p. 378). Also important is what Hill et al. (2008) described as *knowledge of content and teaching*, including an understanding of how to sequence particular content for instruction, and how to evaluate instructional advantages and disadvantages of particular representations and of the knowledge required to make "instructional decisions about which student contributions to pursue and which to ignore or save for a later time" (p. 401).

These perspectives on teacher knowledge also inform decisions on the placement and contribution of tasks to sequences of learning. Decisions on sequences of learning can be informed by what Simon (1995) described as a *hypothetical learning trajectory* (see also Chap. 2) that:

> provides the teacher with a rationale for choosing a particular instructional design; thus, I (as a teacher) make my design decisions based on my best guess of how learning might proceed. This can be seen in the thinking and planning that preceded my instructional interventions … as well as the spontaneous decisions that I make in response to students' thinking. (pp. 135–136)

Simon noted that such a trajectory is made up of three components: the learning goal; the activities to be undertaken; and a hypothetical cognitive process, "a prediction of how the students' thinking and understanding will evolve in the context of the learning activities" (p. 136). These predictions are not related to sequences of explanations but for students to engage in a succession of problem-like tasks, based on recognition that learning is a product of activity that is "individual and personal, and … based on previously constructed knowledge" (Ernest, 1994, p. 2).

In planning and teaching, the role of the teacher is to identify potential and perceived blockages, prompts, supports, challenges, and pathways. In other words, learning occurs as a product of students working on tasks purposefully selected or designed by the teacher and contributing to ongoing interaction with the teacher and their peers on their strategies and products.

3.2.2 Anticipatory Pedagogical Decision-Making

Based on their knowledge of mathematics and pedagogy, teachers make decisions in anticipation of how students will respond to tasks. Gueudet and Trouche (2011), for example, in elaborating the complex factors informing task implementation, noted the potential gap between the availability of resources, in this case the tasks, and the ways that teachers anticipate the tasks for use in classrooms, which we consider as the pedagogy. They described *documentational genesis* as the two-way processes by which tasks are not only interpreted by teachers but also influence the decisions that teachers make (see also Chap. 6). Gueudet and Trouche (2011) described the use of task as a combination of the task as designed and a *scheme of utilization* which "integrates practice (how to use selected resources for teaching a given subject) and knowledge on mathematics, on mathematics teaching, on students, and on technology" (p. 401), i.e., integrating didactics and pedagogy.

We interpret *scheme of utilization* to be similar to the task elements described by Sullivan et al. (2014) who suggested that, when designers communicate with teachers about the intentions and potential of tasks, this can include indications of the mathematical purpose, ways that tasks can be differentiated, and suggestions of actions that can follow the task to implement the learning. In particular, Sullivan and colleagues proposed a scheme of utilization for the type of task and lesson they were describing to include: one or more challenging task(s); one or more additional task(s) that help to consolidate the learning from the earlier ones; preliminary experiences that are prerequisite but which do not detract from the challenge of the tasks; and supplementary tasks that offer the potential for differentiating the experience through the use of enabling prompts and extending prompts (see Sullivan, Mousley, & Jorgensen, 2009). The term *scheme of utilization* emphasizes that advice on anticipated pedagogical actions is not intended as a script but as a prompt to teachers' own decision-making. Another example of a pedagogical scheme is the *five practices* of Smith and Stein (2011) for orchestrating productive mathematics discussions: anticipating, monitoring, selecting, sequencing, and connecting. These five practices are useful in providing a framework for facilitating rich discussions that mathematics teachers may want to see in their classrooms. These schemes suggest a somewhat overlapping boundary between the design and implementation of tasks, lessons, and sequences.

3.3 Some Tasks Presented to Inform Subsequent Discussions

Three tasks are presented in this section to help exemplify and support the discussions throughout the chapter. The first task is an example of the type of task commonly used as part of the Japanese Lesson Study process (see Fernandez & Yoshida, 2004; also Chaps. 2 and 9 in this volume). The second task was described by Bartolini Bussi, Sun, and Ramploud (2013) who reported on its use, initially developed in a Chinese textbook, in Italy. The third task was described by Peled (2008) and subsequently adapted for use as part of a task implementation project (see Sawatzki & Sullivan, 2015).

There are other types of tasks that could have been chosen, such as mathematical investigations intended to be undertaken independently from the teacher, games that illustrate particular mathematical concepts, and matching of different representations of concepts. The particular three tasks we have chosen are intended to be neither exemplary nor representative, but are provided to allow illustration of issues raised in discussions among contributors to this chapter.

3.3.1 The L-Shaped Area: A Lesson from Japan

This *L-Shaped Area* lesson is representative of tasks used as the basis of lessons in Japanese Lesson Study. The intent is to introduce the notion that the number of squares in a rectangular array can be calculated by multiplying the number of rows by the number of columns.

It can be assumed that teachers might establish a context for the area concepts, such as tatami mats, which are traditional rice-straw mats, 90 cm by 180 cm, used commonly as floor covering and sometimes used to describe floor size (area) of rooms or large buildings. One approach is to present students with a worksheet on which there are two copies of the diagram in Fig. 3.1.

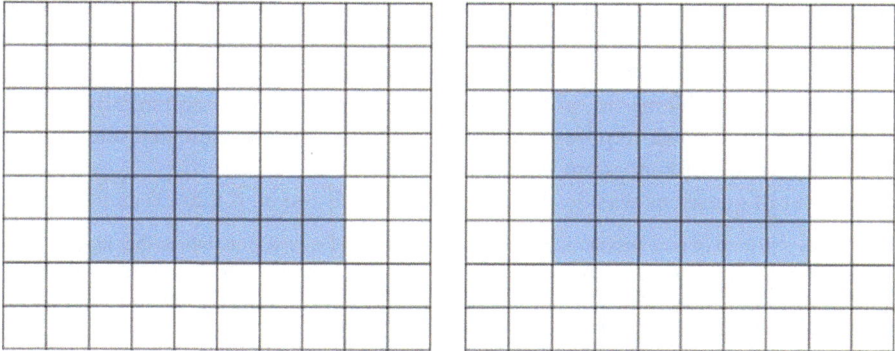

Fig. 3.1 The diagrams used in the L-Shaped Area lesson

The two copies of the diagram on the worksheet communicate to students that, even if they find one solution by counting, a further strategy is expected. Some of the possible solutions include that there are four different ways in which the shape can be rearranged to form a rectangle and the possibility of forming the encompassing rectangle and subtracting the unused portion.

A key phase of such a lesson is the orchestration of selected students' reporting on their strategies, noting that the teacher would have anticipated the types of solutions that students might offer. This would be part of the scheme of utilization of such a task and connects directly to the Smith and Stein (2011) five practices.

There are some interesting characteristics of this lesson: the obvious focus on student-generated strategies; the purposeful choice of students to present and explain their solutions, ensuring that a range of strategies are presented and discussed; and the affordance of a diversity of strategies and representations, allowing students to experience important mathematical ideas.

3.3.2 A Set of Worded Questions

The *Worded Questions* task was described by Bartolini Bussi et al. (2013) who reported on a cross-cultural collaborative project exploring task design and use in both China and Italy (described in more detail below). The set of questions developed in a Chinese textbook are presented in translated form in Fig. 3.2.

The main task is the requested explanation at the top of Fig. 3.2 which may be deduced from consideration of the nine accompanying situations. Note the similarities and differences in the form of the sets of questions and that the numbers are beyond the usual arithmetic range for children in the first years of school in some textbooks. The authors used the task with children aged 8–9, but the task could be presented to older children even in the current format. The framework of questions was similar to that developed as part of the Cognitively Guided Instruction (CGI) project (Carpenter, Fennema, Peterson, Chiang, & Loef, 1989). The CGI framework was developed after a project involving researchers and teachers and adhered to two main tenets: that instruction should focus on problem-solving and that teachers should encourage the use of multiple strategies, listen to student reasoning, and build on what students know. What is different, however, is that, in this case, the collection of problems is given simultaneously and not sequentially. Thus, completion of the task requires a holistic or whole-problem view and attention to similarities and differences among the individual tasks.

The set of questions was described by Bartolini Bussi et al. (2013) as follows:

> This is a system of nine problems concerning addition and subtraction, where the organization in rows refers to … combine, change, compare categorization and the organization in columns refers to the same arithmetic operation (either addition or subtraction, …). In each row there is a problem (in the shaded cell) and two variations.
>
> The task is very complex and requires the students not only to solve each problem but also to explain why the nine problems have been arranged in this way. Each problem is associated with a graphic scheme that models on one or two lines the relationship between quantities. (p. 553)

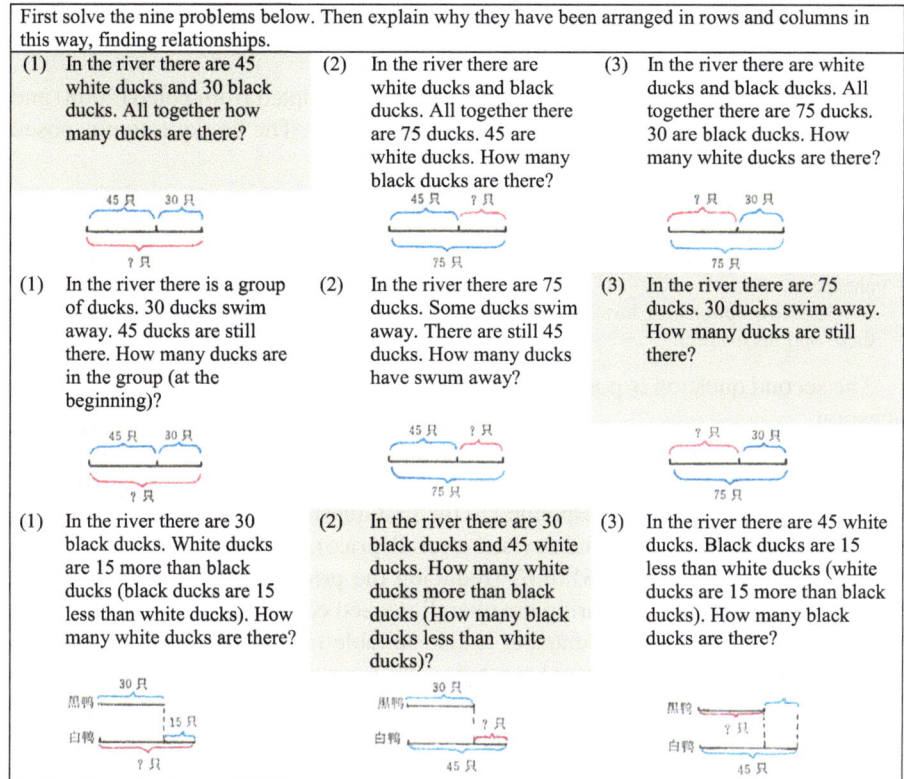

Fig. 3.2 The nine worded questions (Bartolini Bussi, Canalini & Ferri, 2011)

The intention is that the small variations across each row, and also the variations along the columns, allow students to focus on the key elements of the initial problem and the ways that the small variations change the problem and its representation. The nine worded questions construct a three-by-three table and appear as one problem.

From the perspective of variation theory (defined in Chap. 2 and elaborated in Chap. 5), the context of the nine worded questions remains invariant, namely, two groups of ducks in a river, and so is the representation of the part-part-whole relation. In the application of variation theory to task design, mathematics tasks should be designed so that the key idea is varied, allowing learners to experience the effect of the variation in the examples. The known and unknown quantities are varied in each question. The learner's awareness is directed to the pattern of the questions issued in the table. The mathematical aspects, including establishing the relationship between addition and subtraction and also differentiating between forms of subtraction (such as take away and difference), is the focus of the task. Reflection on the individual variations is a more important aspect of this task than the outcomes. Decisions about the use of such a task are a product of teacher knowledge, the scheme of utilization, and the anticipated, hypothetical, learning trajectory.

3.3.3 Shopping for Shoes

The following task based on two related questions is adapted from Peled (2008) and is referred to in the following text as the *Shopping task*. The first question is posed as follows:

> Jenny and Carly go shopping for shoes. Jenny chooses one pair for $110 and another for $100. Carly chooses a pair that costs $160. When they go to pay, the assistant says that there is a sale on and they get 3 pairs of shoes for the price of 2 pairs (the free one is the cheapest).
>
> Give two options for how much Jenny and Carly should each pay. Explain which of these options is fairer.

The second question is posed similarly except that Carly's shoes cost $60 in that question.

Although this task (meaning the two questions together) might seem at first glance to contain little mathematics, the range of arguable solutions is interesting. For example, students have responded to the first question by suggesting that Carly pay $90 (one third of the total revised overall price), or $110 (sharing the $100 overall savings equally), or $126.67 (reducing the price by one third of the $100 overall savings), or $135 (sharing the overall revised cost equally). The task can be used with upper primary students but is also suitable for junior secondary students if solutions based on proportional reasoning are prompted ($116.76 — Carly's fraction of the total value of the 3 pairs of shoes multiplied by the actual total cost). The task for students is not so much to determine one possible answer, but to find a way to resolve the differences between these alternatives. Further, not all of these strategies are applicable to the second task. Some of those solution strategies result in Carly being asked to pay more than her shoes cost.

The tasks raise issues about what constitutes "fairness" (note the parallel with the origins of probability theory) and the ways that social considerations, such as friendship, are integral aspects of the solution. In addition, the task makes it clear to students that they can explain and/or defend a particular answer. It emphasizes argumentation and justification, and because of the degree of ambiguity, it allows consideration of social/cultural mediation in mathematics. Of course, the context is culturally laden, and teachers may choose to adapt the task to a situation and cost familiar to and relevant for the students. Again, the use of the task is dependent on the way that the teacher interprets the mathematical potential and the ways the teachers interpret the intended scheme of utilization.

3.3.4 Determining the Age Range of Students for Which These Tasks Are Appropriately Posed

The potential and appropriateness of these three tasks depend on the prior experiences of the students, pedagogic purpose, and teacher and student expectations. The age range at which these three tasks are best suited also depends on the number combinations used and whether the tasks are posed near the start or end of a relevant sequence of tasks. The L-Shaped Area task has been used with students around age 9, the Worded Questions was used with students of ages 8–9, and the Shopping task is intended for students around age 11 but could be used with older students to motivate the idea of proportional reasoning.

In all cases the tasks are somewhat generic and can be adapted for different levels by minor adjustments of particular task aspects. The appropriateness of the age ranges at which the tasks are posed is also a feature of the expectations for the students. Jaworski, Goodchild, Eriksen, and Daland (2011), for example, describe a task that was simply posed but readily adapted for the use by students anywhere between year 1 and 12. In other words, the grade level for tasks is dependent on a range of contextual factors and is not inherent in the task as designed.

3.4 Design Elements of Tasks

One of the recurring themes in the discussions at the ICMI Study Conference that generated this book was that, while the mathematics exemplified by the task is central, there are many other important considerations in designing tasks, especially when designers wish to anticipate and encourage particular pedagogical choices. Our discussion on task design elements is presented in two parts: first, five task design dilemmas are presented to indicate the range of design considerations; and second, we present six suitability criteria that are intended to facilitate analysis of tasks as well as research on and evaluation of those tasks.

3.4.1 Five Dilemmas

Recognition of inherent tensions is central to the decisions that arise in task design and the associated pedagogies. In delineating decisions on elements of tasks, Barbosa and de Oliveira (2013) focused on various dilemmas associated with designing tasks for groups of learners. They used the dilemmas not only as design considerations but also as ways of evaluating the adequacy of tasks that were designed by teachers in the research project on which they were reporting. There were five dilemmas (or conflicts) identified. In the *Australian Concise Oxford Dictionary*, a dilemma is described as a "situation in which a choice has to be made between two … alternatives". These alternatives represent the extremes of the tensions faced by task designers.

3.4.1.1 Context as a Dilemma

The first dilemma to which Barbosa and de Oliveira (2013) refer arises in the mathematical *context* of tasks, which they describe as ranging from pure mathematics to semi-reality to reality. This dilemma (or more accurately continuum in this case) is, on the one hand, the extent to which tasks are set in a realistic context to maximize engagement of students and, on the other hand, whether the context detracts from the potential of the task to achieve the intended learning. In each of the three tasks previously described, the contexts—the mats representing area as covering, the combinations of ducks on the pond, and the shopping discounts—do not necessarily detract from the mathematics to be learned and in fact help make the potential generalizability of the solutions more accessible. The shoes context has the additional element of raising the social decision-making that is required as well as the discussions about what constitutes "fairness".

It is relevant to note that the use of contexts is far from unambiguous. For example, in a review of the testing system in the United Kingdom, Cooper and Dunne (1998) found that contextualizing mathematics items created particular difficulties for low socioeconomic status (SES) students, so much so that they performed significantly poorer than their middle-class peers, while performance on decontextualized tasks was equivalent. Likewise, Lubienski (2000), in studying the implementation of a curriculum program based on open-ended contextualized problems, found that pupils who preferred the contextualized trial materials and considered them easier all had high SES backgrounds, while most pupils who preferred closed, context-free tasks had low SES. This is a complex issue, and it is not clear whether diminished performance was due to contextualization per se or due to other factors like the particular contexts being unfamiliar and alienating for students in low SES communities, or difficulties in separating contextual knowledge from intended "pure" mathematical actions. In other words, the incorporation of contexts does not necessarily ensure tasks are accessible to all students.

Extending the dilemma on the context of tasks, there is a difference between contexts which can be easily seen to be peripheral and those which are central to the mathematics. For example, the L-Shaped Area task and Worded Questions task can easily be transformed into pure mathematical tasks, or different contexts can be used. In contrast, the context of the Shopping task cannot be minimized, or the task will lose meaning. It is also possible for the context to limit the potential of students to generalize solutions.

3.4.1.2 Language as a Dilemma

The second dilemma is about the *language* of the task and the intended solution. On one hand, mathematical precision is part of the desired learning; on the other hand, clarity for the students is needed to support the learning. The language demand of the L-Shaped Area task is mainly connected to the representation of the potential solutions and so is mathematical. For the Worded Questions, the subtle variations in language exemplify the distinctions between the forms of the question. The language used in the Shopping task may not be clear, and so the task may even need to be modeled or role-played by the teacher, and mathematical and social language is required to explain the "fair" solution. Of course, what constitutes fairness can be context dependent. In each case, it is not the language of the task itself, but the way the language is used and interpreted.

3.4.1.3 Structure as a Dilemma

Barbosa and de Oliveira (2013) described a third dilemma as *structure*, which refers to the degree of openness in tasks. This can be considered as much a function of the task outcome as it is the structure. In this dilemma, the consideration is that specific questions can be posed which, on one hand, scaffold student engagement with a task in a more prescribed way and, on the other hand, allow students greater opportunity to make strategic decisions on pathways and destinations for themselves. Barbosa and de Oliveira (2013) describe this continuum as ranging from more closed to more open. Of course, what constitutes openness is the subject of some debate. For example, Hashimoto and Becker (1999) described three categories of problems: those that use a variety of approaches (that have been described as open-middled—see also Wiliam, 1998); those in which the formulation is open (described as open-started, which is close to problem posing); and those that have a range of solutions (open-ended). The L-Shaped Area task is open-middled in that the focus is on student-devised strategies, and the Shopping task is open-ended in that there is a range of feasible solutions. Although the individual Worded Questions are closed, with just one correct answer, there is openness in the choice of representation and also in the identification of commonalities and differences across the questions in the rows and columns. When presented as a set of problems, the focus for the students is not only in finding the respective answers, but also in identifying, understanding, explaining, and justifying the commonalities and differences.

3.4.1.4 Distribution as a Dilemma

The fourth dilemma, described as *distribution*, refers to selecting content to be focused on in the tasks. This is a function of the cognitive demand of the tasks, described by Smith and Stein (2011) as a hierarchy of classroom tasks that develop from *memorization* to *procedures without connections* to *procedures with connections* to *doing mathematics* tasks. Using this nomenclature, in the L-Shaped Area task, the individual students would be *doing mathematics* when creating their solutions and in considering the solutions of others. When students were answering the individual Worded Questions, they would be performing *procedures with connections*, and when identifying commonalities and differences between the questions, they would be *doing mathematics*. It would be possible to respond to the Shopping task at the level of *procedures without connections*; the extent to which the students engaged in *procedures with connections* or *doing mathematics* would depend on the actions of the teacher.

3.4.1.5 Levels of Interactions as a Dilemma

The fifth dilemma refers to the *levels of interactions* of the participants, meaning between the teachers and the students. This can be interpreted to mean that the task does not exist by itself, but its implementation is influenced by the nature of the intended or anticipated interactions between the teacher and students when they are engaged with the task. This is partly connected to the hypothetical learning trajectory (Simon, 1995) that the teacher has anticipated.

In working on the L-Shaped Area task, the expectation is that students engage with the task first and then discuss the various solutions in small groups, with the teacher and as a class. Similarly, for the Worded Questions, the students would work on the questions, in both the Chinese and Italian contexts, with the teacher leading a critical review of the similarities and differences between and within the rows and the columns. In the Shopping task, the students would formulate their own responses with the essential aspect being the discussion and defense of the various viable solutions.

Designers and teachers confront each of these five dilemmas and make appropriate choices, for each and every task, and teachers may take decisions that were not intended or anticipated by the designer.

3.4.2 Task Suitability Criteria

The dilemmas of task design provide a framework that can be used for analysis of suitability of tasks. Giménez, Font, and Vanegas (2013) provide a suitable framework for analysis of tasks generally.

Giménez et al. (2013) describe *epistemic* suitability as "the extent to which the mathematics taught is 'good mathematics'" (p. 581). Decisions on the mathematics are based on both the local and institutional curriculum and prior experiences of the students. Each of the three tasks previously described addresses important mathematical concepts, although the specific concepts are to some extent dependent on the level at which the tasks are used. This connects directly to the mathematics content knowledge of the teacher, who needs to perceive what mathematics is possible.

Giménez et al. (2013) explain that *cognitive* suitability "reflects the degree to which the teaching objectives and what is actually taught are consistent with the students' developmental potential, as well as the closeness of the match between what is eventually learnt and the original targets" (p. 581). Of course, in our illustrative tasks, the cognitive suitability is mainly a function of the sequencing of these tasks among others and cannot be accurately prescribed out of context, without knowing the objectives of the lesson(s) and what was taught previously. Yet each of the tasks offers a variety of starting points for students. The L-Shaped Area task can be solved by counting methods as well as by more mathematically sophisticated approaches. Students might work on various representations of just one of the Worded Questions, while others might engage in the tasks of comparing and contrasting the questions. The Shopping task is mathematically simple at the level at which it is appropriate, yet contrasting various solutions and arguing which is fair is sophisticated. In other words, all three tasks are adaptable to a level of cognitive demand for which the teacher decides she/he can support the engagement of learners.

Interactional suitability "relates to the extent to which the forms of interaction enable students to identify and resolve conflicts of meaning, and promote independent learning" (p. 581). This is similar to what Barbosa and de Oliveira (2013) described as levels of interaction and can refer to interactions between teacher and students, between the students, and for the student and the task. For the L-Shaped Area task, the nature of the interactions depends on discussions facilitated by the teacher that, for example, compare the solutions. For the Worded Questions, the interactions occur when the teacher encourages the students to contrast the various question forms. The interactional suitability depends on teachers' anticipation of what students could become aware of. For students around age 8 years, the interaction may stop at finding the relationship of addition and subtraction among three quantities (two of them are known and one is unknown). For older students, the interaction may be directed at distinguishing the two different patterns for the subtraction operation. In the second and third columns, one pattern of subtraction is to figure out a partial quantity when the sum and the other partial quantity are known. But the other pattern of subtraction is to compare the difference between a bigger quantity and a smaller quantity. For the Shopping task, the interactions depend on the extent to which the teacher allows and facilitates the consideration of alternatives by the students and prompts discussions about fairness. In other words, each of the tasks has its own scheme of utilization.

Mediational suitability refers to the "availability and adequacy of the material and temporal resources required by the teaching/learning process" (p. 581). The key feature of the L-Shaped Area task is the presentation to students of a worksheet that requires two methods of solution. This is intended to prompt students to offer two different solutions, especially when this has become a normal expectation of both the teacher and the students. The cognitive demand of this task is evidenced by the level and type of engagement. The juxtaposition of the various Worded Questions prompts the students to engage with the questions both one by one (in the Italian implementation) and overall (as in the Chinese model). The context of the Shopping task may require role playing the situation so the nature of the task (not the solution) becomes clear.

Affective suitability reflects the students' degree of involvement (interest, motivation, etc.) in the task. As Middleton (1995) argued, a key task characteristic that influences affective responses is the degree of control, which is better described as the opportunity for student decision-making, meaning the choices that the students can make. Middleton (1995) also suggested that interest and arousal are important determinants of student motivation. In the L-Shaped Area task, the choice is the mode of solution, while the interest in the task is motivated by the use of a familiar context. Although there is limited choice in the individual Worded Questions, students make decisions on ways they interpret the similarities and differences between the questions; in the Shopping task, the choices are the decisions on what is "fair" and imagining themselves in a related situation.

Giménez et al. (2013) considered *ecological* suitability as "the degree of compatibility between the study process and the school's educational policies, the curricular guidelines and the characteristics of the social context, etc." (p. 581). The L-Shaped Area task and the Worded Questions attempt to address multiple solutions to a problem at different cognitive levels often explicit in the mathematics curriculum. The link between the Shopping task and a conventional mathematics curriculum is, however, more tenuous and requires intervention by the teacher for this to become explicit. Often, the aims of mathematics curricula are difficult to discern. In Chap. 5, the complexity of such curricula is described in more detail, with an attempt to categorize the different purposes afforded by text-based tasks.

Overall, this section summarizes two frameworks that describe the elements of, and design considerations for, tasks. They illustrate not only that task design is multidimensional, but also that there are tensions to be considered at each phase of design. The tensions are present for task designers and teacher adaptation, whether they are designing tasks for themselves or for others. The next section focuses on consideration of the mathematical content of tasks.

3.5 The Nature of the Mathematics that is the Focus of the Tasks

Perhaps the most critical element of task design is the potential for the task to prompt the learning of the intended mathematical concepts. But there are different perspectives of mathematics that can be considered. On one hand, Ernest (2010) described the goals of a *practical* perspective of mathematics as students learning the mathematics adequate for general employment and functioning in society, drawing on the mathematics used by various professional and industry groups. Ernest included in this perspective the types of calculations one does as part of everyday living, including best-buy comparisons, time management, budgeting, planning home maintenance projects, choosing routes to travel, interpreting data in the newspapers, and so on.

On the other hand, Ernest described a *specialized* perspective as that mathematical understanding which forms the basis of university studies in science, technology, and engineering. He argued that this includes an ability to pose and solve problems, appreciate the contributions of mathematics to culture, the nature of reasoning, and intuitive appreciation of mathematical ideas such as "pattern, symmetry, structure, proof, paradox, recursion, randomness, chaos, and infinity" (Ernest, 2010, p. 24).

Both perspectives are directly connected to the teachers' mathematical knowledge for teaching and clearly inform task design. In taking a specialized perspective, the following subsection elaborates considerations for tasks that prioritize explicit mathematical goals. In taking a practical perspective, the subsequent subsection explores issues associated with tasks that focus on mathematical literacy, described here as numeracy. As with other design elements, these two can be in tension, in that a focus on one can detract from the goals associated with the other.

3.5.1 Tasks that Address Specialized Mathematical Goals

Connected to the teacher knowledge that informs task design and implementation are two aspects: the conceptual ideas represented by a specialized perspective and the mathematical processes in which the students might be expected to engage.

From this perspective, it is assumed that teachers will be explicit about the nature of the expected mathematical goals for the students, not only as part of their planning but also in their ongoing interactions with students. There is, however, an inherent tension here between articulating a mathematical goal to students and having students discover or investigate a mathematical concept or idea in a lesson. In the latter case, the articulation of the goal needs to be rather general, so as to not reveal the concept to be discovered or investigated. Further, Smith and Stein (2011) argue that articulating the mathematical goals (at the design phase, especially if the design is done other than by the class teacher) can support furthering teacher knowledge of the specialized mathematics.

In the case of the L-Shaped Area task, the mathematical goals include the array model of multiplication, conservation of area, and seeing other ways of calculating area other than counting squares one by one. Experience with the task also lays a foundation for study of later concepts such as area conservation (that is useful in the process of calculating the area of parallelograms), breaking a composite shape into parts (that can inform the calculation of the area of trapezoids), and subtracting areas (that may be used in calculating the area of paths around shapes). The task would be entirely different if the area formula, $A = l \times w$, was made explicit to students as the goal of the task.

In the Worded Questions, the mathematical concepts include the relationship between addition, subtraction, and their representation and the different forms of subtraction (e.g., take away, difference), with generality in recognizing reciprocal relationships between addition and subtraction. Such aspects commonly are emphasized in curriculum statements. In Australia, for example, the content of relevant aspects of the curriculum is presented through statements such as:

> Represent and solve simple addition and subtraction problems using a range of strategies including counting on, partitioning and rearranging parts.

The Shopping task does not focus on specific mathematical concepts, unless it is posed in the context of proportionality, in which case the extent to which the proportional allocation of the costs represents fairness in the different tasks can be the focus of discussion.

The specialized perspective, as described by Ernest, also addresses the process goals associated with the tasks. Examples of such process goals for students are:

- Making connections between intuitive knowledge and formal mathematical principles/conventions/ideas
- Developing mathematical modeling and problem-solving skills
- Developing algebraic thinking/the ability to express generality

- Learning that jumping with the first idea that comes to mind is not always a good strategy
- Learning the value of examining multiple solutions to a problem and building connections between those solutions

These process aspects are implied in the L-Shaped Area task by the invitation to students to provide two solution methods. In Worded Questions, the potential for generality is in recognizing reciprocal relationships as well as in the interaction between the question forms. In the Shopping task, the mathematical processes include the justifications of a "fair" solution, comparing solution options, and the applicability of the same solution method to both of the questions.

Such process goals are evident in four strands of mathematical proficiency described by Kilpatrick, Swafford, and Findell (2001). The first strand, *conceptual understanding*, includes the comprehension of mathematical concepts, operations, and relations. The second strand, *procedural fluency*, refers to carrying out procedures flexibly, accurately, efficiently, and appropriately and, in addition to these procedures, having factual knowledge and concepts that come to mind readily. The third strand, *strategic competence*, includes the ability to formulate, represent, and solve mathematical problems. The fourth strand, *adaptive reasoning*, includes the capacity for logical thought, reflection, explanation, and justification.

3.5.2 Designing Tasks that Address a Practical Perspective

Taking a different stance, Goos, Geiger, and Dole (2010) use a model of mathematics focusing on real-life contexts, application of mathematical knowledge, use of representational, physical, and digital tools and that emphasizes cultivation of positive dispositions toward mathematics. Their model, shown in Fig. 3.3, illustrates the considerations associated with tasks.

This model connects various aspects informing task design. Goos, Geiger, and Dole (2013) used the term *mathematical knowledge* to include not only fluency with accessing concepts and skills, but also problem-solving strategies and the ability to make sensible estimations. Such knowledge is accessed in solving the L-Shaped Area task and the Worded Questions. On one level, the mathematical demand of the Shopping task is limited, although it is noted that in junior secondary levels it can be anticipated that some students might propose a solution based on proportionality which would represent that mathematical knowledge.

Goos et al. also proposed *positive dispositions*, "a willingness and confidence to engage with tasks and apply their mathematical knowledge flexibly and adaptively" (p. 591), as part of their model (note that this is also the fifth proficiency from Kilpatrick et al., 2001). One of the elements of disposition is related to the opportunities for students to make decisions on the nature of the solution and the pathway to the solution. This relates to the notions of control and student decision-making. The L-Shaped Area task and the Shopping task both allow such opportunities.

3 Task Design, Pedagogies and Learning

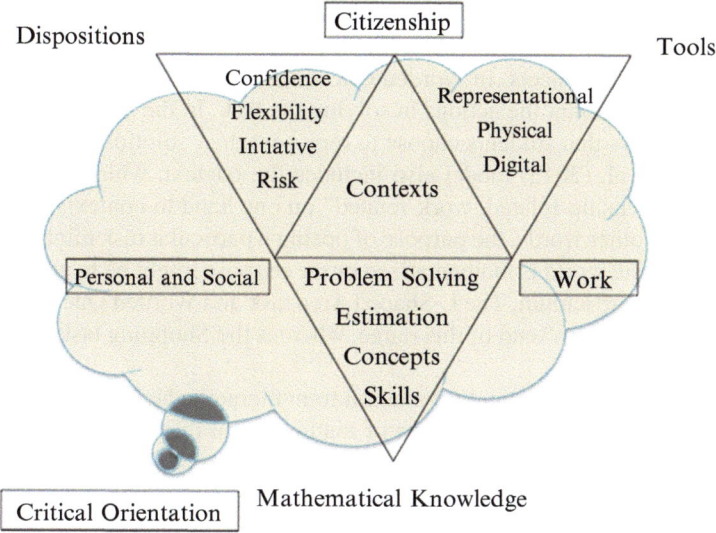

Fig. 3.3 The model proposed by Goos et al. (2010)

Worded Questions has this element when the students are seeking to describe similarities and differences between the questions.

Another element that can foster a positive disposition is if the task has what is described as a "low floor, but high ceiling". Recognizing that it is not clear whether it is the task that fosters the positive disposition or the disposition that facilitates the engagement with the task, all three of the tasks have the potential to foster improved disposition. The "floor" (which in this case refers to the level at which students might initially engage with the task) in the L-Shaped Area task is represented by a solution in which the square units are counted individually, in the individual Worded Questions by making a physical model of the ducks or a number line segment, and in the Shopping task by finding a single possible cost breakdown. The "ceiling" in the L-Shaped Area task is represented by any of the solutions expressed in general form and also perhaps by the articulation of a general solution strategy; in Worded Questions by explaining the similarities and differences between the rows and columns, respectively; and in the Shopping task by contrasting the solutions for the two forms and considering which approaches can apply to both problems and which do not.

A third element presented by Goos et al. (2013) is the tools, which they elaborate as follows:

> In school and workplace contexts, tools may be representational (symbol systems, graphs, maps, diagrams, drawings, tables), physical (models, measuring instruments), and digital (computers, software, calculators, internet). (p. 591)

None of the three tasks require the use of digital technology to assist in formulating solutions, although calculators may be useful for students who might otherwise not be able to engage with the tasks. In the L-Shaped Area task, the two diagrams provide a tool with which to represent the solutions and could be presented as manipulable screen objects. In Worded Questions, one tool could be a number line that is used to represent the actions in solving the task. In the Shopping task, a tool could be the ways that students choose to represent their solutions.

The Goos et al. (2013) model also includes the context, which can range from "personal, citizenship-related, work-related" on one hand to contexts related to the curriculum. In other words, the purpose of posing a particular task might be to explicate ways that the world is mathematized or the purpose might be to address aspects of the intended curriculum. The L-Shaped Area task and Worded Questions are situated at the "curriculum" end of this range, whereas the Shopping task is more at the "personal" end.

A further aspect is a critical orientation to numeracy which Goos et al. describe as appropriate and inappropriate uses of mathematical thinking. A critical orientation is most evident in the Shopping task. Given that arguable solutions range from $90 to $135 (ignoring the solution in which the discount is not shared) in the first situation, the considerations that might inform choices about what the shoppers might pay come to the fore. Indeed, not only explaining the respective solutions but also listening to the explanations of others is connected to developing such a critical orientation. This might include consideration of friendships and the perspective on fairness.

Although it could be argued that tasks designed for supporting the learning of mathematics are more likely to achieve their goal if the mathematics is both important and explicit, this section illustrates that there is tension in resolving the balance between purely mathematical goals and those which are more social or personal or related to illustrating the usefulness of mathematics in making everyday decisions.

3.6 Task Design Processes

Chapter 2 presents an overview of frameworks and principles for task design, analyzing them on different frame levels: grand, intermediate, and domain-specific frames. Clear throughout this discussion is the inextricable relationship between the design of the task and that of the learning environment. These two must be considered simultaneously.

Ron, Zaslavsky, and Zodik (2013) described a three-stage, backward-design process that includes:

- Stating goal(s) and connecting the task to the goal(s)
- Designing a generic task that addresses these goals; and then (when applicable)
- Carefully choosing the specific examples to "plug in" the generic task (p. 641)

One critical aspect of this design process is to provide well-thought-out starting points for teachers. Another aspect is to explore the role of tasks in fostering awareness of the role of tools, in this case meaning the mathematical routines or procedures that can be enlisted in the solution of problems. They argued, "From a design perspective, with respect to tools, we would like teachers to be able to design tasks that foster discussion of the merits and limitation of existing as well as new tools" (p. 642). They continued:

> This constitutes a challenge for teachers and teacher-educators, because on the one hand, we often want to point to the limitation of existing tools for a particular purpose, while we need to maintain the usefulness and merits of the existing tools for other purposes (otherwise students will believe that much of what they learn will need to be abandoned in the future). ... Thus, the progression from existing tools/concepts to new tools/concepts should lead to an extended mathematical 'toolbox'. (p. 642)

In their perspective, the focus of the task design was to emphasize to students that they were acquiring new tools for future use, such as maybe learning a formula for area of a rectangle in the L-Shaped Area task or number line use in Worded Questions. An associated perspective was outlined by Chu (2013) who described a process of task design with two components: the task design itself and consideration of a particular context in which students were learning mathematics in a second language (English). The process he outlined started with specific academic and linguistic goals, selection of inputs to tasks, specifying the conditions constraining those inputs, clarifying procedures needed, and predicting outcomes as both products (artifacts produced) and processes (ways of engaging). He argued that:

> This framework shapes activities built around mathematical practices to scaffold student engagement in interactive tasks that foster their emerging autonomy. ... Results suggest trajectories for teachers' shifting understanding of conceptual, academic, and linguistic goals as they appropriate a pedagogy of promise that fully develops the potential of all (English language learners). (p. 559)

As Chu explained, one aspect is the design of tasks, and another aspect is the design of the instruction that draws upon and connects those tasks. Chu articulated five principles that guide the design of instructional experiences for students: academic rigor, high challenge/high support, quality interactions, language focus, and well-constructed curriculum.

Knott, Olson, Adams, and Ely (2013) describe a process that focuses on adapting suggestions from texts to inform instruction. They suggested that this process, which they refer to as *turning a lesson upside down*, involves both components—design of the task and consideration of the learning environment:

- Selecting a lesson from a text and identifying the key mathematical understanding or idea
- Writing the mathematical idea as a generalization
- Deciding whether the key understandings entail justification
- Finding or designing a task or sequence of tasks that promotes exploration of the key idea
- Writing questions for students that can prompt them to generalize the key idea

It is important to consider further the process of converting textbook examples to classroom tasks. Particularly in the United States, teachers often take text resources and the associated teachers' guides as the curriculum and plan their teaching from there.

For example, Remillard, Herbel-Eisenmann, and Lloyd (2009) described planning processes as "transforming curriculum ideals, captured in the form of mathematical tasks, lesson plans, and pedagogical recommendations into real classroom events" (p. 1). Stein et al. (1996) described the initial phase of such planning as the teacher taking the mathematical task as presented in instructional materials, before the transformation process.

Peled and Suzan (2013) describe a different process of task design. They focus on what they described as "simple" tasks, meaning tasks that are parsimoniously posed in order to support a shift toward a model of teaching based on problem-solving approaches. They argued that such tasks are preferable to complex modeling tasks because of the resistance of teachers to using such tasks initially. Peled and Suzan (2013) suggested that their tasks serve a double purpose: both to create learning opportunities and to serve as a model for future task creation by teachers. In offering examples of their simple tasks, they wrote:

> One of the tasks involved cutting greeting cards ... and the second involved pouring beer from a container into cans. The main and relevant difference between the problems involves the rigidity of the cardboard versus the "flexibility" of liquid. This feature results in different types of "remainders", as the rigid material does not allow remaining scraps to be put together (unlike a situation such as cutting cookies with "flexible" dough). This difference leads to fitting very different mathematical models. (p. 633)

In other words, the process of task design can focus on affective, practical, and/or mathematical aspects, and these foci can be specific or implied. Further, the focus might be on tools and sequences, and this design process influences both the task itself and the learning experiences constructed around the task, including the learning of teachers.

3.6.1 The Role of the Authority and Autonomy of the Teacher in Designing and Implementing Tasks

A further influence on the design and implementation of tasks in classrooms is the role of the teacher either in adapting a task developed by others (as previously indicated) or in designing the task in the first place. Although there is substantial evidence that the implementation of tasks by teachers can subvert the aims of the task's designer, such as by reducing or increasing the demand of the tasks on the students (see, e.g., Desforges & Cockburn, 1987; Prestage & Perks, 2007; Stein et al., 1996; Tzur, 2008), it seems also that involving teachers in consideration of design issues can affect the potential of the tasks. This aspect of task design and implementation is further elaborated on in Chaps. 2, 5, and 6. Here we note two aspects connected directly with task design.

Askew and Canty (2013) collaborated on a task design and teacher learning project. They introduced teachers to broad principles underpinning the tasks and argued that the tasks were:

> ...appropriated in ways that may not have matched with the intentions of the designer, but rather than this being an obstacle it led to rich discussions around the nature of teaching

and learning. ... Rather than trying to construct tasks that are 'teacher-proof' or at least supported by materials that explicate in detail the intended style of implementation, that working with the fuzziness of appropriation can be a strength. (p. 531)

Kullberg, Runesson, and Mårtensson (2013, 2014) describe a project in which teacher adaptations of a task enhanced student learning. Rather than seeking to limit teacher adaptations, they fostered and celebrated them. In their learning study approach, Kullberg et al. (2013) focused on the principles of variation (for more on Variation Theory, see Chap. 2) to structure a lesson focused on division by a number between 0 and 1. In their analysis they concluded:

> The case study is an example of a design project where teachers' reflection on their teaching and the learners' responses can lead to a refinement of the task design ..., but also to a greater accuracy and clarity about what to point out and make discernible to the learners. (p. 617)

Each of these examples recognizes the central role of the teacher and the teacher's knowledge and learning in the (re)design of tasks and their implementation. Rather than fearing that teacher adaptations may limit the potential of the task, as is assumed by some designers, involving teachers as far as possible in the intentions of the designer can enhance the implementation of the task.

Authorship is considered further in Chap. 5. Here we have not said much about teachers creating tasks for their own use, but in Chap. 4 the notion of emergent task design in response to what takes place in lessons is an important related idea.

3.6.2 Problematic Aspects of Converting Tasks from One Culture to Another

An issue about anticipated pedagogical intentions is the adaptation of a task designed for one culture for the use in a different culture. Culture here is taken in both its broad interpretation as being associated with a different geographic location and language and at times in its narrower interpretation to mean the prevalent cultural context of the subject (mathematics) and of the classroom, including the social and sociomathematical norms in place. This section addresses the former of these, while the latter issues are considered in the subsequent section.

There are a number of key connections between task formulation and culture. These include: the cultural specificity of task context, the relationship between cultural considerations and the types of solution prompted, the precision of the available language and its relationship to mathematics concepts, and the compatibility between the cultural background of the teacher and the students, national traditions, and classroom constraints.

It is not a simple task to take curriculum from one language and culture to use in another. If it were merely a matter of translation rather than transformation, one could seamlessly appropriate curricular materials. But much of the cultural context, especially in mathematics, is implicit, nestled within the sequencing of tasks and activities, the choice of context, and the social and mathematical norms of the specific classroom and of the broader society. Further, both across and within cultures, socioeconomic differences can influence the contexts that are meaningful to students. The culturally and contextually dependent relationships among different mathematical concepts are complex. In short, the successful implementation of specific curriculum and/or tasks depends to a large extent on the teacher as a cultural interpreter. It is the teacher's role to understand and interpret the task as it is contextualized in one culture and re-present it as a culturally relevant and appropriate task in her own context. The mathematical goals of a set of tasks or a piece of curriculum may be quite different across cultures.

To illustrate this point, the L-Shaped Area lesson described earlier used Japanese tatami mats to introduce area as covering; these are not square. In a Western context, the teacher may choose to reinterpret the task to involve carpet or tile squares or some other contextually relevant material. Alternatively, the teacher might use the tatami mats with part of the intention being to include that international dimension in the student experience. The Shopping task, as posed, is specific to higher-income groups, but the notion of "two-for-one" discounts are common, and so it can be expected that teachers adapt contexts to suit their students while preserving the essential elements of the tasks.

In the study that generated the Worded Questions example, Bartolini Bussi et al. (2013) compared and contrasted the cultural contexts of a piece of mathematics curriculum and described the differences in approach, context, sequencing, and understanding between Chinese and Italian teaching cultures. They conducted a study in which Italian teachers reinterpreted the mathematics inherent in the task from the Chinese curricular approach to make it culturally accessible for teaching in Italy. Bartolini Bussi et al. (2013) described how they used a complex task from a Chinese textbook that emphasized the connectedness and complementarity of addition and subtraction and transformed it into several separate tasks, some involving addition and some subtraction. In the Italian curriculum, as in many western societies, the mathematical concepts of addition and subtraction are sometimes taught as separate mathematical concepts, each with its own set of rules. In contrast, in the Chinese tradition those operations are seen as inextricably intertwined representations of the additive relationship, as yin and yang and warp and weft, with understanding developing only by considering the whole fabric. Bartolini Bussi et al. (2013) illustrated a fundamental principle of the Chinese curriculum: "one problem, multiple changes", which emphasizes varying conditions and conclusions. This stands in stark contrast to the western approach of sequencing learning from one concept to a single subsequent concept, with limited emphasis on connections. Bartolini Bussi et al. (2013) described how practicing teachers in their project "re-designed it to tailor it to the Italian tradition and to their individual teaching styles and systems of beliefs" (p. 554). They reported:

> Three main changes were introduced: (1) the single task has been transformed into a set of several tasks; (2) classroom work was organized according to a sequence inspired by the theoretical framework of semiotic mediation after a Vygotskian approach …; (a) individual or small group solution of each row of problems followed by the invention of three problems similar to the given ones, to foster the awareness of the problem structure; (b) collective discussion of the findings, with teacher's orchestration. (p. 554)

In other words, rather than presenting the entire set of questions simultaneously, they were presented sequentially, with the variations emerging progressively. These changes are indicative of the expected ways of teaching in the respective countries. This was also evident in the decisions about representations in the Italian context:

> Moreover the solving graphic schemes (at the beginning) were removed and introduced later, after thorough exploration and solution of the problems, as students were not familiar with such schemes. In this way the use of a graphic scheme was acknowledged by students as meaningful and not perceived as an automatic answer to a given task. (p. 554)

In the Chinese classes, the number line was proposed as a prompt to an alternate representation of a solution. In the Italian classes, the line was removed to avoid it predetermining the solution path chosen by the students.

It is important to note that it is the particular features of the task, deemed important and emphasized by the teacher, that are culturally dependent. In one culture, the emphasis might be on *solving the task and getting an answer*. In another culture, the emphasis might be on the *process or processes by which the task is solved*, and in yet another context, it may be *the connections and patterns that are observed over a set of problems* that are emphasized. The classroom work then would focus on the processes by which different solutions were obtained, and much less emphasis would be placed on the answer itself.

Interestingly, the notion of perspectives on teaching and learning and task design being connected to particular cultures and languages is not restricted to the transfer of tasks across national boundaries. For example, in designing tasks for Australian Indigenous students, not only can the familiarity of the context be considered but also mathematical strengths of the students. Indigenous Australian students have well-developed conceptions of location that can be used in the teaching of more formal geometrical concepts. Further, where the composition of classrooms includes a mix of ethnic, racial, language, and socioeconomic student backgrounds, the differences between the experience and orientation of the respective groups are important design and pedagogical considerations.

3.6.3 Classroom Culture and Anticipated Pedagogies

Student practices and expectations in the classroom depend on the establishment of social and sociomathematical norms. The prevailing classroom culture can have a significant impact on anticipated pedagogies. If, for example, a teacher intends that students replicate routines that have been explicitly demonstrated, then a teacher-directed lesson structure supported by classrooms in which students attend to accuracy and detail is important. If, on the other hand, teachers seek to transfer some responsibility for learning to the students, then different processes and ways of communicating are needed. This is in part a function of the classroom culture and processes that are established over time.

In an important meta-analysis of 49 research studies on classroom culture between 1991 and 2011, Rollard (2012) described three significant and relevant findings that inform the connections between task design and pedagogies. Firstly, the meta-analysis found that the middle years of schooling (ages 9–14 years) are critical for connecting classroom goal structures and the formation of student attitudes because it is in these years that parents and teachers become more interested in assessment of success, and there is more overt competition between students. Students in these years may be more reluctant to engage with tasks that they have not been shown how to do and sometimes avoid the perception of trying hard to avoid censure from other students (see Sullivan, Tobias, & McDonough, 2006). Particularly at these levels, establishing a positive classroom culture is a prerequisite to effective use of some types of tasks.

Secondly, Rollard (2012) concluded that classrooms that promote mastery, meaning those that focus on the learning of the content rather than competitive performance, are more likely to foster positive student attitudes to learning. This is similar to the findings of Dweck (2000) who explained that students who seek mastery of content are more willing to make learning decisions for themselves and are less dependent on the affirmation of others. Such students tend to develop a growth mindset approach to learning, believing that hard work pays off. Rollard (2012) suggested that teachers can actively promote a mastery orientation in the students, in part by paying attention to the type of tasks that are posed and by emphasizing the process rather than the answer in the classroom. Dweck suggests that an emphasis be placed on hard work rather than on intelligence (we describe her ideas more fully below). Thirdly, Rollard concluded from the meta-analysis that classrooms in which teachers actively support the learning of all students promote high achievement and effort.

It is interesting to consider the similarities and differences in Rollard's conclusions and other models of classroom culture. For example, Cobb and McClain (1999) argued that students should have opportunities for "personally experienced mathematical problems ... (which) would constitute opportunities for them to learn" (p. 12). They also described the importance of classroom social norms, such as "explaining and justifying solutions, attempting to make sense of explanations given by others ... and questioning alternatives when a conflict of interpretations had become apparent" (p. 10). For Rollard and also Cobb and McClain, classroom culture is not created by establishing rules in advance but through the structure of lessons, the types of tasks that are posed, the ongoing interactions between teachers and students during lessons, and the relationship of the students with the teacher.

In another study, Brown and Coles (2013) explored specific ways in which teachers took into account their established classroom cultures when designing tasks, so as to effectively establish or reinforce a desired classroom culture. Teachers considered the student age group, the classroom culture, their prior knowledge and skills, and the place of the task in the larger curriculum. In turn, the design of the task impacted the classroom culture, the students' knowledge and skills, and often the larger curriculum.

Brown and Coles used the term *relentless consistency* to describe the desired teaching orientation needed for supporting student learning. For example, if the desire is to create a classroom environment where children are comfortable struggling with complex, open-ended problems, then time must be spent in establishing this comfort level. Once children are used to this, their expectation that they work in this way persists. Rather than seeking to design tasks that can be implemented as written, teachers make choices that require them to be "relentlessly consistent" about not telling students what to do. This is best achieved by designing and/or using a very familiar task, freeing up the teacher to attend to the consistency of what she values in the students' work.

Of course, classroom culture is also a characteristic of context and the community culture. Examples of relevant factors include the size of the class groups, the flexibility of the furniture, whether the language of instruction is the first language of the students, the classroom resources, and the processes of selecting students for the class. There are also factors related to the overall national cultural context. For example, there are Japanese technical terms that describe the purpose and enactment of various aspects of lessons (see Chap. 2). Such terminology would no doubt assist teachers in establishing classroom ways of working.

Elaborating on this notion of classroom culture, Brown and Coles (2013) described part of the teachers' role as being to create classrooms in which persistence, consideration of alternatives, and justification of reasoning are the norm. Brown and Coles (2013) also argued that establishing such a classroom culture and routines takes time to foster.

> … in designing and implementing tasks, teachers have, as a base for decision-making, the classroom cultures they have already established with their students. These cultures are developed over time from the first lessons with a new group. (p. 623)

Similarly, Chu (2013) addressed pedagogical features of the specialized learning environment for learners of English on which he focused:

> … an architecture of three moments assists teachers in deconstructing broad goals into connected intermediate objectives that flow together smoothly. (p. 561)

These three "moments" are specific phases of a lesson: preparing the learners, interacting with the concept, and extending understanding.

A further perspective on classroom culture is described by GEMAD (2013). Their approach includes an expectation that students will:

> develop their own cognitive strategies, manage different representations of the mathematical concepts, choose the best solution strategies, argue about their decisions and communicate fluently their thinking processes. (p. 570)

GEMAD described the teacher's role in fostering this classroom culture to include making learning goals explicit, prompting groups of students to create their own solutions and to present these solutions to the class, and focusing on "argumentation and justification" (p. 572). An interesting aspect of the classroom culture that the GEMAD project was seeking is their explicit intention that students be grouped heterogeneously.

Effective implementation of the L-Shaped Area task is a product of a classroom culture which had been established earlier using the Japanese Lesson Study process. Although the Worded Questions task might be representative of a conventional approach to teaching, the task of synthesizing across the tasks requires students to take a meta-view of the set of tasks and to explicitly look for and make connections among different but related procedures—a higher level of cognitive demand. This way of working is not established overnight. Similarly, the requirement in the Shopping task for students to consider the options suggested by others and to articulate their own preferences is a product of the classroom norms that have been already established.

If the designer of the task is not the classroom teacher, there is a need to anticipate what the culture of the classroom might bear, at least to suggest what kind of classroom culture and instructional approach might appropriately support the implementation of the designed task or task sequence. Taking the reverse perspective, the teacher can hypothesize the classroom approaches that might best support the implementation of a task. Indeed this emphasizes that tasks cannot be "teacher-proofed" and teachers must make active decisions on the implementation of tasks. In any case, there is always interaction between the task itself and its classroom realization.

3.7 Considering the Students' Responses in Anticipating the Pedagogies

Common to the three illustrative tasks are expectations that students will create mathematical knowledge by engaging with the task with thoughtful support from the teacher. The starting point is generally a task that is appropriately challenging for those who will engage with it and with the potential to be supportive of various mathematical product and process goals and to positively influence affective dimensions of student engagement with the task.

Although many aspects of pedagogy have been addressed in earlier sections, the following seeks to describe some initiating aspects of pedagogy that have not so far been considered. There are three issues: the motivation of the students, the introduction of a task, and differentiating the task to ensure it is accessible to all students.

3.7.1 Student Motivation

The first issue associated and discussed in this section is motivation; goals might include:

- Students enjoy the mathematics they are learning.
- Students see the usefulness of the mathematics to them.
- Students be able to interpret the world mathematically.
- Students see the connection between mathematics learning and their future study and career options.
- Students know that they can learn mathematics if they persist.

Recognizing that all of these are important, and also acknowledging that different readers will have different preferences, it is arguable that the most critical goal is that students come to know that they can learn mathematics. If they do learn, then there is the potential that such learning becomes a lifelong endeavor and not merely a pathway to some possibly unrelated goal. Whether any of these goals are achieved depends on the implementation of the task and the response of the students. Dweck (2000) argues that finding ways to support students is as much connected to their orientation to learning as it is to cognitive approaches. She categorizes students' orientation to learning in terms of whether they hold either *mastery* goals or *performance* goals.

Students with mastery goals, according to Dweck, seek to "master" the content and self-evaluate in terms of whether they feel they can transfer their knowledge to other situations. They remain focused on mastery especially when challenged. Such students do not see failure as a negative reflection on themselves, and they connect effort with success. In contrast, students with performance goals are interested merely in whether their answers are correct. Such students want to learn but are more comfortable on tasks with which they are familiar. They give up quickly when challenged; they evaluate their achievements based on positive feedback from a teacher.

Another motivational factor is the mathematical intention behind the task. For example, the task designer might intend that students will learn particular mathematical concepts, they might apply the mathematics to a social situation, or the goal might be simply to elicit positive motivation of the students by increasing their interest in the result. All three of the illustrative tasks incorporate a mix of such factors. The L-Shaped Area task offers students experience with concepts which have the potential for future use rather than immediate benefits for learning. Similarly, delayed usefulness can claimed for the Worded Questions task. The Shopping task has a potential immediate utility and only if the teacher is able to elicit an effective discussion about processes of determining fairness would the longer-term utility become evident.

3.7.2 Introducing the Task to the Students

A second issue is considering ways of introducing tasks to students. On one hand, teachers want students to be able to interpret the task demands. On the other hand, it is assumed that teachers will not give so much direction to students that it becomes impossible for them to create their own mathematics through working on the task.

Several studies find teachers who somehow reduce the challenge of tasks. Stein et al. (1996), in a classroom-based study of task implementation, noted a tendency for teachers to reduce the potential demand of tasks. Tzur (2008) argued that teachers modify tasks when planning if they anticipate that students might not engage with the tasks without assistance. Charalambous (2008) argued that the mathematical knowledge of teachers is a factor in determining whether they reduce the mathematical demand of tasks based on their expectations for the students. Another factor that places pressure on teachers is the reluctance of some students to take risks in their learning. Desforges and Cockburn (1987), for example, reported a detailed study of primary classrooms in the United Kingdom and found that students and teachers conspired to reduce the level of risk for the students. Desforges and Cockburn argued that teachers can sometimes avoid the challenge of dealing with students who have given up by reducing the demand of the task rather than reflecting on what might be causing them to give up. Teachers who increase the challenge of tasks have not been so systematically studied, but strategies for doing so have been reported by Knott et al. (2013) (see above), Lee, and Lee and Park (2013) (see Chap. 5), Prestage and Perks (2007).

Of course, many of the decisions on how to introduce tasks are made during the process of the introduction itself. For example, teachers explore what prerequisite language is known by the students and what they understand about the context in which the task is being posed. Again involving teachers more closely in the intentions of the task designer, or the design process itself, may help to inform the task introduction process.

3.7.3 Access to Tasks by All Students

A third pedagogical anticipation is that if a task is appropriately challenging for most students, it can be anticipated that some will find it too difficult and may not engage with the task or rely too heavily on prompts from the teacher. The metaphor of Vygotsky's (1978) Zone of Proximal Development defines the work of the class as going beyond tasks that students can solve independently, so that the students are working on challenges for which they need support. It seems that one approach is for teachers to plan variations to the original task that are more accessible for those students experiencing difficulty or to plan tasks with multiple entry points, providing access for all students.

This notion of planned task variations is a consistent theme in advice to teachers. For example, a working group of teachers identified 34 different strategies they used when intervening while students are working (Association of Teachers of Mathematics (ATM), 1988). The strategies were then grouped under headings that describe the major decisions teachers have to make about interventions, such as whether or not to intervene, why intervention is advisable, how to initiate an intervention, whether to withdraw or proceed with the intervention, how to end an intervention, and so on. The level of detail was fine-grained; for example, there were 14 specific intervention suggestions about supporting students experiencing difficulty, about half of which relate to task differentiation. Although it makes no sense to assume that a teacher can adopt all such strategies successfully, articulating teachers' practices in this way does make them available for others to use.

Christiansen and Walther (1986), in describing the nature of student engagement in their learning, argued that:

> Through various means, actions are envisaged, discussed and developed in a co-operation between the teacher and the students. One of the many aims of the teacher is here to

differentiate according to the different needs for support but to ensure that all learners recognise that these processes of actions are created deliberately and with specific purposes. (p. 261)

It is assumed that such approaches involve teachers inviting students who experience difficulty to work on tasks that are similar to the ones undertaken by other class members, but differentiated in some way to increase the accessibility of the task without reducing conceptual content. The design of these alternate tasks can be undertaken by the original designer or by the teacher, either in anticipation or interactively during the lesson.

It is perhaps in consideration and anticipation of students' responses to tasks that the teacher's role becomes critical. As described in this section, the teacher considers the motivation of the students, the level of prerequisite knowledge to engage with the task, the prevailing classroom mathematical culture, and the extent to which the task can be differentiated to allow all students to engage effectively.

3.8 Summary and Conclusion

This chapter described factors influencing task design and features of task design that inform and are informed by teachers' decisions about mathematical goals and anticipated pedagogies. By analyzing three typical tasks as examples, attention was paid to five dilemmas (context, language, structure, distribution, levels of interaction) and six tensions (epistemic, cognitive, interactional, mediational, affective, ecological). Designers and teachers need to consider these multiple dimensions to address different aspects of the task and pedagogic design, based on their anticipation of classroom implementation and students' learning.

The process of task design could focus on either or both the specialized and practical aspects of mathematics, formal and natural language, and this focus can be specific or implied. We recognized and described the central role of the teacher in design/redesign of tasks and their implementation.

Teachers and designers might be aware of the cultural assumptions of a task in their (re)design process. Especially in the implementation in the classroom, students' understanding and activities are influenced by social and sociomathematical norms, and it is necessary to consider what the culture of the classroom might bear and at least suggest what kind of classroom culture and instructional approach might appropriately support the implementation of the designed task or task sequence.

Designers may seek to either limit the decision-making of teachers or augment it, either as part of the design process or by direct collaboration. Teachers in turn anticipate the pedagogies through the creation of compatible classroom cultures and consideration of hypothetical learning trajectories. Both designers and teachers may consider affective issues of task design, including the motivational responses of students and the need to maximize the engagement of all students.

References

Askew, M., & Canty, L. (2013).Teachers and researchers collaborating to develop teaching through problem solving in primary mathematics. In C. Margolinas (Ed.), *Task design in mathematics education* (Proceedings of the International Commission on Mathematical Instruction Study 22, pp. 531–540), Oxford, UK. Available from http://hal.archives-ouvertes.fr/hal-00834054

Association of Teachers of Mathematics (ATM). (1988). *Reflections on teacher intervention.* Derby: ATM.

Barbosa, J. C., & de Oliveira, A. M. (2013). Collaborative groups and their conflicts in designing tasks. In C. Margolinas (Ed.), *Task design in mathematics education* (Proceedings of the International Commission on Mathematical Instruction Study 22, pp. 541–548), Oxford, UK. Available from http://hal.archives-ouvertes.fr/hal-00834054

Bartolini Bussi Maria, G., Canalini, R., & Ferri, F. (2011). Towards cultural analysis of content: problems with variation in primary school. In J. Novotna & H. Moraova (Eds.), *Proceedings of SEMT'11, International Symposium Elementary Maths Teaching: The mathematical knowledge needed for teaching in Elementary School* (pp. 9–20). Prague: Faculty of Education, Charles University.

Bartolini Bussi, M., Sun, X., & Ramploud, A. (2013). A dialogue between cultures about task design for primary school. In C. Margolinas (Ed.), *Task design in mathematics education* (Proceedings of the International Commission on Mathematical Instruction Study 22, pp. 549–558), Oxford, UK. Available from http://hal.archives-ouvertes.fr/hal-00834054

Brown, L., & Coles, A. (2013).On doing the same problem – first lessons and relentless consistency. In C. Margolinas (Ed.), *Task design in mathematics education* (Proceedings of the International Commission on Mathematical Instruction Study 22, pp. 617–626), Oxford, UK. Available from http://hal.archives-ouvertes.fr/hal-00834054

Carpenter, T. P., Fennema, E., Peterson, P. L., Chiang, C.-P., & Loef, M. (1989). Using knowledge of children's mathematics thinking in classroom teaching: An experimental study. *American Educational Research Journal, 26*(4), 499–531.

Charalambous, C. Y. (2008). Mathematical knowledge for teaching and the unfolding of tasks in mathematics lessons: Integrating two lines of research. In O. Figueras, J. L. Cortina, S. Alatorre, T. Rojano, & A. Sepulveda (Eds.), *Proceedings of the 32nd Annual Conference of the International Group for the Psychology of Mathematics Education* (Vol. 2, pp. 281–288). Morelia: PME.

Christiansen, B., & Walther, G. (1986). Task and activity. In B. Christiansen, A. G. Howson, & M. Otte (Eds.), *Perspectives on mathematics education* (pp. 243–307). Dordrecht, The Netherlands: Reidel.

Chu, H. (2013). Scaffolding tasks for the professional development of mathematics teachers of English language learners. In C. Margolinas (Ed.) *Task design in mathematics education* (Proceedings of the International Commission on Mathematical Instruction Study 22, pp. 559–568), Oxford, UK. Available from http://hal.archives-ouvertes.fr/hal-00834054

Cobb, P., & McClain, K. (1999). Supporting teachers' learning in social and institutional contexts. In F.-L. Lin (Ed.), *Proceedings of the 1999 International Conference on Mathematics Teacher Education* (pp. 7–77). Taipei: National Taiwan Normal University.

Cooper, B., & Dunne, M. (1998). Anyone for tennis? Social class differences in children's responses to national curriculum mathematics testing. *The Sociological Review*, (Jan), 115–148.

Desforges, C., & Cockburn, A. (1987). *Understanding the mathematics teacher: A study of practice in first schools*. London: The Palmer Press.

Dweck, C. S. (2000). *Self theories: Their role in motivation, personality, and development.* Philadelphia, VA: Psychology Press.

Ernest, P. (1994). Varieties of constructivism: Their metaphors, epistemologies and pedagogical implications. *Hiroshima Journal of Mathematics Education, 2*, 1–14.

Ernest, P. (2010). The social outcomes of learning mathematics: Standard, unintended or visionary? In *Make it count: What research tells us about effective mathematics teaching and learning* (pp. 21–26). Camberwell: Australian Council for Educational Research.

Fernandez, C., & Yoshida, M. (2004). *Lesson study: A Japanese approach to improving mathematics teaching and learning*. London: Routledge.

GEMAD. (2013). An experience of teacher education on task design in Colombia. In C. Margolinas (Ed.), *Task design in mathematics education* (Proceedings of the International Commission on Mathematical Instruction Study 22, pp. 569–578), Oxford, UK. Available from http://hal.archives-ouvertes.fr/hal-00834054

Giménez, J., Font, V., & Vanegas, Y. (2013). Designing professional tasks for didactical analysis as a research process. In C. Margolinas (Ed.), *Task design in mathematics education* (Proceedings of the International Commission on Mathematical Instruction Study 22, pp. 579–588), Oxford, UK. Available from http://hal.archives-ouvertes.fr/hal-00834054

Goos, M., Geiger, V., & Dole, S. (2010). Auditing the numeracy demands of the middle years curriculum. In L. Sparrow, B. Kissane, & C. Hurst (Eds.), *Shaping the future of mathematics education* (Proceedings of the 33rd Annual Conference of the Mathematics Education Research Group of Australasia, pp. 210–217). Fremantle: MERGA.

Goos, M., Geiger, V., & Dole, S. (2013).Designing rich numeracy tasks. In C. Margolinas (Ed.), *Task design in mathematics education* (Proceedings of the International Commission on Mathematical Instruction Study 22, pp. 589–598), Oxford, UK. Available from http://hal.archives-ouvertes.fr/hal-00834054

Gueudet, G., & Trouche, L. (2011). Communities, documents and professional geneses: Interrelated stories. In G. Gueudet, B. Pepin, & L. Trouche (Eds.), *Mathematics curriculum material and teacher development* (pp. 305–322). New York: Springer.

Hashimoto, Y., & Becker, J. (1999). The open approach to teaching mathematics – creating a culture of mathematics in the classroom: Japan. In L. Sheffield (Ed.), *Developing mathematically promising students* (pp. 101–120). Reston, VA: National Council of Teachers of Mathematics.

Hiebert, J., & Wearne, D. (1997). Instructional tasks, classroom discourse and student learning in second grade arithmetic. *American Educational Research Journal, 30*, 393–425.

Hill, H., Ball, D., & Schilling, S. (2008). Unpacking pedagogical content knowledge: Conceptualising and measuring teachers' topic-specific knowledge of students. *Journal for Research in Mathematics Education, 39*(4), 372–400.

Jaworski, B. (2014). Mathematics education development. Research in teaching Learning in practice. In J. Anderson, M. Cavanagh, & A. Prescott (Eds.), *Curriculum in focus: Research guided practice* (Proceedings of the 37th Annual Conference of the Mathematics Education Research Group of Australasia, pp. 2–23). Sydney: MERGA.

Jaworski, B., Goodchild, S., Eriksen, S., & Daland, E. (2011). Inquiry, mediation and development: Use of tasks in developing mathematics learning and teaching. In O. Zaslavsky & P. Sullivan (Eds.), *Constructing knowledge for teaching secondary mathematics: Tasks to enhance prospective and practicing teacher learning* (pp. 143–160). Norwell, MA: Springer.

Kilpatrick, J., Swafford, J., & Findell, B. (Eds.). (2001). *Adding it up: Helping children learn mathematics*. Washington, D.C.: National Academy Press.

Knott, L., Olson, J., Adams, A., & Ely, R. (2013). Task design: Supporting teachers to independently create rich tasks. In C. Margolinas (Ed.), *Task design in mathematics education* (Proceedings of the International Commission on Mathematical Instruction Study 22, pp. 599–608), Oxford, UK. Available from http://hal.archives-ouvertes.fr/hal-00834054

Kullberg, A., Runesson, U., & Mårtensson, P. (2013). The same task? – Different learning possibilities. In C. Margolinas (Ed.), *Task design in mathematics education* (Proceedings of the International Commission on Mathematical Instruction Study 22, pp. 609–616), Oxford, UK. Available from http://hal.archives-ouvertes.fr/hal-00834054

Kullberg, A., Runesson, U., & Mårtensson, P. (2014). Different possibilities to learn from the same task. *PNA, 8*(4), 139–150.

Lee, K., Lee, E., & Park, M. (2013). Task modification and knowledge utilization by Korean prospective mathematics teachers. In C. Margolinas (Ed.), *Task design in mathematics education* (Proceedings of the International Commission on Mathematical Instruction Study 22, pp. 347–356), Oxford, UK. Available from http://hal.archives-ouvertes.fr/hal-00834054

Lubienski, S. T. (2000). Problem solving as a means toward mathematics for all: An exploratory look through a class lens. *Journal for Research in Mathematics Education, 31*, 454–482.

Middleton, J. A. (1995). A study of intrinsic motivation in the mathematics classroom: A personal construct approach. *Journal for Research in Mathematics Education, 26*(3), 254–279.

Peled, I. (2008). Who is the boss? The roles of mathematics and reality in problem solving. In J. Vincent, R. Pierce, & J. Dowsey (Eds.), *Connected maths* (pp. 274–283). Melbourne: Mathematical Association of Victoria.

Peled, I., & Suzan, A. (2013). Designed to facilitate learning: Simple problems that run deep. In C. Margolinas (Ed.), *Task design in mathematics education* (Proceedings of the International Commission on Mathematical Instruction Study 22, pp. 633–640), Oxford, UK. Available from http://hal.archives-ouvertes.fr/hal-00834054

Prestage, S., & Perks, P. (2007). Developing teacher knowledge using a tool for creating tasks for the classroom. *Journal of Mathematics Teacher Education, 10*(4–6), 381–390.

Remillard, J. T., Herbel-Eisenmann, B. A., & Lloyd, G. M. (Eds.). (2009). *Mathematics teachers at work: Connecting curriculum materials and classroom instruction*. New York: Routledge.

Rollard, R. G. (2012). Synthesizing the evidence on classroom goal structures in middle and secondary schools: A meta analysis and narrative review. *Review of Educational Research, 82*(4), 396–435.

Ron, G., Zaslavsky, O. & Zodik, I. (2013). Engaging teachers in the web of considerations underlying the design of tasks that foster the need for new mathematical concept tools. In C. Margolinas (Ed.), *Task design in mathematics education* (Proceedings of the International Commission on Mathematical Instruction Study 22, pp. 641–647), Oxford, UK. Available from http://hal.archives-ouvertes.fr/hal-00834054

Sawatzki, C., & Sullivan, P. (2015). Situating mathematics in "real life" contexts: Using a financial dilemma to connect students to social and mathematical reasoning. *Mathematics Education Research Journal*.

Simon, M. (1995). Reconstructing mathematics pedagogy from a constructivist perspective. *Journal for Research in Mathematics Education, 26*, 114–145.

Smith, M. S., & Stein, M. K. (2011). *Five practices for orchestrating productive mathematics discussions*. Reston, VA/Thousand Oaks, CA: National Council of Teachers of Mathematics/Corwin Press.

Stein, M. K., Grover, B. W., & Henningsen, M. (1996). Building student capacity for mathematical thinking and reasoning: An analysis of mathematical tasks used in reform classrooms. *American Educational Research Journal, 33*(2), 455–488.

Sullivan, P., Askew, M., Cheeseman, J., Clarke, D., Mornane, A., Roche, A., et al. (2014). Supporting teachers in structuring mathematics lessons involving challenging tasks. *Journal of Mathematics Teacher Education*, April. doi:10.1007/s10857-014-9279-2

Sullivan, P., Mousley, J., & Jorgensen, R. (2009). Tasks and pedagogies that facilitate mathematical problem solving. In B. Kaur (Ed.), *Mathematical problem solving* (pp. 17–42). Association of Mathematics Educators. Singapore/USA/UK: World Scientific Publishing.

Sullivan, P., Tobias, S., & McDonough, A. (2006). Perhaps the decision of some students not to engage in learning mathematics in school is deliberate. *Educational Studies in Mathematics, 62*, 81–99.

Tzur, R. (2008). Profound awareness of the learning paradox. In B. Jaworski & T. Wood (Eds.), *The mathematics teacher educator as a developing professional* (pp. 137–156). Sense: Rotterdam.

Vygotsky, V. (1978). *Mind in society*. Cambridge, MA: Harvard University Press.

Watson, A., & Sullivan, P. (2008). Teachers learning about tasks and lessons. In D. Tirosh & T. Wood (Eds.), *Tools and resources in mathematics teacher education* (pp. 109–135). Rotterdam: Sense Publishers.

Wiliam, D. (1998, July). *Open beginnings and open ends*. Paper distributed at the open-ended questions Discussion Group, International Conference for the Psychology of Mathematics Education, Stellenbosch, South Africa.

Chapter 4
Accounting for Student Perspectives in Task Design

Janet Ainley and Claire Margolinas
with *Berta Barquero, Angelika Bikner-Ahsbahs, James Calleja, David Clarke, Alf Coles, Kimberly Gardner, Joaquim Gimenez, Heather L. Johnson, Pi-Jen Lin, Johan Lithner, Heidi Strømskag Måsøval, Peter Radonich, Annie Savard*
and additional contributions from *Laurinda Brown, Viktor Freiman, Claudine Gervais, Carina Granberg, Thomas Janßen, Bert Jonsson, Yvonne Liljekvist, Carmel Mesiti, Mathias Norqvist, Jan Olsson, Pedro Palhares, Elena Polotskaia, Wen-Huan Tsai, Leonel Vieira, Caroline Yoon*

4.1 Introduction

Mathematical tasks or sequences of tasks are, we may assume, designed to embody mathematical knowledge in ways that are accessible to students and to improve students' mathematical thinking. However, if we look beyond the intentions of those who design and select tasks and focus on the impact of students' perceptions of tasks on their mathematical learning, some important questions are raised. One of the aims of this chapter is to gain insights into students' perspectives about the meanings and purposes of mathematical tasks and to better understand how appropriate task design might help to minimize the gap between teacher intentions and student mathematical activity.

The title of the chapter is deliberately ambiguous; we attempt both to explore *accounts of* how students understand the meaning and purpose of the mathematical activity they undertake and to discuss how task design might *take account of* what we know about these perspectives. In Sect. 4.2 we explore research that indicates ways in which the perceptions of students may differ from the intentions of teachers and task designers and attempt to articulate more clearly the nature of those differences. Such research raises both theoretical and methodological challenges concerning how an observer can appreciate the student's point of view. In Sect. 4.3 we explore ways

---

J. Ainley (✉)
School of Education, University of Leicester, Leicester LE1 7RH, UK
e-mail: janet.ainley@le.ac.uk

C. Margolinas
Laboratoire ACTé, Université Blaise Pascal, Clermont Université, Aubière, France

This chapter has been made open access under a CC BY-NC-ND 4.0 license. For details on rights and licenses please read the Correction https://doi.org/10.1007/978-3-319-09629-2_13

© The Author(s) 2021
A. Watson, M. Ohtani (eds.), *Task Design In Mathematics Education*,
New ICMI Study Series, https://doi.org/10.1007/978-3-319-09629-2_4

in which task design that takes account of students' responses might reduce the discrepancies between the intentions of designers and/or teachers and students' perceptions of their activity and achievements. Finally, in Sect. 4.4 we raise some questions for further research.

4.2 Articulating the Gap Between Teachers' Intentions and Students' Perceptions and Responses

4.2.1 Students' Responses to Word Problems

One area in which research has focused on learners' perspectives is the common practice of setting mathematical tasks within "everyday" contexts. The use of contextualized word problems has a long history and indeed is so deeply embedded in school mathematics that such problems have become stereotypical of the experience of learning, and perhaps more significantly being assessed in, mathematics. Contexts are often used by teachers or designers in order to make learning easier by giving meaning to mathematical ideas and showing their usefulness. However, there is an ambiguity in the use of word problems, as they are also traditionally used in assessment, which suggests they are seen as more challenging than straightforward calculations; the lack of attention sometimes paid to the realism of the contexts chosen suggests that meaning and usefulness are lower priorities than the mathematical content (Ainley, 2012). In this section we try to understand the difficulties that arise when students are dealing with word problems; in Sect. 4.2.2 we return to the question of "meaning".

Research about learners' perceptions of the use of contexts in mathematical tasks has suggested that these can differ considerably from the intentions of those who designed them (Cooper & Dunne, 2000). Although designers may choose contexts to offer real-world models to think with or to illustrate the usefulness of mathematical concepts in real life, pedagogic practice may lead students to adopt "tricks" to bypass the contextual elements (e.g., Gerofsky, 1996; Verschaffel, Greer, & Torbeyns, 2006) or fail to appreciate the extent to which everyday knowledge is intended to be utilized in the mathematical task (Cooper & Dunne, 2000). Thus, despite the intentions of task designers, the use of contexts can serve to distract attention from the mathematics ideas. Tasks or task sequences which draw on real-world contexts, but which do not reflect the purposes for which mathematics is used in the real world, may be perceived by students as evidence of the gap between school mathematics and relevance to their everyday lives.

Although the peculiar conventions of word problems as a genre have been explored (Gerofsky, 1996) and the relative cognitive challenges posed by different styles and formats of problems extensively researched (e.g., Csikos, Szitànyi, & Kelemen, 2012; Patkin & Gazit, 2011), the pedagogic value of such problems is rarely questioned (Ainley, 2012). Much of the research in this field has demonstrated a considerable gap between the intentions of the teacher (or the task designer) and the activity of pupils. Many studies (e.g., Hershkovitz & Nesher, 1999; Verschaffel, 2002) have revealed that the strategies used by pupils (and sometimes encouraged by teachers) to answer word problems successfully involve ignoring the context in which the problem has been set. They involve identifying key features of the problem, particularly the numbers and words such as *altogether* which signal the operation, and moving as quickly as possible to a numerical calculation; the story context of the problem is seen as a distraction. Gerofsky (1996) paraphrases this approach as follows:

> I am to ignore ... any story elements of this problem, use the math we have just learned to transform ... [it]... into correct arithmetic or algebraic form, solve the problem to find one correct answer.... (p. 39)

This problem-solving strategy, using a "translation" of key words into arithmetic operations (Hegarty, Mayer, & Monk, 1995) can be highly effective in terms of obtaining correct answers and achieving good test scores, even though it may bypass the intentions of the teacher by failing to give consideration to either the mathematical structure of the problem or the context.

It is unsurprising that pupils "perceive school word problems as artificial, routine-based tasks which are unrelated to the real world" (Verschaffel et al., 2006, p. 60), particularly given the lack of attention given to realistic content by some writers of word problems (Gerofsky, 1996). Pupils' recognition of this lack of realism is demonstrated vividly in studies which have engaged pupils in creating their own problems, where the examples include inappropriate uses of decimals (for specifying numbers of sweets) or ownership of unrealistic numbers of items, such as household irons (Pimm, 1987).

Verschaffel (2002) argues that this tendency to disregard everyday knowledge arises from pupils' experience of the culture and practice of mathematics classrooms. A more nuanced view is presented in a wide ranging study by Cooper and Dunne (2000) which provides evidence of a difficulty some children, and particularly those from lower socioeconomic groups, have in understanding the implicit rules of the mathematics classroom. They report observations of pupils who approached solving word problems by drawing on aspects of their everyday knowledge in ways which were not intended by the teacher or task designer. For example, when given a problem designed to be solved using simultaneous equations, set in the context of buying drinks and popcorn at the cinema, some pupils used the actual price that they have recently paid for a canned drink rather than using information given in the problem to work it out.

It appears that the use of problems which contextualize mathematics in "real-world" situations may serve to extend rather than to reduce the gap between teachers' intentions and students' responses. We now introduce two concepts which offer a model of the classroom context with the potential to shed light on the reasons for this gap.

4.2.2 The Student's Situation

In this section, we turn to the question of the "meaning" which can be constructed by the students in the situations that they encounter at school. We thus discuss the conditions in which it might be useful to use a material or "real" context. When the teacher is planning a task or managing this task during a lesson, he/she has some specific objectives; he/she knows more or less in advance what the goals are. In contrast, students try to adjust to what they perceive as the teacher's objectives.

In order to better understand what the point of view of the students is, two concepts seem useful within the framework of the Theory of Didactical Situations in mathematics (Brousseau, 1997; Brousseau, Brousseau, & Warfield, 2014). *Didactic contract* points to the implicit interpretations that have been shared between students and teachers about a specific type of task or knowledge (see Chap. 2). *Milieu* refers to what the student is actually dealing with: concrete objects and elementary mathematical objects. The student's situation, and hence their understanding of the task, is based on these two aspects (see Barquero & Bosch, 2015).

One may think that all this is quite straightforward; the teacher sets the contract and the student's milieu, and thus he/she makes his/her intention clear for the students. But it is not so simple! The didactic contract is mostly *implicit* and its features are never really fully decided by a teacher. Pupils become aware of some recurrent aspects relative to mathematical tasks, and these regular features evolve into some kind of implicit rules. The situation described by Cooper and Dunne in relation to contextualized problems might be seen as an example. The implicit contract adopted by the teacher or the task designer is that the real-world context provides a frame within which the mathematical problem is given meaning. It is not important that this context is genuinely realistic in its details (such as the actual prices of items) because it is the mathematics that is really important. Many pupils will derive from this the rule described by Gerofsky (1996) (see Sect. 4.2.1), and indeed this may be reinforced by aspects of pedagogy. However, others may not recognize the implicit didactic contract and focus attention on the context rather than the mathematics.

Often teachers are not aware of these "rules", particularly when they are not mathematically correct. For instance, pupils may think that all equations always have a unique solution, because all the equations they have solved previously did have only one solution (e.g., $ax + b = 0$; $a \neq 0$). Thus, the didactic contract is at the core of implicit understandings between pupils and teachers and also plays a part in some gaps of understanding between teachers and pupils.

Other important differences in points of view may be generated by the *milieu* of the task itself. The teacher, when he/she plans a task, has her teaching goal in view, whereas the student has only the *milieu* and the contract in order to understand what the task really is about. One important point is that the *milieu* is never only material, even when pupils work with objects. For instance, in infant school a teacher may show a picture with three identical toy bears and ask: "how many bears are in the picture?" and a pupil may say "mine is different". This seems not to be relevant for the task, because for the teacher, the bears are not interesting in themselves and only their quantity is interesting. But for this pupil, bears are the important things and the quantity is not relevant since they are all the same! This kind of gap may originate from the fact that "how many" was only related to the contract "when the teacher says *how many* you have to count and say the last number". *How many* does not trigger any feedback from the situation; only the teacher and some of the more advanced pupils know if the answer is correct or not. There is no reason inherent in the task for wanting to know how many bears there are. In contrast, if the question was "put in a basket the exact quantity of caps needed to give one cap to each bear", counting the bears now has a purpose. The interest of the pupil may be on the bears, but even if she is distracted by the context, she might realize that in order to give the exact quantity of caps and make the distribution of the caps she has to do something, to develop a strategy. One of these strategies is to count the bears and count the caps until the last number pronounced is the same in both. In this case, the activity of counting is not only a response to the contract but prompted by the existence of feedback from the milieu. Thus, the design of the task itself creates the possibility for students and teachers to have the same interpretation of the task.

We can now turn back to the question of "meaning", using the example of the bears. Often teachers use contexts which are emotionally important for young pupils because they want to trigger their interest, and sometimes this is wrongly understood as the way to give "meaning". As we mentioned in Sect. 4.2.1, this may lead to serious misunderstanding. But material objects are necessary in order to build tasks which allow students to try their own procedures. Brousseau (1997) describes the *milieu* as the non-intentional part of the situation; objects have no intention, but they have properties which may give feedback if the situation is built to permit this feedback. That is how we analyze the difference between "how many bears?" and "put the exact quantity of caps in order to give one cap to each bear". In the first situation, if the pupil does not know how to count or gives the wrong number, the teacher has to tell her that this is not the right number (or asks other pupils to do so). To give an explanation, the teacher has to count the bears, that is, to do the action in place of the child herself. In the second situation, if the pupil has put the wrong number of caps in her basket, she will realize that this is so when she is allowed to try the caps on. This feedback is given by the *milieu*, it cannot be confused with a moral judgment, as right or wrong are sometimes considered. The role of the teacher is thus completely different: he/she can interact with the student in order to prompt her to try another solution or to explain what she thinks might be the problem. The student can play the same "game" again and learn from this replay.

We now use the notions of contract and milieu to shed light on the development of our main questions.

4.2.3 When Student's Milieu and Teacher's Planned Milieu Are Not the Same: An Example

Different interpretations of the classroom situation by students have been discussed in Sect. 4.2.1 in the context of word problems, where a "real-world" interpretation of the problem by some students may lead to different views of what is required in the same task. In this section, we develop an example of the same phenomenon of misunderstanding which does not originate from a word problem. This analysis (Comiti, Grenier, & Margolinas, 1995) has triggered a lot of French research about similar phenomena.

The question "is −1 the square of a number?" was part of a set of ten preliminary questions during the first lesson of a module on square roots for 15-year-old students. The (very experienced and competent) teacher expected no major difficulty for the students in answering this question.

The path of reasoning by the students, which was anticipated by the teacher, was:

- Try some possible candidates: +1, −1 calculate the square and obtain +1.
- Connect this fact to the known rule: minus by minus is plus.
- Answer: it is not possible, there is no number of which the square is −1.

Thus, the teacher was quite surprised when Michael told her that he had a solution, which was "the negative square". Because she didn't understand fully what Michael had in mind, she asked him to go to the blackboard. Michael wrote thus:

$$-(1)^2 = -1$$

She was even more surprised when a lot of students declared that they agreed with Michael. In her expected view of the whole lesson, unexpected phenomena had occurred and the planned progression did not go smoothly.

Our analysis, which was based on the structure of the *milieu* (Brousseau, 1986, 1990; Margolinas & Steinbring, 1994), offers a way to understand Michael's situation. Like every student, Michael was introduced to whole numbers in close relationship to their written form; "2" and "two" were in this sense *the same thing*, which means that "2" was not distinguished as a particular notation for the concept "two". This is certainly normal for small integers, when you want children to fluently link * *, "two stars," and 2. When decimal and rational numbers were introduced, Michael might have been told that 2, 2.0, and 4/2 represent the same number, but he might have understood that 2 was the result (the "true" result) of 2.0 and 4/2. What we want to stress is that Michael might not have had the occasion to *differentiate the number and its written signs*.

If we keep that in mind, we might understand Michael's answer. We can infer his possible reasoning as follows:

- Write the possible well-formed expressions using the signs: "−", "1", "2", and the brackets "()"; you can obtain $(-1)^2$, $-(1)^2$, and -1^2 (but not, e.g., $(-)1^2$ which is not well formed)
- Calculate the result and see if some expressions are equal to −1. You find two expressions: $-(1)^2$ and -1^2 which are the same if you consider that the brackets in the case of (1) are not useful
- Tell the teacher your answer

The objects of Michael's reasoning are not numbers but signs. The *milieu* Michael interacts with is totally different from the *milieu* which was anticipated by the teacher. Furthermore, this *milieu* is more familiar to other students who agree with him.

During the following part of the lesson, the students resisted the teacher's tentative attempts to reestablish the lesson on the basis she needed. For instance, one of the following questions was "is it possible for two different numbers to have the same square?"; 4 and −4 are proposed but some students considered these numbers as opposites but not different. If you refer to notation and not numbers, this is not so strange, for instance, "lion" and "Lion" are the same words, even if their writing is slightly different, which will be taken into account in certain circumstances (e.g., if the word "lion" is at the beginning of a sentence).

Furthermore, at the end of the lesson, the teacher asked the students to write the following sentences:

- a^2 is the square of a.
- a^2 is the square of −a.

The students contradicted the teacher and asked for some brackets to be added in the last sentence:

- a^2 is the square of (−a).

We interpret the students' response as follows: if you consider "the square" as "the sign ^{2}" and apply this rule in writing, the teacher's second statement should be *−a^2 is the square of −a*, and since they know that a^2 is not equal to $-a^2$, the students want the teacher to add the brackets, giving $(-a)^2$ as the square, which they believe is equal to a^2.

This case shows that it is difficult to define "a task". In fact, if the teacher had phrased her question differently, for instance, if the question had been "is it possible to multiply a number by itself and obtain the result −1?" the absence of any reference to the written signs for a square number might not have triggered the same misunderstanding. For the teacher, both questions are the same mathematical question because she considers numbers as theoretical objects, but for Michael and other students, these questions are not the same.

This case was the first description of a phenomenon which has been named as a *situation's bifurcation* (Margolinas, 2005) and investigated by different French and francophone researchers (e.g., Bloch, 1999; Clivaz, 2012). In this case study, we show that prior knowledge (and not only a lack of prior knowledge), when it is not what the teacher is expecting, can lead to a deep epistemological misunderstanding between student and teacher. But another phenomenon is interesting here, which is related to teacher's knowledge. In fact the task, which was considered very straightforward for the teacher, is ambiguous if you take into account students' knowledge about numbers. In this sense the task is ambiguous, but this could have been a wonderful occasion for the teacher to explain to the students what a square is (multiplication of a number by itself) and to open a mathematical discussion about writing and numbers. In a sense, this task might be considered as an example of a *missed*

learning opportunity (Bikner-Ahsbahs & Janßen, 2013, pp. 156–157) and a missed *emergent task*. The conditions described for a successful emergent task are that:
The teacher must:

- Have mathematical knowledge that extends the content of the lesson.
- Show interest in the students' learning processes.
- Be open for unusual ways on the part of the students. She must be willing to abstain from the planned course (Bikner-Ahsbahs & Janßen, 2013, p. 160).

In our observation of the "square of minus one" case, the teacher really showed interest in the students' learning processes and was open to deviate from the planned course (in fact the first part of the lesson took much longer than she had expected, and she accepted that), but she did not know that students of this level might confuse number and written signs. Thus, what she lacked was not exactly "mathematical knowledge" but "mathematical knowledge for teaching" (Ball, Hill, & Bass, 2005; Ball, Thames, & Phelps, 2008).

We link this with our previous discussions, because word problems might appear as a particular case of a more general phenomenon, which takes into account students' previous knowledge about the situation. When everyday knowledge is engaged, as it may be in the case of word problems, the risk of different interpretations of the problem is certainly higher, which might explain why this kind of phenomenon has been documented more frequently in "real-life"-based problems. Thus, the student's perspective changes the very definition of task.

4.2.4 Influence of the Didactical Contract on the Definition of Task

What we call a *task* has different possible definitions (see Chap. 2). Speaking solely about *a task* is a reduction of what the actual involvement of student and teacher implies, which is dealing with a situation which comprises a *milieu* and a *contract*. The situation which has been set up according to a particular design is constantly changing during class interaction: "the actions of teacher and student are mutually informing during the performance of a task" (Clarke & Mesiti, 2013, p. 173).

We have seen earlier that the way pupils interpret the *milieu* of the task can engage some pupils in a totally unexpected situation. If we now consider the *contract*, an exercise which seems to be a very straightforward task might instead develop into a completely open task depending on the *contract*. Clarke and Mesiti (*ibid.*, p. 178) offer an example from their observation in a Japanese school, in which "the seemingly simple pair of simultaneous equations $5x+2y=9$ and $5x+3y=1$ engaged the class for a 50-min lesson" (*ibid.*, p. 178). Discussing the lesson, the teacher emphasized prompting students' reflections on "what solving equations is all about" (*ibid.*, p. 178) and not only on the actual result. Clarke and Mesiti argue for a focus on "well-taught" mathematics rather than "good tasks", taking into account the way students may involve themselves in mathematical tasks and the way mathematical tasks are employed in order to maximize students' voice and agency (*ibid*, p. 181).

Moreover, tasks cannot be dissociated from the ways of working. As Coles and Brown (2013) claim: "In any task, as well as learning some mathematics, students are learning about *what learning mathematics is like* in this classroom; for us, the choice to use any task cannot be dissociated from a choice about *ways* of working" (p. 184).

The didactical contract being mostly implicit, it is not mainly shaped by what the teacher says explicitly but by what the teacher regularly encourages in the classroom. The implicit nature of the didactical contract implies that explanations from the teacher are not sufficient to generate a change of contract.

For instance, Johnson (2013) designed a sequence of four tasks for 7th grade pre-algebra students by adapting the well-known bottle problem developed by Swan (Swan, 1985). "Given the context of a bottle filling with liquid being dispensed into the bottle at a constant rate and a picture of a bottle, the bottle problem requires students to sketch a graph of the changing height of the liquid as a function of the changing volume" (Johnson, 2013, p. 212). To adapt this basic task, Johnson first reversed the activity by providing a graph and asking students to sketch the appropriate bottle and later by providing a computer environment which linked dynamic sketches of filling shapes to graphs as in Fig. 4.1.

The dynamic sketches were intended to foster students' consideration of relationships between covarying quantities. "The task sequence is designed to support students' progression in using nonnumerical quantitative reasoning to coordinate covarying quantities" (*ibid.*, p. 213).

However, the importance of calculation during a mathematics lesson, acquired during previous schooling, might have been too strong for some students:

> The responses of two students, Navarro and Myra (who participated in different interview pairs), provide insight into the kind of difficulty students might have. When Navarro and Myra were presented with the filling triangle task, both of them attempted to determine

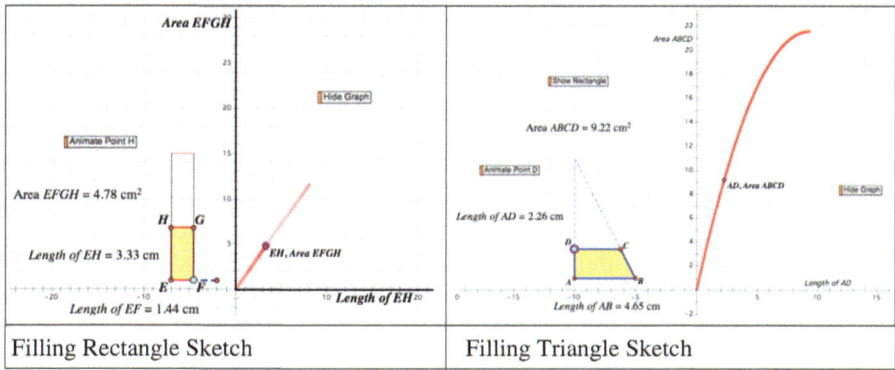

Fig. 4.1 Different sketches used by Johnson (Johnson, 2013, p. 213). (**a**) Filling *rectangle sketch*. (**b**) Filling *triangle sketch*

amounts of area. Even after prompting to not worry about making calculations, Navarro's persistence in trying to calculate amounts of area made it seem as if he depended on calculating amounts of areas to make such predictions. Unlike Navarro, after my prompt to not worry about how to calculate the area, Myra smiled and exclaimed "Oh, I get you now!" When I asked her to explain, she said "the area is getting bigger, but how much it increases is getting smaller." By no longer attempting to determine amounts of area, Myra was able to describe variation in how the area was increasing. (*ibid.*, pp. 217–218)

The prompting to not worry about calculation proved sufficient to trigger another kind of reasoning from Myra, but was not sufficient to trigger the same response from Navarro.

More generally, given that the epistemological choices which have been made throughout the mathematics curriculum are generally built on deductive theory, it may be difficult, not only for students but also for teachers, to interpret a way of dealing with mathematics based on the modeling aspect of mathematics (Job & Schneider, 2013). This is particularly problematic when it is not possible for students to understand the deductive aspects of some mathematical knowledge at the beginning of their studies, which is the case in particular for calculus in high school.

These reflections lead to the consideration of extreme complexity in what a task is when we consider how the student might understand the task within a didactical contract. It also challenges researchers' analysis when their observations are too rapid; if they are not aware of the prevalent contract in the class, they might not understand fully the task-student-teacher interactions.

4.2.5 Student's Perception of the Meaning and Purpose of the Task

The way in which a student perceives the meaning and purpose of a task will have an impact on the aspects of the task he/she focuses on and the activity he/she undertakes in response to it. As has already been discussed in previous sections, the student's perception of the purpose of the task may be rather different from that of the teacher or the task designer. In the example in Sect. 4.2.2, although the teacher had designed a task about counting, the child saw the purpose of the task as talking about bears. We also see differences in the ways in which the meaning of a task may be interpreted by students and by teachers. Cooper and Dunne (2000) report students misinterpreting the role of everyday context within tasks, and thus failing to see the purpose as being about mathematical content. Further, given the same task, in the same classroom, within the same didactical contract, students may have different interpretations of the meaning of the task depending on their previous knowledge and experience, as illustrated in the case study in Sect. 4.2.3, and what they think is important for them when attending a mathematics course.

Accessing students' perceptions poses considerable methodological challenges. Gardner (2013, p. 194) uses a phenomenological approach to categorizing students'

Table 4.1 Outcome space for conceptions of statistics (Gardner, 2013, pp. 194–195)

Conception 1: Statistics as facts or algorithms	
Definition	Statistics is a class in which one states terms, evaluates expressions and formulas, solves equations, and makes and describes graphs
Approach	Write and study examples or facts the teacher presents, memorize formulas and procedures, manipulate a calculator, solve problems the way they are done in class
Capabilities	Do well on a statistics test, remember formulas and facts after a long period of time
Conception 2: Concepts about and procedures for handling data	
Definition	Statistics is the study of contextualized techniques for collecting, representing, and analyzing data
Approach	Write or state a contextual interpretation of graphs and numerical summaries, execute procedures with and without technology, relate personal experience and knowledge to statistical concepts, determine the appropriate statistical method for a given scenario
Capabilities	Explain or teach statistics to another person, read and understand statistics in media, use technology, know when it is appropriate to use a particular procedure or method
Conception 3: Summarize, estimate, infer, and predict	
Definition	Statistics is the study of processes used to estimate population attributes and to generalize or predict trends
Approach	Use multiple approaches; utilize technology to differentiate or discover trends; recognize when data need to be collected; explain assumptions, procedures, and results to others; assess the reliability of results; provide support for conclusions drawn or estimates made
Capabilities	Write or present a detailed analysis of an inference, estimate, or prediction that includes an assessment of assumptions, interpret statistical output from software, appreciate the practicality of statistics
Conception 4: Adapting, restructuring, changing viewpoint	
Definition	Statistics is a way to acquire knowledge about a population and illuminate trends to improve the quality of life, inform decisions, and change one's outlook of the world. It also comes with the responsibility to use and monitor ethical practices
Approach	Adapt to the variable nature of statistics, question the ethical treatment of subjects in studies involving humans or animals, employ the highest ethical standards and design principles, disseminate results of studies to illuminate attributes and inform decisions
Capabilities	Devise a plan of action to change policies or perceptions based on reliable study results, redefine one's understanding of statistics as new processes are learned, formulate theories, restructure one's view of the world

written responses, which she argues communicate their perceptions of the task's purpose. This approach is based on the determination of different possible conceptions about the chosen subject, in this case statistics (Table 4.1):

> Data were collected from one section of a graduate course in data analysis and probability for preservice and inservice teachers. The task in Fig. 4.2 is an item from the course midsemester examination. The item assessed the student's performance level on analyzing and reporting summarized data. (Gardner, 2013, p. 196)

27. Class data were collected on scores students made when playing the 1996 Bop It game. The statistical summaries are below. Write a short report summarizing the results.

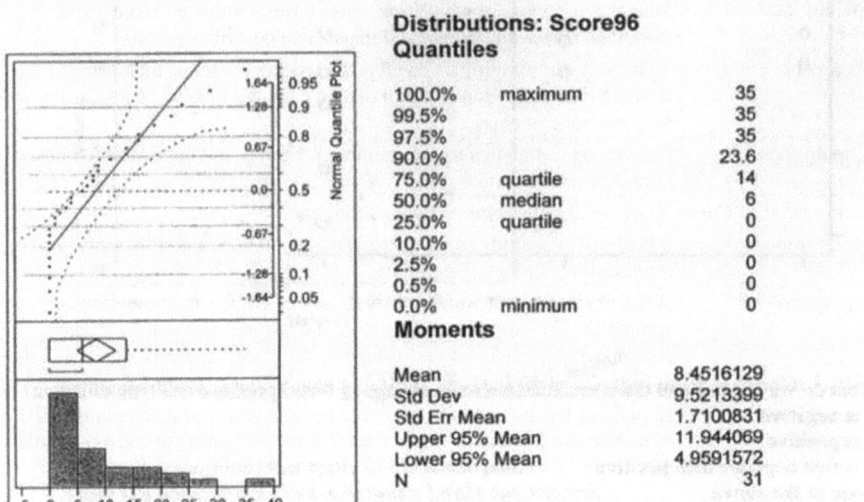

Fig. 4.2 Assessment task on descriptive statistics (Gardner, 2013, p. 196. Copyright Gardner 2007)

Gardner interprets the responses of three students as communicating their interpretation of the purpose of the task.

For one student, Anna, the purpose of the task might be described as "to determine whether she can recall facts about statistical summaries" (Gardner, 2013, p. 195). For another student, Byron, it might be described as "a means for him to communicate his understanding of concepts about data" (Gardner, 2013, p. 196). For a third one, Charles, the purpose was "a means for him to demonstrate his ability to summarize data and support the conclusion drawn" (Gardner, 2013, p. 197). Thus, in response to the same task, students not only answer differently but may be engaging in quite different kinds of activity.

4.3 Taking Account of Student Perspectives: How Task Design Might Reduce the Gap

In this section, we turn to the second interpretation of our chapter title and consider approaches to task design which might take account of the issues discussed in the previous section, concerning the gap which may arise between teachers' (and task designers') intentions and the responses of students. A theme which threads through this section is the challenge in moving beyond the idea that successful completion of a task is the end point and a valid proxy for mathematical learning. A number of the studies we refer to draw on the work of Brousseau (1997) who described a pattern of interactions in which pupils draw on indications from the teacher's behavior, and other aspects of the situation, to find out what is required to complete a given task, and the teacher accepts this as evidence of learning. For example, Strømskag Måsøval (2013, 2015) describes a teacher posing easier and easier questions to students who were struggling to understand the requirements of a task, leading to completely different knowledge. This has been called the *Topaze Effect* by Brousseau (Brousseau, 1982) and *funneling* by Bauersfeld (1995) and reported widely (e.g., Mason, 2002). In the process of funneling questions toward simple answers, it is not always clear whether it is the teacher's desire for an answer or the students' lack of response that leads the process. Some authors (e.g., Wood, 1998) contrast this with *focusing* which is a deliberate interactive act by the teacher.

4.3.1 Students' Expectations

As we discussed in Sect. 4.2, a gap may occur between the intentions of the teacher and the perceptions of the students when the milieu can be interpreted in several ways and/or the didactic contract is disturbed, that is, when students have expectations which are not aligned to the teacher's expectations or intentions. In Sect. 4.2.3, we discussed an example in which students' previous learning led them to overgeneralize in a way that the teacher had not anticipated. Deciding when it is, or is not, appropriate to generalize a mathematical idea can be challenging for students, and this may be a source of disconnection between teachers' and students' expectations.

Rote learning and algorithmic reasoning are very common at the core of the didactic contract, leading students to expect that this is the response that will be required. This poses a challenge for teachers who want to focus on developing creative reasoning.

Lithner et al. (2013) report a study concerned with introducing creative mathematical reasoning (see also Jonsson, Norqvist, Liljekvist, & Lithner, 2014). Despite carefully designed tasks, the students' expectation was that an algorithm would be provided. They also describe a typical response to the frustrations expressed by students in this situation in which "the teacher lets the teaching act collapse" (*ibid.*, p. 225), taking back the responsibility for the students' work by simplifying aspects of the task.

A similar phenomenon is reported by Calleja (2013), in an action research study which also aimed to move students toward more open problem solving. Calleja describes an interview with a student who frequently asked for help. The student commented that Calleja (as her teacher) responded in ways that made it easier for her to complete the tasks. Calleja reflects that he "evidently avoided the student's frustration and speeded up the task completion" (*ibid.*, p. 169), but at the expense of the deeper understanding, the tasks had been designed to support. Similar responses are reported in other studies which address student perspectives in relation to task design.

In selecting or designing tasks, it can be challenging for teachers to take account of students' expectations which differ from their own, and have the potential to present real obstacles to learning. Strømskag Måsøval (2013, p. 235) describes a rather confused conversation between a teacher and a group of students about a particular diagram in a numerical investigation designed by the teacher.

	(a) If this shape were part of a sequence of shapes, what would the next one look like?
	(b) What kinds of figurate numbers do you find in the bright and the dark areas and in the shape as a whole?
	(c) Express what the shape tells you about these numbers in terms of a mathematical statement.

Paul [is] insecure about what the teacher asks for; he wonders whether it is only the first element or it is the sequence of elements they are supposed to consider:

> 598. Paul: If we are supposed to see the connection, it is only this very shape we shall look at now? [Draws a curve with his pencil around the element given in the task.] It is not the next shapes we have made [points at the succeeding elements drawn in his notebook when he says "next"]?
> 599. Teacher: You may well look at it as it stands there [Pause 1–3 s] uh [Pause 1–3 s] [indecipherable]
> 600. Paul: Not further, ok.

The teacher's response in turn 599 I interpret as confirming that it is satisfactory that the students look at the element given in the task (a 5×5 square) as a basis for finding answers to Tasks 4b and 4c. It is plausible that the teacher takes this stance as a consequence of seeing the 5×5 square as a generic example. […]

These general properties are however not addressed in the classroom situation. The teacher does not express to the students that he uses the 5×5 square in the sense of a generic example, nor does he use the term "generic". What I interpret as the teacher's implicit utilization of a generic example contributes to vagueness in the discourse: The stance taken by the teacher about the sufficiency of looking at one element of the shape pattern (genericity of the 5×5 square) is consistent with the formulation [of the following tasks], a correspondence which may be expected since the task is designed by the same teacher. Application of *singular* number in the noun "the shape" indicates that the shape presented in the task is seen as generic:

> What kinds of figurate numbers do you find in the bright and the dark areas, and in *the shape* as a whole? [Task 4b, emphasis added] Express what *the shape* tells you about these numbers in terms of a mathematical statement. [Task 4c, emphasis added]
> (Strømskag Måsøval (2013, p. 235)

The teacher's intention, which was interpreted by Strømskag Måsøval as to offer a single diagram as a generic example, was not apparent to the students. This leads to the students being unaware of the teacher's aim for the task. Mason and Pimm (1984) describe this process as an inherent implicit feature of a didactical situation; if the teacher offers an example, the learner is supposed to appreciate the general, and if the teacher offers a generality, the learner is supposed to be able to apply it to examples. Strømskag Måsøval exemplifies this phenomenon, suggesting that the teacher's focus on the mathematical content of the task, and his familiarity with the subject matter, led him to pay insufficient attention to the wording of the task, which also required students to "express what the shape tells you about these numbers in terms of a mathematical statement" (Strømskag Måsøval, 2013, p. 233), without any clear explanation about what "mathematical statement" may mean. Acknowledging the difficulties which may be presented when students' expectations lead them to look for algorithmic solutions rather than engaging in more open problem solving, a number of studies have developed task design approaches in deliberate contrast to traditional formats. Savard, Polotskaia, Frieman, and Gervais (2013) set out design principles for tasks which aim to promote holistic reasoning about mathematical structures. In particular, they state that "The task should not contain any explicit and immediate questions that could be answered by finding one particular number […] However the task should include an intriguing element" (*ibid.*, p. 272).

The example they present, designed for young children, shares many of the features of a traditional word problem, consisting of simple statements about the numbers of marbles three boys say that they own. However, the statements offered are incompatible, and the challenge of the task is to provide an argument for which statement is incorrect:

> This is an example of a text proposed to students.
> *Peter, Gabriel and Daniel are playing marbles. Peter says, "I have 5 marbles." Gabriel says, "I have 8 marbles." Daniel says, "Peter has 4 marbles less than Gabriel."*
> We introduce this text as a strange situation or as a situation where one of the persons made a mistake. Students are invited to explain why the text is unrealistic and how it can be corrected considering different quantities involved. (*ibid.*, p. 273)

This situation is thus not one in which the students have to act in order to get the right feedback from the milieu, but a validation situation (Brousseau, 1997; Brousseau et al., 2014) where the students exchange arguments which can be tested through the milieu (in this case through the manipulation of marbles). Reviewing the outcomes of the study, Savard et al. (2013) comment on the progress made by the children in solving problems which required holistic analysis and also on the challenges for teachers in resisting the pressure to revert to more traditional formats with a single "answer".

One response to the problem of mismatch between students' expectations and the teacher's intentions may be for the teacher to state more explicitly the kinds of behaviors that are expected in response to the task. But there is a tension here: if the teacher's aim is for students to develop creative, flexible, and independent mathematical thinking, specifying particular desirable behaviors may be counterproductive.

Coles and Brown (2013) capture this tension by saying that "the more the desired behaviors in students are specified, the less these behaviours are likely to emanate from the students' own awareness" (p. 184). Drawing on theoretical perspectives

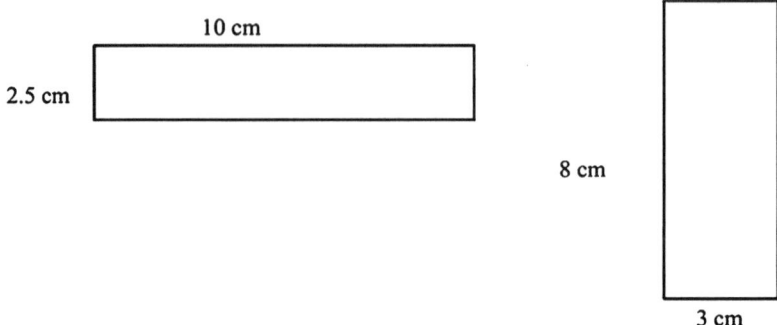

Fig. 4.3 Two contrasting examples (Coles & Brown, 2013, p. 187)

from enactivism and the development of mathematical thinking, Coles and Brown describe and exemplify design principles which underpin an approach to helping students develop patterns of thinking over an extended period:

> These shapes [See Fig. 4.3] are 'two contrasting examples' […]. With this image on the board, the teacher asks students, 'what is the same and what is different' […]. (*ibid.*, p.187)

Coles and Brown's description encompasses both the reflective task design, which teachers and task designers undertake in planning for teaching, and emergent design which takes place "in the moment" in the classroom, in response to the activity of students. Research addressing student perspectives in task design focuses on both of these forms of design, which we discuss in the next two sections.

4.3.2 Reflective Task Design

In this section, we discuss research studies focusing on ways to take account of student perspectives in task design which takes place in a *reflective* space away from the classroom (see also Chap. 9). In these studies, the design may be led by researchers, task designers, teacher educators, or teachers and often by a team combining individuals with different or dual roles (e.g., Coles & Brown, 2013; Lin & Tsai, 2013; Radonich & Yoon, 2013). Many of these studies draw explicitly or implicitly on the methodology of design-based research (Design-Based Research Collaborative, 2003), in which each iteration of task design is based on conjectures drawn from the analysis of student responses (Calleja, 2013; Lithner et al., 2013).

The development of hypothetical learning trajectories (Simon, 1995) or thought experiments in which the designer anticipates student learning is an important tool in such design (see Chap. 2 of this volume). Calleja (2013) and Palhares, Vieira, and Gimenez (2013) report on studies in which a priori analysis of the mathematical knowledge is involved, and students' likely responses are used to develop sequences of tasks. Palhares et al., however, conclude that "[c]ognitive analysis seems not to be enough to decide about ordering [of tasks]" (p. 247), in this case of tasks designed to promote algebraic thinking in young students. They found that students who started with a set of sequential tasks seemed to be more capable of establishing distant generalization than a group who started with structural tasks, and to retain their performance more. Therefore, they highlight the need for further cycles of design development in the light of the children's responses in order to improve the design of task sequences rather than single tasks.

A common theme in studies about reflective task design is the use of aspects of design to direct students' attention to certain mathematical features, particularly in situations in which previous research has indicated students' tendencies to respond in other, less productive, ways. A simple change in a mathematical question can make a big difference in the cognitive demand for the students. For instance, Sullivan and Lilburn (2004) argue that you can transform a straightforward and closed question, like "731−256=?", into what they consider a *good question*: "Arrange the digits so that the difference is between 100 and 200" (*ibid.*, p. 4). Thus, the transformed task is now less about the procedure of subtraction and more about estimation. There are (at least) 6 possible answers without moving digits from one number to the other in the initial question. The revised question emphasizes the point that a small change in the question can prompt quite different thinking from the students.

Another example is given by Johnson (2013) who describes the principles she used in the design of a sequence of covariation tasks. In previous studies (Johnson, 2012a, 2012b), she had found that students tended to focus on change in variables as though they were independent, and so she included the use of an animation with prompt questions such as "what changes and what stays the same?" and the requirement for students to predict characteristics of graphs and other representations in order to focus their attention on relationships rather than the results of calculations (Johnson, 2013, p. 214). In Chap. 3 of this volume, further ways in which teachers can adapt cognitive demand are described, particularly focusing on moving learners toward appreciating generalities rather than following procedures (Knott, Olson, Adams, & Ely, 2013).

The use of a nonstandard version of a familiar task is also at the center of Lin and Tsai's (2013) study which aimed to develop conjecturing by primary school students:

> The task designed by the teacher was to ask students to make a conjecture and verify whether it is true. The statement is that "*In any two figures, if the area of one figure is bigger than the other, then the perimeter of the figure is greater than the other, too. Do you agree? Why? Show your work on the grid paper.*" The task for conjecturing is initiated from a false statement. (*ibid.*, p. 252)

Here, the use of a false conjecture about the relationship between the area and perimeter of a pair of figures was used to focus students' attention on the ways in which conjectures can be explored and tested rather than on the calculation of specific values.

Coles and Brown (2013) highlight the importance of focusing students' attention on making distinctions and include the use of two contrasting examples, with prompt questions similar to those used by Johnson, and the introduction of appropriate mathematical language and notation to record the distinctions identified by students, within their list of task design principles. The starting point for one of the examples they offer, also concerning area and perimeter, is shown earlier in Fig. 4.3. Pupils are asked to consider what is the same and what is different about the rectangles. When attention is focused on the area of the rectangles, pupils notice that in one rectangle the values of the area and the perimeter are the same. This is designated as an *equable* shape, providing pupils with a new distinction with which to work in their exploration. The rationale for this approach is not based specifically on evidence of difficulties arising in the study of area and perimeter but rather on the *enactivist* approach which equates learning with the ability to act differently based on changing perceptions of distinctions in a particular sphere of action (Maturana & Varela, 1987). Coles and Brown offer the metaphor of a wine taster to illustrate this view. As the wine taster's palate develops to make finer distinctions, he/she is able to act differently on the basis of this perception.

Potential negative consequences of the unintentional direction of students' attention are identified by Calleja (2013) in his discussion of one iteration of a task sequence, designed on the basis of hypothetical learning trajectories. He notes that the activity of some students appears to have been influenced by the title assigned to a task (*Investigate Pythagoras' Theorem*), which may have focused their attention on a limited range of mathematical content. This leads him to speculate that a more open title (*Investigate Right-Angled Triangles*) may encourage a more open approach, and, incidentally, avoid discouraging those who are as yet unfamiliar with Pythagoras.

A somewhat different perspective on the direction of students' attention through task design is at the heart of a design framework developed by Ainley and Pratt (Ainley, 2008; Ainley, Pratt, & Hansen, 2006) which also relies on detailed analysis of mathematical content to develop learning trajectories. This analysis focuses on the *utility* of mathematical ideas, that is, how and why the ideas are useful. They argue that understanding utility is a key component of mathematical thinking which is often overlooked. For example, teaching about measures of average may include both the procedures needed to calculate the mean and median and conceptual exploration of the nature of the measures, but fail to address how and why such measures can be used in solving problems. In order to create opportunities to experience the utility of mathematical ideas, and to focus students' attention on the power of using them, Ainley and Pratt design tasks which have a clear and immediate purpose for students within the context of the lesson. This might be designing a product, such as an efficient paper spinner or a computer-based model to generate data, or solving an engaging problem, in which the mathematical idea (i.e., the teacher's intended content for the task) is used in a meaningful way. This design framework offers a way in which the teacher's intentions (including a focus on the utility of the mathematical content) can be aligned with the students' activity, which is driven by the purposeful nature of the task, even though the two elements of purpose and utility remain quite distinct.

4.3.3 Emergent Task Design

An area of research which is complementary to the studies of reflective task design discussed in the previous section concerns the ways in which teachers develop tasks during the flow of classroom activity, in response to the actions of students. Bikner-Ahsbahs and Janßen (2013) use the term *emergent tasks* to refer specifically to situations in which "the teacher conceives the mathematical potential of a learning opportunity and translates it into a task" (p. 154) in such a way that students' interest is maintained. To understand how to build on students' questions in order to create emergent tasks in the moment is challenging for teachers, and in this section we discuss studies which both explore emergent task design and consider how teachers' skills can be developed.

Bikner-Ahsbahs and Janßen (2013) build on previous research into the creation of interest-dense situations in mathematics lessons, that is, situations in which learners are deeply engaged with mathematical questions, developing successively deeper meanings and coming to see the importance of a mathematical object (Bikner-Ahsbahs, 2003). They explore how teachers exploit such situations by aligning emergent tasks to what they perceived as the students' epistemic needs. The challenge for the teacher is, therefore, to understand a mathematical problem students have encountered within the interest-dense situation and to translate it into a task for the class:

> Previous to the scene presented here the students had worked on the question how to divide a round licorice stick evenly among three persons. […]
>
> 149. Rahel: yes Mister Kramer once more a stupid question, how does one GET the central point how did they GET that because that is so small.
> 150. S: that is just
> 151. T: that's another problem right. that's a practical problem (..) oh no, how does one even find the central point in such a small circle right' (.) exactly. those are questions'
> 152. Anji: a very small compass right'
> 153. T: yes one can find out with the compass, only when one is just drawing the circle' one has the central point. but when one has the circle already right'
> 154. Rahel: yes
> 155. L: that's exactly what geometry works with.
> 156. Rahel: I know that
> 157. L: there are possibilities to find out a-n-d you can puzzle at home maybe someone finds a possibility'
>
> […] the teacher probably notices already that the missing central point poses an additional problem and wants to postpone it to an exercise. But the students insist on an immediate clarification by asking "but how" (can we exercise that). Rahel reacts by naming the difficulty in dividing the circle without knowing the central point (149). She grasps the epistemic gap and thus *sees* the mathematical *structure*. Now the teacher summarizes the two problems: How does one even find the central point in such a small circle? Commenting "those are questions" he documents wonder about the deep involvement of the students that he tries to take up. (*ibid.*, p.158)

From observations and analysis of occasions on which teachers are, or are not, able to identify suitable situations and develop emergent tasks, Bikner-Ahsbahs and Janßen identify the three requirements for the teacher to meet this challenge successfully (previous noted in Sect. 4.2.3): sufficient mathematical knowledge to extend the content of the lesson, a genuine interest in students' learning, and a willingness to deviate from the planned lesson to follow unexpected directions in students' activity.

Emergent task design can build flexibly and effectively on student responses to support their engagement with mathematical thinking, but makes considerable demands on the teacher's ability to act "in the moment". It is, of course, not clearly separated from the more reflective design discussed in the previous section, but involves adaptation and adjustment of an initial task during the progress of a lesson. This performance of a task (Clarke & Mesiti, 2013) is shaped by interactions between the teacher and students but guided by the teacher's intentions. Presenting evidence from three lessons which form part of a wider international study, Clarke and Mesiti draw attention to the extended time and attention given to relatively straightforward mathematical tasks by teachers in Japan and China who construct their lessons around opportunities for students to discuss and report their reasoning. Clarke and Mesiti do not address the question of how teachers develop these skills, but this issue is made explicit by Coles and Brown (2013) in their discussion of the development of generic design principles within a school. These principles are used in reflective task design undertaken collaboratively among colleagues and, thus, come to inform the more spontaneous actions in the classroom: "Creating opportunities for students to make distinctions within mathematics can also become a habit for teachers and a normal way of both planning activity and informing decisions in the classroom" (Coles & Brown, 2013, pp. 191–192). Coles and Brown illustrate this through analysis of an example of emergent design in a lesson about area and perimeter (see Sect. 4.3.1).

In addition to the challenge for teachers, there is a potential threat to the coherence of the curriculum itself within emergent tasks. This difficulty might lead to very different didactical contracts in the class during the completion of these tasks. One contract includes covering curriculum; the knowledge to be learned is therefore determined in advance or at least has to be compatible with the curriculum and the coherent epistemological foundation of the knowledge. However, the mathematical content of emergent tasks, which follow the students' responses and offer an opportunity for students to engage in rich mathematical activity, may be less predictable.

4.3.4 Open Tasks: Voice and Agency

Underpinning much of the research we have discussed in this chapter is the recognition not just of the importance of accounting for student perspectives in task design, in order to reduce the gap between the intentions of teachers and the activity of students, but also of the significant role of student agency and voice in the development of mathematical thinking. In this respect, our approach contrasts strongly with some other perspectives on learning in which the student is positioned as a rather passive recipient of information that is then processed into mathematical meaning. For example, cognitive load theory (e.g., Sweller, 1994) characterizes tasks according to the number of variables that learners have to manipulate in order to be successful and analyzes those that are essential for completion and those that are extraneous. The implication for design could be that only essential variables should be given, to simplify the load. This idea has to undergo considerable expansion to include the development of mathematical thinking, and students' capabilities in mathematizing situations, so that *load* has to include variables that, while not being essential in a mathematical sense, are germane to the situation.

Within the Theory of Didactical Situations, which positions the student as actively mathematizing, agency has been modeled through the concept of *adidactical* situations. This concept derives from the idea that in ordinary life, in *non-didactical* situations (e.g., trying to float in the water in a swimming pool), we acquire some implicit knowledge in the interaction with the milieu. This kind of knowledge is directly useful (you float or not) and meaningful in the situation. The idea of Brousseau was to study a sort of image of these non-didactical situations, which he named *adidactical*, designed to allow the acquisition of predetermined knowledge. Thanks to this design, the student is offered the possibility of trying her own procedure in order to succeed in dealing with the adidactical situation and to encounter the determined knowledge in a situation that becomes meaningful for the student. Therefore, Brousseau's idea is clearly inserted within a contract where the knowledge to be learned is entirely determined in advance and where the a priori analysis, which includes the careful study of the hypothetical learning trajectory, is the basis of the design. The epistemological aspects of this kind of design are crucial; the study of the knowledge and the situation in which this knowledge is useful is at the core of anticipating learning through adidactical situations.

Design approaches that aim to encourage student agency may vary considerably. Clarke and Mesiti (2013) describe lessons which start from relatively closed tasks but which are developed and extended around students' responses. The development of reasoning and argument may be stimulated by presenting students with situations which contain faulty mathematical statements or conjectures (Lin & Tsai, 2013, Savard et al., 2013). The design principles described by Coles and Brown (2013) start from offering a closed task, which is then developed through inviting students to focus on distinctions. Bikner-Ahsbahs and Janßen (2013) base their account of an example of an emergent task on the (apparently simple) initial task of dividing a strip of paper into three equal pieces—an open-ended task intended to initiate discussion about fractions. Other studies have focused on the use of more open, contextualized tasks in order to encourage and value students' independent creative activity.

In a design-based research study, Calleja (2013) developed a design framework which included structured, semi-structured, and unstructured (open) tasks in order to support students to move from the experience of traditional teaching to progressively become familiar with "the social experiences of mathematical inquiry, discussion and communication" (p. 166). This framework acknowledges both the importance of student agency and the challenge that this may present to their expectations.

Radonich and Yoon (2013) utilize a design framework based on model-eliciting activities (Lesh, Hoover, & Kelly, 1993), in which real-world problems are presented to students as a stimulus for mathematical modeling. The work of Lesh and his colleagues has generated many examples of this kind of work, as has the Realistic Mathematics Education tradition (see de Lange, 2015). The example offered by Radonich and Yoon is particularly interesting:

> The problem begins with a comic that tells how the Renaissance artist, Giotto, gained the pope's attention by drawing a perfect freehand circle [...]. After reading the comic, students are asked to draw their own freehand circles and choose the best among them. Next, they watch a short YouTube video of a mathematics teacher who professes to be the world's freehand circle drawing champion and appears to draw a perfect freehand circle.
>
> [...] Students then meet the problem statement, which introduces them to a client, Bonnie, who is holding a circle drawing competition at the local Pancake House. The students are asked to work in teams of three to develop a method for ranking circle attempts from most circular to the least circular, which Bonnie can use to judge the circle drawing attempts on the night of the competition. Students are asked to test their method on some examples of circle attempts [...], but their method must also work for any circle attempt that could be drawn on the night of the competition. The student teams write their final method in the form of a letter to Bonnie. (*ibid.*, p. 261)

Although clearly located in the mathematical topic of circle properties, the openness of the task allows for students to develop and justify a wide range of models, which may draw on different theorems, challenging the expectation that there will be a single correct approach. The main focus of the project that Radonich and Yoon report is, however, not the initial performance of the task, but how the teacher might effectively build on the wide range of activity students have engaged in to ensure some curriculum coherence. Their approach involves presenting back to the class an example of the work from a group of students carefully chosen to present a good, but incomplete, model. The challenge for the class is then to test and improve this model, allowing the teacher to focus on reinforcing particular mathematical content. The use of students' work in this context is a deliberate challenge to students' expectations that teacher-presented ideas are better than their own, but Radonich and Yoon acknowledge the potential sensitivities in doing this, emphasizing that "effort will be required to create a culture where discussing student work is a natural and safe part of the teaching and learning process" (p. 266).

4.4 Conclusion: Some Topics for Future Research

In this chapter, we have attempted to both account for the gap which can open up between the intentions of teachers and task designers and the experiences of students tackling mathematical tasks and to reflect research which attempts to minimize this gap through innovative task design. What is revealed is a complex picture, which we believe is relatively under-researched. Although some evidence is presented within this chapter, there is relatively little research which addresses directly the ways in which students perceive the meaning and purpose of tasks. Most studies that do address this issue do so by inferring students' perceptions from their actions. There is a need for further research which uses alternative methods to understand student perspectives more fully, particularly in the context of innovative task design. We suggest that the areas for further research in aspects of task design identified elsewhere in this volume might also be enhanced by a parallel study of student perspectives.

In our chapter, we have highlighted the importance of the context (in particular the *contract*) in the different impacts that similar tasks might have on students. Word problems have been thoroughly investigated because these kinds of tasks, which are often linked to "real-world" contexts, are more susceptible to alternative interpretations from the students. This has contributed to deep interrogations of the validity of such kinds of tasks. However, the intention of any tasks which include some "real" setting might be wrongly understood, and even tasks which are only mathematical tasks are not immune to misunderstanding. It seems thus very important *to collect more data and to develop analysis which is focused on the effect of the contract on the variability of student's understanding of the tasks* (see Sarrazy, 2002 for an example of such a study). This topic might be very interesting to develop in intercultural research, because the kinds of contracts that develop in different countries might be different and thus reveal the nature of the contract itself.

Research in mathematics education has often been developed without any special *interest in the differential effect of the tasks on students*: who is benefiting from a certain kind of task and who is not? Qualitative methodology has been reported in our chapter, which helps us to understand that students might interpret tasks very differently. However, it would be useful if this question were to be addressed in quantitative methodology-based research. The statistical analysis needed to compare the progress of different students has to be specific and some has already been developed in some publications (e.g., Sarrazy & Chopin, 2010).

More generally, particularly helpful might be research that addresses the question: *are some tasks more robust than others?* The *robustness* of tasks might be intended as *resistant to changes from the teacher* but also *understandable and useful for all the students*.

Of course, the student's perspective, which is the purpose of this chapter, cannot be detached from the teacher's ability to develop and implement "good tasks" or "good teaching". We suggest that *emergent tasks* (Bikner-Ahsbahs & Janßen, 2013) *might be developed into a general concept*. In fact, interest-dense tasks might emerge from a lot of tasks; it is not yet clear if this is mostly due to the intrinsic quality of the initial task or due to efficient mathematical knowledge for teaching or both. The ability of teachers to observe students' procedures and to develop the task accordingly might need special attention.

References

Ainley, J. (2008). Task design based on *purpose* and *utility*. *ICMI 11*. Available from http://tsg.icme11.org/document/get/291

Ainley, J. (2012). Developing purposeful mathematical thinking: A curious tale of apple trees. *PNA, 6*(3), 85–103. Available from http://www.pna.es/Numeros2/pdf/Ainley2012PNA6%283%29Developing.pdf

Ainley, J., Pratt, D., & Hansen, A. (2006). Connecting engagement and focus in pedagogic task design. *British Educational Research Journal, 32*(1), 23–38.

Ball, D. L., Hill, H. C., & Bass, H. (2005). Knowing mathematics for teaching. Who knows mathematics well enough to teach third grade, and how can we decide? *American Educator, 2005*, 14–46.

Ball, D. L., Thames, M. H., & Phelps, G. (2008). Content knowledge for teaching: What makes it special? *Journal of Teacher Education, 59*(5), 389–407.

Bauersfeld, H. (1995). "Language Games" in the mathematics classroom: Their function and their effects. In P. Cobb & H. Bauersfeld (Eds.), *The emergence of mathematical meaning: Interaction in classroom cultures* (pp. 271–291). Hillsdale, NJ: Lawrence Erlbaum Associates.

Barquero, B., & Bosch, M. (2015). Didactic engineering as a research methodology: From fundamental situations to study and research paths. In A. Watson & M. Ohtani (Eds.), *Task design in mathematics education: An ICMI Study 22*. Heidelberg: Springer.

Bikner-Ahsbahs, A. (2003). Social extension of a psychological interest theory. In N. A. Pateman, B. J. Dougherty, & J. T. Zilliox (Eds.), *Proceedings of the 27th Conference of the International Group for the Psychology of Mathematics Education and PMENA* (Vol. 2, pp. 97–104). Honolulu, Hawaii: PME.

Bikner-Ahsbahs, A.-A., & Janßen, T. (2013). Emergent tasks—spontaneous design supporting in-depth learning. In C. Margolinas (Ed.), *Task design in mathematics education*. (Proceedings of the International Commission on Mathematical Instruction Study 22, pp. 153–162), Oxford, UK. Available from http://hal.archives-ouvertes.fr/hal-00834054

Bloch, I. (1999). L'articulation du travail mathématique du professeur et de l'élève dans l'enseignement de l'analyse en Première scientifique. *Recherches en Didactique des Mathématiques, 19*(2), 135–193.

Brousseau, G. (1982). *Les "effets" du contrat didactique*. Deuxième école d'été de didactique des mathématiques, Olivet. Available from http://guy-brousseau.com/2315/les-%C2%AB-effets-%C2%BB-du-%C2%AB-contrat-didactique-%C2%BB-1982/

Brousseau, G. (1986). La relation didactique: le milieu *4e école d'été de didactique des mathématiques* (pp. 54–68): IREM de Paris 7. Available from http://math.unipa.it/~grim/brousseau_03_milieu.pdf

Brousseau, G. (1990). Le contrat didactique: le milieu. *Recherches en Didactique des Mathématiques, 9*(3), 309–336.

Brousseau, G. (1997). *Theory of didactical situations in mathematics*. Dordrecht: Kluwer Academic.

Brousseau, G., Brousseau, N., & Warfield, G. (2014). *Teaching fractions through situations: A fundamental experiment*. Dordrecht/Heidelberg/New York/London: Springer.

Calleja, J. (2013). Mathematical investigations: The impact of students' enacted activity on design, development, evaluation and implementation In C. Margolinas (Ed.), *Task design in mathematics education* (Proceedings of the International Commission on Mathematical Instruction Study 22, pp. 163–172), Oxford, UK. Available from http://hal.archives-ouvertes.fr/hal-00834054

Clarke, D., & Mesiti, C. (2013). Writing the student into the task: Agency and voice. In C. Margolinas (Ed.), *Task design in mathematics education* (Proceedings of the International Commission on Mathematical Instruction Study 22, pp. 173–182), Oxford, UK. Available from http://hal.archives-ouvertes.fr/hal-00834054

Clivaz, S. (2012). Connaissances didactiques de l'enseignant et bifurcations didactiques: analyse d'un épisode. *Recherches en didactique, 14*, 29–46.

Coles, A., & Brown, L. (2013). Making distinctions in task design and student activity. In C. Margolinas (Ed.), *Task design in mathematics education* (Proceedings of the International Commission on Mathematical Instruction Study 22, pp. 183–192), Oxford, UK. Available from http://hal.archives-ouvertes.fr/hal-00834054

Collaborative, D.-B. R. (2003). Design-based research: An emerging paradigm for educational enquiry. *Educational Researcher, 32*(1), 5–8.

Comiti, C., Grenier, D., & Margolinas, C. (1995). Niveaux de connaissances en jeu lors d'interactions en situation de classe et modélisation de phénomènes didactiques. In G. Arsac, J. Gréa, D. Grenier, & A. Tiberghien (Eds.), *Différents types de savoirs et leur articulation* (pp. 92–113). Grenoble: La Pensée Sauvage. Available from http://halshs.archives-ouvertes.fr/halshs-00421007

Cooper, C., & Dunne, M. (2000). *Assessing children's mathematical knowledge*. Buckingham, UK: Open University Press.

Csikos, C., Szitànyi, J., & Kelemen, R. (2012). The effects of using drawings in developing young children's mathematical word problem solving: A design experiment with third-grade Hungarian students. *Educational Studies in Mathematics, 81*, 47–65.

de Lange, J. (2015). There is, probably, no need for this presentation. In A. Watson & M. Ohtani (Eds.), *Task design in mathematics education: An ICMI Study 22*. Heidelberg: Springer.

Gardner, K. (2007). Investigating secondary school student's experience of learning statistics. *Dissertation Abstracts International* (UMI No. 3301002).

Gardner, K. (2013). Applying the phenomenographic approach to students' conceptions of tasks. In C. Margolinas (Ed.), *Task design in mathematics education* (Proceedings of the International Commission on Mathematical Instruction Study 22, pp. 193–202), Oxford, UK. Available from http://hal.archives-ouvertes.fr/hal-00834054

Gerofsky, S. (1996). A linguistic and narrative view of word problems in mathematics education. *For the Learning of Mathematics, 16*(2), 36–45.

Hegarty, M., Mayer, R., & Monk, C. (1995). Comprehension of arithmetic word problems: A comparison of successful and unsuccessful problem solvers. *Journal of Educational Psychology, 87*(1), 18–32.

Hershkovitz, S., & Nesher, P. (1999). Tools to think with: Detecting different strategies in solving arithmetic word problems. *Journal of Computers for Mathematical Learning, 3*, 255–273.

Job, P., & Schneider, M. (2013). On what epistemological thinking brings (or does not bring) to the analysis of tasks in terms of potentialities for mathematical learning. In C. Margolinas (Ed.), *Task design in mathematics education* (Proceedings of the International Commission on Mathematical Instruction Study 22, pp. 203–210), Oxford, UK. Available from http://hal.archives-ouvertes.fr/hal-00834054

Johnson, H. L. (2012a). Reasoning about quantities involved in rate of change as varying simultaneously and independently. In R. Mayes & L. L. Hatfield (Eds.), *Quantitative reasoning and mathematical modeling: A driver for STEM integrated education and teaching in context* (Vol. 2, pp. 39–53). Laramie, WY: University of Wyoming College of Education.

Johnson, H. L. (2012b). Reasoning about variation in the intensity of change in covarying quantities involved in rate of change. *Journal of Mathematical Behavior, 31*(3), 313–330.

Johnson, H. L. (2013). Designing covariation tasks to support students' reasoning about quantities involved in rate of change In C. Margolinas (Ed.), *Task design in mathematics education* (Proceedings of the International Commission on Mathematical Instruction Study 22, pp. 211–220), Oxford, UK. Available from http://hal.archives-ouvertes.fr/hal-00834054

Jonsson, B., Norqvist, M., Liljekvist, Y., & Lithner, J. (2014). Learning mathematics through algorithmic and creative reasoning. *The Journal of Mathematical Behavior, 36*, 20–32.

Knott, L., Olson, J., Adams, A., & Ely, R. (2013). Task design: Supporting teachers to independently create rich tasks. In C. Margolinas (Ed.), *Task design in mathematics education* (Proceedings of the International Commission on Mathematical Instruction Study 22, pp. 599–608), Oxford, UK. Available from http://hal.archives-ouvertes.fr/hal-00834054

Lesh, R., Hoover, M., & Kelly, A. E. (1993). Equity, assessment, and thinking mathematically: Principles for the design of model-eliciting activities. In *Developments in school mathematics education around the world* (Vol. 3, pp. 104–130). Reston, VA: National Council of Teachers of Mathematics.

Lin, P.-J., & Tsai, W.-H. (2013). A task design for conjecturing in primary classroom contexts. In C. Margolinas (Ed.), *Task design in mathematics education* (Proceedings of the International Commission on Mathematical Instruction Study 22, pp. 249–258), Oxford, UK. Available from http://hal.archives-ouvertes.fr/hal-00834054

Lithner, J., Jonsson, B., Granberg, C., Liljekvist, Y., Norqvist, M., & Olsson, J. (2013). Designing tasks that enhance mathematics learning through creative reasoning. In C. Margolinas (Ed.), *Task design in mathematics education* (Proceedings of the International Commission on Mathematical Instruction Study 22, pp. 221–230), Oxford, UK. Available from http://hal.archives-ouvertes.fr/hal-00834054

Margolinas, C. (2005). Les situations à bifurcations multiples: indices de dysfonctionnement ou de cohérence. In A. Mercier & C. Margolinas (Eds.), *Balises en didactique des mathématiques* (pp. Cédérom). Grenoble: La Pensée Sauvage. Available from http://halshs.archives-ouvertes.fr/halshs-00432229/fr/

Margolinas, C., & Steinbring, H. (1994). Double analyse d'un épisode: cercle épistémologique et structuration du milieu. In M. Artigue, R. Gras, C. Laborde, P. Tavignot, & N. Balacheff (Eds.), *Vingt ans de didactique des mathématiques en France. Hommage à Guy Brousseau et Gérard Vergnaud* (pp. 250–258). Grenoble: La pensée sauvage.

Mason, J. (2002). Minding your Qs and Rs: Effective questioning and responding in the mathematics classroom. In L. Haggarty (Ed.), *Aspects of teaching secondary mathematics: Perspectives on practice* (pp. 248–258). London: RoutledgeFalmer.

Mason, J., & Pimm, D. (1984). Generic examples: Seeing the general in the particular. *Educational Studies in Mathematics, 15*(3), 277–289.

Maturana, H. R., & Varela, F. J. (1987). *The tree of knowledge: The biological roots of human understanding*. Boston: Shambala.

Palhares, P., Vieira, L., & Gimenez, J. (2013). Order of tasks in sequences of early algebra. In C. Margolinas (Ed.), *Task design in mathematics education* (Proceedings of the International Commission on Mathematical Instruction Study 22, pp. 241–248), Oxford, UK. Available from http://hal.archives-ouvertes.fr/hal-00834054

Patkin, D., & Gazit, A. (2011). Effect of difference in word formulation and mathematical characteristics of story problems on mathematics preservice teachers and practising teachers. *International Journal of Mathematical Education in Science and Technology, 42*(1), 75–87.

Pimm, D. (1987). *Speaking mathematically*. London: Routledge and Kegan Paul.

Radonich, P., & Yoon, C. (2013). Using student solutions to design follow-up tasks to model-eliciting activities. In C. Margolinas (Ed.), *Task design in mathematics education* (Proceedings of the International Commission on Mathematical Instruction Study 22, pp. 259–268), Oxford, UK. Available from http://hal.archives-ouvertes.fr/hal-00834054

Sarrazy, B. (2002). Effects of variability of teaching on responsiveness to the didactic contract in arithmetic problem-solving among pupils of 9–10 years. *European Journal of Psychology of Education, 17*(4), 321–341.

Sarrazy, B., & Chopin, M.-P. (2010). Anthropo-didactical approach to teacher-pupil interactions in teaching mathematics at elementary school. *Scientia in Educatione, 1*(1), 73–85. Available from http://www.scied.cz/index.php/scied/article/view/55.

Savard, A., Polotskaia, E., Freiman, V., & Gervais, C. (2013). Tasks to promote holistic flexible reasoning about simple additive structure. In C. Margolinas (Ed.), *Task design in mathematics education* (Proceedings of the International Commission on Mathematical Instruction Study 22, pp. 269–278), Oxford, UK. Available from http://hal.archives-ouvertes.fr/hal-00834054

Simon, M.-A. (1995). Reconstructing mathematics pedagogy from a constructivist perspective. *Journal for Research in Mathematics Education, 26*, 114–145.

Strømskag Måsøval, H. (2013). Shortcomings in the milieu for algebraic generalisation arising from task design and vagueness in mathematical discourse. In C. Margolinas (Ed.), *Task design in mathematics education* (Proceedings of the International Commission on Mathematical Instruction Study 22, pp. 231–240), Oxford, UK. Available from http://hal.archives-ouvertes.fr/hal-00834054

Strømskag Måsøval, H. (2015). Students' mathematical activity constrained by the *milieu*: A case of algebra. In *Proceedings of NORMA 14, The Seventh Nordic Conference on Mathematics Education*. Turku, Finland: University of Turku.

Sullivan, P., & Lilburn, P. (2004). *Open-ended Maths activities: Using "good" questions to enhance learning in mathematics*. Melbourne: Oxford University Press.

Swan, M. (1985). *The language of functions and graphs*. Nottingham: Shell Centre.

Sweller, J. (1994). Cognitive load theory, learning difficulty, and instructional design. *Learning and Instruction, 4*(4), 295–312.

Verschaffel, L. (2002). Taking the modelling perspective seriously at elementary school level: Promises and pitfalls. In A. Cockburn & E. Nardi (Eds.), *Proceedings of the Twenty Sixth Annual Conference of the International Group for the Psychology of Mathematics Education* (Vol. 1, pp. 64–80). Norwich, UK: PME.

Verschaffel, L., Greer, B., & Torbeyns, J. (2006). Numerical thinking. In A. Gutiérez & P. Boero (Eds.), *Handbook of research on the psychology of mathematics education* (pp. 51–82). Rotterdam: Sense Publishers.

Wood, T. (1998). Alternative patterns of communication in mathematics classes: Funneling or focusing? In H. Steinbring, M. Bartolini Bussi, & A. Sierpinska (Eds.), *Language and communication in the mathematics classroom* (pp. 167–178). Reston, VA: National Council of Teachers of Mathematics.

Chapter 5
Design Issues Related to Text-Based Tasks

Anne Watson and Denisse R. Thompson
with *Marita Barabash, Bärbel Barzel, Yu-Ping Chan, Jaguthsing Dindyal, Patricia Hunsader, Kyeong-Hwa Lee, Anna Lundberg, Katja Maass, Birgit Pepin, Susan Staats, Geoff Wake, Sun Xuhua*
and additional contributions from *Toh Pee Choon, Yael Edri, Francisco Javier Garcia, Tay Eng Guan, Ghislaine Gueudet, Raisa Guberman, Eric Hart, Ho Foo Him, Leong Yew Hoong, Stephan Hußmann, Jason Johnson, Cecilia Kilhamn, Toh Tin Lam, Eun-Jung Lee, Timo Leuders, Fou-Lai Lin, Nitsa Movshovitz-Hadar, Nicholas Mousoulides, Teresa Neto, Loudes Ordóñez, Min-Sun Park, Susanne Prediger, Kristina Reiss, Quek Khiok Seng, Luc Trouche, Barbara Zorin*

5.1 Introduction

In this chapter, we focus on design issues related to written tasks and prompts for mathematical action and the sequencing and norms of collections of tasks, such as textbooks, that shape the expected action. We take a *task* to be the written presentation of a planned mathematical experience for a learner, which could be one action or a sequence of actions that form an overall experience. Thus, a task could consist of anything from a single problem, or a textbook exercise, to a complex interdisciplinary exploration. The design process for such tasks is not necessarily long or cyclic, but we are interested in particular issues that might or should be considered when designing tasks to be presented in text.

A *text*-based task is intended to create mathematical action through prepared and published inert written and visual images, in worksheets, textbooks, screen images, video, assessment instruments, digital interactive textbooks, and other digital technologies. We emphasize that text-based tasks are any such inert, static tasks with which learners interact and do not refer just to tasks in textbooks. In collections of text-based materials, such as textbooks, tasks do not exist on their own, but as components of the whole collection. Hence, in this chapter we also consider how the overall principles and aims of a collection (e.g., a textbook, a set of worksheets) can be embodied in different kinds of tasks. When we talk about prepared, published

---

A. Watson (✉)
University of Oxford, Oxford, UK
e-mail: anne.watson@education.ox.ac.uk

D.R. Thompson
University of South Florida, Tampa, FL 33620, USA

This chapter has been made open access under a CC BY-NC-ND 4.0 license. For details on rights and licenses please read the Correction https://doi.org/10.1007/978-3-319-09629-2_13

© The Author(s) 2021
A. Watson, M. Ohtani (eds.), *Task Design In Mathematics Education*,
New ICMI Study Series, https://doi.org/10.1007/978-3-319-09629-2_5

collections of tasks, we focus solely on the school-level curriculum and learners in educational environments. We also consider freestanding tasks that might be used by learners of any age and in varied phases of learning as relevant. Chapters 3 and 4 look at how teachers and learners, respectively, work with tasks, including digital tasks. Further insights into dynamic digital technologies are found in Chap. 6.

Throughout this chapter, we highlight design principles and research perspectives that are salient whether the tasks are freestanding or part of textbook collections. Although the design of text-based tasks and the design of textbooks share many common features, we believe that text-based task design is worthy of study in its own right, separate from but related to design and research on textbooks.

5.1.1 Research on Text-Based Tasks Within Textbooks

Some work on task design has delineated how different interpretations of curriculum aims and standards influence the design of collections of tasks embodied in textbooks. For instance, in the USA a number of large-scale curriculum projects were developed in the 1990s in response to the publication of the *Curriculum and Evaluation Standards* by their National Council of Teachers of Mathematics (1989). As Hirsch (2007) indicated, this Standards document provided a "basic design framework" that influenced authors regarding the nature and scope of content, integration of technology, embedded assessments, professional development for teachers, and active engagement of learners through explorations within cooperative groups. Thus, a design issue facing the curriculum developers was to provide text-based tasks or materials that embodied both intention and implementation (see Chap. 2). Other research on textbooks has focused on how the principles of design of the textbook tasks might have influenced teachers' enactment of those tasks and how such enactment influenced learner achievement (see Chap. 3; also see multiple perspectives on enactment in Remillard, Herbel-Eisenmann, and Lloyd, 2009). Still others have focused on design principles through comparisons of the tasks within textbooks, sometimes from a neutral stance of simply highlighting differences (e.g., Pepin and Haggarty, 2001) and other times from a more critical stance that highlights differences in tasks and their affordances for student learning (Huntley and Terrell, 2014). A summary of textbook research presented by Fan, Zhu, and Miao (2013) points to the shortage of research about the design process itself and a shortage of research into relationships between textbooks, teaching, and learning.

Our focus is on text-based tasks, and there is a growing body of research that approaches textbook analysis through the opportunities afforded by task content. In these approaches, it is presumed that some affordances are desirable according to some theoretical frame or curriculum aims, and research generally reveals the presence or lack of a particular feature. For example, research into some textbooks has shown a lack of explanations (Dole and Shield, 2008; Stacey and Vincent, 2009), of higher-order thinking (Nicely, 1985; Nicely, Fiber, and Bobango, 1986), of worked-out examples (Mayer, Sims, and Tajika, 1995), of opportunities for reasoning (Stylianides, 2009, 2014; Thompson, Senk, and Johnson, 2012), of problem variation (Stigler, Fuson, Ham, and Kim, 1986), of word problems (Xin, 2008), of conceptual connections (Sun, 2011), or of conceptual robustness (Harel, 2009).

Detailed examination of such textbook analysis research is beyond the scope of our chapter as we are interested in individual and sequential task design and how design embodies curriculum principles and theories of learning, either implicitly or explicitly. However, the assumptions behind many of these studies are that (a) the textbook defines the learners' mathematical experience because of the prevalence of textbooks in schools around the world and (b) textbooks could or should provide all these experiences. With both of these assumptions, learners' experiences with the tasks are intertwined with the associated teaching and how the teacher enacts the tasks as part of instruction (see Thompson and Usiskin (2014) for more insights on the enactment of curriculum). Thus, while we consider presence or lack of particular textual features, we cannot claim that presence ensures experience or absence implies teaching deficits. Rather, we consider design features that underlie the development of tasks, generally irrespective of the implementation of those tasks by either teachers or learners. In other words, we examine the relationship between authors' intentions for the task and the affordances and opportunities that the task provides; that is, we are interested in the bridge between curriculum intention and pedagogic implementation that can be provided through static text. This requires imagination and anticipation about learners' and teachers' engagement with tasks. Our approach is to draw on existing research where possible and also to take a scholarly perspective to task design issues by drawing on professional experience and published tasks. Many of the tasks presented in this chapter were shared and discussed by participants during the conference for ICMI Study 22 on which this book is based.

5.1.2 Shape of the Chapter

We consider three interrelated aspects of the design issues of text-based tasks: (1) nature and structure of such tasks, (2) pedagogic/didactic purpose of their design (i.e., intentions), and (3) intended/implemented mathematical activity as embedded in them (i.e., affordances and opportunities for learning). Although the chapter considers these headings, the relationships between the task designer, a teacher, a task, the mathematics within the task, and the learner are important at every stage. In particular, the relationship between task and teaching is like two sides of a coin because both contribute to the context of the learners' mathematical activity.

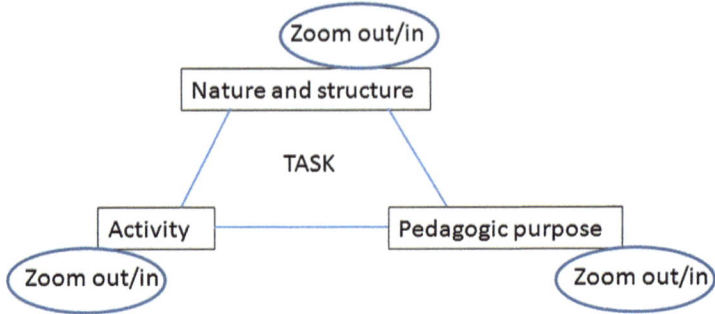

Fig. 5.1 Task design intention triangle

The three interrelated perspectives can be seen as a triangular structure with nodes (Fig. 5.1). Each node can be considered by zooming out and thinking about the overall educational context and how this affects task design and also by zooming in to the imagined interaction between one learner and the task.

Throughout the chapter, we present and compare examples of tasks to illustrate variation within each node and how principles of design play out within each node.

5.2 Nature and Structure of Tasks

We consider design principles related to three aspects of the nature and structure of tasks: (1) the types of text materials in which tasks are found, (2) the authorship, authority, or voice of the task, and (3) the mathematical content of the task. Within each of these aspects, particular issues related to design are evident.

5.2.1 Different Kinds of Text Materials

We start by making distinctions among different kinds of text-based materials. *Learning management systems* are those in which learning is presumed to be managed either by sequencing (such as in a traditional textbook series) or by a planned interplay between formative assessment and instructional tasks of various kinds. In some online curriculum packages (e.g., *I Can Learn* in the USA), the learner is essentially in his/her own private classroom using a software system that provides tasks and then gives mechanistic feedback to both the learner and the teacher about the learner's interaction with the task. Based on that feedback, the learner might move forward to new tasks on new concepts or might engage in tasks designed to offer remediation. The overall topic sequence is thus managed by a (sometimes virtual) teacher and/or possibly learners themselves.

In textbooks, the sequence is fixed according to a designed narrative suggested by the authors, with tasks potentially designed to build on each other and with careful consideration to necessary prerequisites. Teachers often choose to modify the textbook sequence based on perceived needs of their learners or mandated curriculum goals and must consider what assumed knowledge their students may not possess in a revised sequence and adjust tasks accordingly. In an online system, sequence might be varied according to learners' responses but a designed narrative controls those variations. In *task banks*, collections of varied tasks are published for which the teacher (or even the learner) is the effective learning manager and makes decisions about who does what and in what order; the individual tasks themselves may not be linked by a narrative (e.g., Yerushalmy (2015, Chap. 7, this volume); SMILE, n.d.). *Freestanding* tasks are those that do not form part of a curriculum package but are supplementary or fulfill a special purpose. For example, the NRich website provides extension tasks accessible by students, teachers, and parents that are intended as curriculum supplements (nrich.maths.org); the COMPASS (Common Problem Solving Strategies as Links Between Mathematics and Science) (2013) project (Maaβ, Garcia, Mousoulides, and Wake, 2013) provides interdisciplinary tasks within a European setting that can be used within mathematics and science classrooms. Freestanding tasks are typically designed to be self-contained, without relying on completion of previous tasks; if prerequisite knowledge is needed, then that information would need to be provided to a potential task user.

Task collections might exist in printed form as banks or books for learners with or without teacher guidance, in multimedia form such as paper and digital and/or physical materials for learners with or without teacher guidance, or in the form of guidance for teachers with materials for tasks, but no text for learners (e.g., Numicon at: global.oup.com/education/content/primary/series/numicon/). We do not consider the latter type in detail here as there are no given text-based tasks for learners unless the teacher constructs one, but our remarks about construction and use of text-based tasks apply also to teacher-made text-based tasks.

There are general observations about task design that apply across these multiple kinds of text. However, we also acknowledge that tasks designed to be included in mandated curricula material or those adopted by a governing body (e.g., school, district, ministry of education) may be designed under content and pedagogy constraints that do not exist for designers of supplementary materials or for teachers who design tasks for use with their own learners (Gueudet, Pepin, and Trouche, 2013). For instance, in tasks within mandated curricula, there may be a focus on problems addressing particular processes (e.g., reasoning, graphical representations) or particular solution approaches (e.g., written explanations); tasks may be designed to be facilitated by a teacher, with appropriate teacher guidance provided for implementation of the task. In contrast, in collections of tasks for supplementary use, there may be more of a focus on inquiry or exploratory approaches or multiple solution pathways. Similarly, a task designed for a specific purpose, such as introducing learners to engineering as a career choice, would adhere to different principles than a teacher-designed task to help a class learn a particular mathematical idea. The first requires a zoomed-out view of the design intention triangle, perhaps focusing on the value of engineering or the types of problems engineers solve, while the second will be a zoomed-in view at the classroom level, perhaps focusing on skills and understandings learners need for an assessment or to provide evidence of meeting an established curricular goal.

5.2.2 Authorship, Authority, and Voice

As designers plan and develop tasks, several issues come into play: (1) how will designers interact with each other and with the ultimate users of their tasks, namely, teachers and learners; (2) how will they position authority for evaluating the correctness or completeness of a task, namely, within the task or within the user of the task; and (3) what voice is used, namely, whether the task is addressed to a teacher or to a learner. We consider each of these issues in turn.

5.2.2.1 Authorship

Task designers work together in various formats to author mathematical text-based tasks:

- Substantial teams working with a long development process that includes field trials to obtain teacher input. Large-scale curriculum development projects, such as the School Mathematics Project in the UK, the many Standards-based curriculum projects in the 1990s in the USA, or the Chinese government teams developing national curriculum texts, have used this format. Also this format is used for many projects with more specific aims, such as numeracy recovery. The Canada-based project JumpMaths includes information about the effects of experience on its genesis (https://jumpmath.org/cms/).
- Author teams working on short time frames using design principles imposed by publishers or authorities. For example, official textbook production in China in 1960 took place within a 1-year cycle; there were serious learning problems within the textbooks that had to be changed (Li, Zhang, and Ma, 2009). Anecdotally, we know of one US state which requested a new official textbook in 6 months.
- Project teams working within particular agreed principles for pedagogical and epistemological coherence. For instance, in the COMPASS project, a large team works across multiple European countries to develop interdisciplinary tasks, with specific principles, such as a project-centered approach, an inquiry-oriented pedagogy, and an integration of information technology (Maaβ et al., 2013).
- Individuals or teams developing innovative or idiosyncratic materials with a specific focus or for use under specific conditions. For instance, Staats and Johnson (2013) created specific interdisciplinary tasks for use in college algebra and Movshovitz-Hadar and Edri (2013) developed social justice tasks to focus on values education in Israeli classrooms.

5   Design Issues Related to Text-Based Tasks   149

- Teams of teachers producing editable materials or task banks in a dynamic process. For instance, a group of lower secondary teachers in France are working together to produce a text easily adapted by all (*Sesamath* as described in Gueudet et al., 2013); a group of Israeli teachers are designing tasks using *Wikitext* (Even and Olsher, 2012).
- Individual teachers or small local teams disseminating their ideas. The proliferation of Internet resources has made it possible for teachers to post lessons and tasks for use by teachers anywhere in the world, at no cost or for a nominal fee (e.g., http://www.teachmathematics.net/; http://www.teacherspayteachers.com/).

A team which has come together because of an underlying shared belief and agreed-upon design principles, such as a team designing tasks that use a particular software (e.g., Geogebra, Cabri) or have a particular curriculum aim (e.g., Realistic Mathematics Education in the Netherlands), is presumed to have epistemological and conceptual coherence in their work. In contrast, a Wiki-type team might produce materials with variable principles (e.g., the French team for *Sesamath*). Even when materials have initially been developed by teams with specific design principles, the move from the design stage to more widespread use and adoption through commercial publication can create constraints or pressures that force adaptations or modifications in tasks to satisfy publishing needs.

An example of innovative text-based task development was the Resources for Learning and Development Unit (RLDU, n.d.) in which teams of about 10 teachers worked together to design tasks which they trialed and adapted. The final tasks were published in a format that implied certain pedagogic principles, namely, that tasks would be presented in a learner-friendly format and that learners would engage with mathematics through exploration and inquiry without being told exactly what to do or how to do it. Authorship for the text of the task rested with the teacher team, but authority for the mathematics rested in the explorations of the students (Llinares, Krainer, and Brown, 2014). Further information can be obtained from Laurinda Brown who coordinated and edited the resources (laurinda.brown@bristol.ac.uk).

Consider the RLDU task in Fig. 5.2. Although there is a sequence of questions to facilitate engagement with the task, there are no numbers or measures within the task. Thus, as written, there is not enough information for learners to answer the questions; rather there is the comment, "Bring a bicycle into the classroom!" This comment suggests the task designer wanted teachers to use a physical bicycle to facilitate inquiry; learners were expected to use the bicycle to explore mathematics and possibly to consider differences in answers to the questions for bicycles of different sizes. The questions should be relatively easy for learners to understand, but the mathematics is only accessible through exploration.

5.2.2.2   Authority for Mathematics

As task developers design a task or a collection of tasks to be included in a textbook, the manner in which the task is written can determine where authority lies for evaluating the mathematics that is the product of the task. In considering textbook

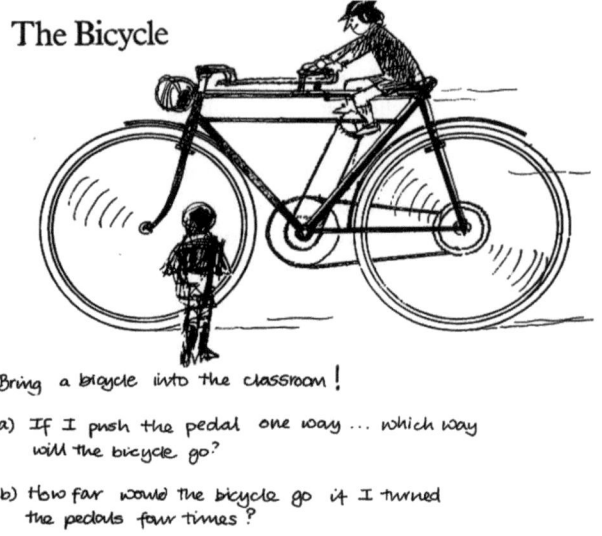

Fig. 5.2 An RLDU (Resources for Learning and Development Unit) worksheet (n.d.)

authority, Herbel-Eisenmann (2009) cites research indicating that textbooks often derive their authority from the structure of the text itself as well as the pedagogy that results or the political and cultural context. Authority for mathematics is ultimately within mathematics itself at the school level: most results can be checked by working backward or using different tools or searching for implications or contradictions, so long as learners are working within the usual conventions. However, mathematical authority is often ceded to the textbook authors who provide an answer book or the teacher who checks answers. Rarely do text-based tasks support the development of the self-checking learner, using mathematics to verify solutions. Developing self-checking learners could be accomplished by including regular explicit self-checking strategies or implicit strategies in which a later action modifies an earlier action or tasks that embed immediate feedback or tasks with a solution obtained from multiple approaches. These are possible using digital resources that show implications of incorrect reasoning and interactive software that allows for adjustment and self-correction as an integral activity. Some curriculum materials have attempted to build

5   Design Issues Related to Text-Based Tasks   151

in features to facilitate this self-monitoring. For instance, curriculum materials for secondary learners developed by the University of Chicago School Mathematics Project have Quiz Yourself questions at different points within a lesson for learners to stop and check their comprehension, with answers to these short tasks at the end of the lesson. Throughout recent decades, various teams have developed programmed learning suites, either in hard copy or digitally, which provide multiple pathways, triggered according to diagnostic evaluation of learners' success so far. Responsibility for diagnosing common errors might be with the designers (Anderson and Schunn, 2000) but could be valuable for the learners themselves. For example, the German textbook series *Mathewerkstatt* contains tasks that enable learners to self-check their work and diagnose potential errors, ensuring that learners do not proceed too far without feedback relative to their mathematical progress (Hußmann, Leuders, Prediger, and Barzel, 2011a). In some of Swan's tasks, learners diagnose errors in the work of other anonymous learners.

5.2.2.3   Voice of a Task

In designing tasks, developers must consider how to address the ultimate user of the task and what messages are conveyed through different usages of language. When analyzing *voice* of text-based tasks, differences might be found between text which is directed at the teacher or at learners in the classroom or at learners who are assumed to be studying on their own. Zooming in to the learner's perspective, the main voice used in tasks is imperative. Learners are told what to do: *look at*, *write*, *solve*, *measure*, *find out*, and so on. Instructions may be supplemented with questions, some of which trigger action or application (such as *how many …?*) and others which trigger reflection, conjecture, and generalization (such as *what do you think would happen if…?*). There is evidence from a study of 400 learners that those who are used to imperative texts will scan the text looking for *what to do* (Shuard and Rothery, 1984, p. 114). When examining textbooks, we have found that the imperative tone dominates; when the teacher then refers to the textbook during instruction, such as "what does the textbook say?", the authority of the textbook as a means to resolve issues or questions is reinforced (Herbel-Eisenmann, 2009).

In some texts, there is no direct instruction for tasks but an assumption that some action will be carried out, triggered by a format to be completed or some objects to be contemplated. Such tasks may be used for young children who may be unable to read, but also be used with older students to encourage inquiry and exploration. The implied instruction may be to *complete* or *fill in the gaps*. In all texts, we might ask whether the learner is positioned as a compliant learner or as a co-creator of knowledge. Typical worksheets or workbook pages for very young learners provide a printed writing frame in which the learner merely fills in answers or uses color, arrows, and so on to indicate connections or classifications. In these cases, dialogue with a teacher shapes whether the experience is merely compliant or seen as the creation of meaning. Another way to frame this issue is to identify whether the overall mathematical narrative is delivered by the text or is created by the learner.

Fig. 5.3 Worksheet from RLDU (n.d.)

RATIO AND PROPORTION

For example, consider the *worksheet* from the RLDU in Fig. 5.3. There are no questions or instructions provided for the task. However, learners who have previously worked with tasks from this resource know they are to create their own questions and explorations from the figure. Thus, the task positions the learner as creator of the narrative, possibly working with others and in dialogue with the teacher.

Another example of how voice can shift authority is demonstrated by Wagner (2012) who analyzes two of his own tasks using Rotman's (1988) distinction between imperatives that require learners to write (*scribble*) and those that require them to think. The first of his tasks is traditionally imperative; the first two parts tell the learner what to do and the final part invites some thinking.

This polygon is drawn on 1 cm dot paper.

1. Find its area by dividing it into a rectangle and two triangles.
2. Find its area by dividing it into two triangles.
3. Show another way you can divide the polygon into two triangles.

The second task uses the first person, as if a real person is talking and sharing a solution but in a problem-solving sequence used to find the area of an unnamed shape as if it is the only method. So, authorship of the second task is more overt than in the first task but still focuses mainly on *scribbling* with a little attention to thinking and no room for another learner's own ideas.

- I knew it was a trapezoid because the arrow marks showed that it had exactly two parallel sides.
- I identified the bases and the height. I noticed that the 6.8 cm side length was extra information that I didn't need.
- I used the formula. (Wagner cites from Small, Connelly, Hamilton, Sterenberg, and Wagner, 2008, pp. 144–145.)

In an example from Korea, a solution uses the phrases "Let's think" and "Let's find" to suggest that the learner work alongside an unknown person (Lee, Lee, and Park, 2013). Thus, the voice invites the learner to be a co-creator of the mathematical solution.

Fig. 5.4 Collection of sequences to be completed

a. 7, ..., 25, 34, ..., 52, 61, ...,

b. ..., 1.6, 2.5, 3.4, ..., 5.2, ..., 7.0,

c. 1.7, 2.6, ..., ..., 5.3, ..., 7.1, ...,

d. ..., 0.16, ..., 0.34, 0.43, 0.52, 0.61, ...,

e. -7, 2, 11, 20, 29, ..., ..., ...,

5.2.3 Nature of Mathematics

The examples in Figs. 5.2 and 5.3 strongly imply that mathematics provides tools for posing questions and understanding phenomena and also that mathematics can emerge from enquiry. By contrast, the teacher-designed sequence task in Fig. 5.4 implies that mathematics consists of symbolic objects that are acted on according to some rules and that mathematical activity consists of mental reasoning (in this case calculation) and writing and then seeking and expressing patterns. Task design influences the nature of mathematical activity and therefore the learners' perception of what it means to do mathematics. If the task in Fig. 5.4 is left at the stage of *filling in the blanks* with no reflection on connections between the five sequences, some opportunities to learn will have been missed as learners may simply think of mathematics as doing computations. However, if learners look for similarities and patterns among the sequences, then they are able to develop understanding of the structure and regularity within numbers.

Comparison of text-based tasks from different educational cultures can be of value in highlighting deeper implications about the nature of mathematics as presented in the task and text. Although ontology is important, in the context of this chapter we cannot separate this from epistemology and, hence, the intended or assumed nature of mathematical activity. For example, in the Resources for Learning and Development Unit (RLDU), an overall intention was to develop a style of mathematics that focused on mathematization, posing mathematical questions, collaboration, and problem-solving; zooming in, the learner was encouraged to experiment, make conjectures, and search for or create mathematical procedures to carry out their enquiries. The mathematical activity that can be imagined being initiated by these tasks would involve practical materials, physical activity, discussion, application of techniques, and definitions. In addition, the nature of the tasks naturally lends itself to learners working collaboratively in small groups. Thus, pedagogy is not only implied but also structured by the materials, as in the note to bring a bicycle to class.

To explore how different views of the nature of mathematics play out in task collections, we look at three examples of comparative studies. The first example shows that different perspectives on the nature of mathematics can be related to the notion of authority previously discussed. Gueudet et al. (2013) compared features of two different French textbooks. Differences between them can be considered to create, rather than to reflect, differences in school mathematics cultures. One textbook, *Helice* (Chesné, Le Yaouanq, Coulange, and Grapin, 2009), was of a traditional type written by four experts; the other, *Sesamath* (Sesamath, 2009), was developed collaboratively by a group of 57 teachers using digital open-access tools. As a result of the manner in which the books and their related tasks are created, *Helice* has more overall consistency in its presentation of mathematics and mathematical activity and more coherence of conceptual development. *Sesamath*, while possibly empowering teachers, offers a fragmented and atomistic approach to concepts. For instance, *Sesamath* presents a counting or enumeration approach to finding area; in contrast, *Helice* presents a conceptual approach to understanding area as a conserved property of 2-d shapes by having students decompose and recompose shapes. In addition, *Helice* offers problems with a variety of solutions but *Sesamath* offers one expert solution. Learners therefore experience mathematics either as a connected whole with several possibilities for action or a fragmented collection of limited actions.

In the second example, the focus of tasks might inculcate different ways of reasoning. Chang, Lin, and Reiss (2013) compared a Taiwanese and a German textbook series according to principles of continuity, accessibility, and contextualization and the ways that content was structured. They took different approaches to proof and, more importantly for our purposes, the tasks that followed the proof. Figure 5.5 illustrates the Taiwanese visual-algebraic approach which was followed by several visual-algebraic tasks that used the theorem; the German approach was deductive and followed by tasks with a focus on area.

There is a subtle but important difference when we zoom in to the learner's perspective. In the Taiwanese approach, learners extract the relationships from a diagram directly but must have prior knowledge of how to determine the area of a triangle and the area of a square; in the German approach, learners apply a priori formal knowledge to a diagram using knowledge of similar triangles embedded

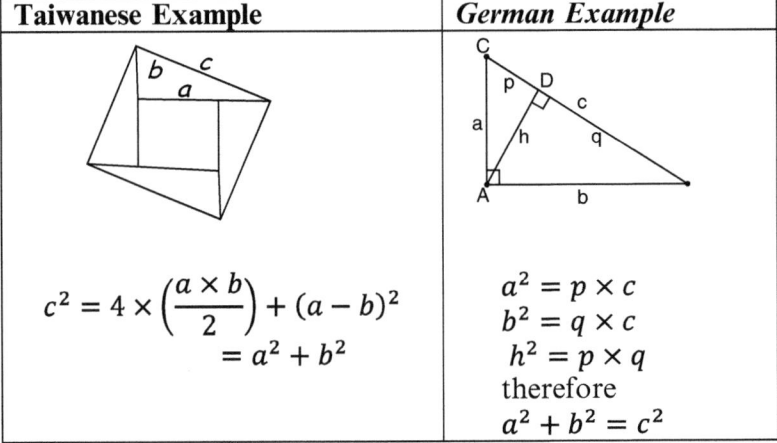

Fig. 5.5 Comparison of Taiwanese and German typical approaches to Pythagoras' theorem (Adapted from Chang et al., 2013, p. 308)

5 Design Issues Related to Text-Based Tasks

within the large right triangle. Although the study offered these as examples of differences in overall mathematics pedagogy, this contrast highlights how differences in task design can engender different learner experiences of geometric reasoning through different kinds of mental activity.

The third comparison we offer contrasts the treatment of the additive relations between small whole numbers for young children in the Chinese textbook and a Portuguese textbook (Sun, Neto, and Ordóñez, 2013). Analysis of control and variation in the tasks for this concept shows significant differences in opportunities to learn. In the Chinese textbook (Fig. 5.6), the focus on each page is representing a part-part-whole relation visually, physically, and in alternative equivalent symbolic forms. Thus, the Chinese textbook presents mathematics as a variety of formal representations of some actions, and this implies that mathematical activity consists of carrying out physical actions, forming mental images and expressing them in numerical instantiations.

In the Portuguese textbook, which we cannot show here, the focus was on active methods (such as "doubling plus 1") applied to several sums. If we look at one

Fig. 5.6 Page from Chinese textbook (Mathematics Textbook Developer Group for Elementary School, 2005, p.68)

page as a text-based task, in the Chinese approach the object of learning and the underlying concept is the additive relation, with connections made between addition and subtraction and reasoning with tens and ones. In the Portuguese textbook, the object of learning appeared to be a different procedure applied to different problems on every page.

To understand what can be learned from one task, it is useful to consider the immediate sequence. In the Chinese textbook, the subsequent tasks also focus on additive relations, while in the Portuguese textbook a variety of methods which can be used in different cases are presented, leaving the teacher to make the connections. Sun also shows this difference applies to some US textbooks (Sun, 2011).

These three comparative examples suggest three issues related to task design in terms of how learners may view what it means to do mathematics:

- Is the learner encouraged to explore and compare different solution methods, or must the learner apply one given method?
- Is the learner expected to apply a priori knowledge or to apply mathematical reasoning (such as expressing relationships) to access new ideas?
- Do the choice and sequence of examples prioritize conceptual understanding (such as through relating actions to symbols or comparing representations) or prioritize methods for reaching a solution?

5.2.4 Summary

In this section, we have discussed the nature of text-based tasks, the view of mathematics they imply through their structure and expectations, and the authority they assume based on the nature of their authorship or the voice they employ. We have not intended to imply that any particular set of design principles in these aspects is inherently better than any other. Text-based tasks are prepared and static. A major question to be considered is: "How can tasks shape an experience of mathematics that is dynamic and dialogic and sees the learner as a sensemaking creator of connections, insights, and solutions?" Some of the tasks already presented could provide the opportunity for such dynamic dialogue, but only if the associated pedagogy supports this. In the next section, we consider the pedagogic issues in task design.

5.3 Pedagogic/Didactic Purpose of Text-Based Tasks

The pedagogic intent of a task also influences how that task might be designed by its developers. In this section, we consider how cultural differences in purpose influence design, how learners are made aware of purpose, how developers ensure a coherent purpose within a collection of tasks, and how new knowledge is integrated with existing knowledge.

5.3.1 Cultural Differences in Purpose

Throughout this chapter, we assume the aims of mathematics education are multifaceted, so that learners become knowledgeable about concepts, competent with procedures, and capable and willing to select, adapt, and use mathematics in a variety of familiar and unfamiliar contexts and problems. An overarching question is whether and how text-based tasks can contribute to all these aims. The relative importance of these aims is to some extent cultural, and there is some evidence of cultural differences in how aims are translated into text-based tasks and used to promote learning.

In several countries, lessons typically have three levels: the first level is basic facts and formula; the second level is to make connections between these; and the third level is for learners to apply some higher-level thinking to problems. The third level of classroom mathematical activity is hardest to achieve and depends on progression embedded within task sequences. One example of a three-level task lesson from Taiwan follows:

1. Have a triangle with three angles and put the angles together to form a straight line with 180°.
2. Show that the exterior angle sum is 360°.
3. Then use the exterior angle sum to prove the sum of the interior angles (National Academy for Educational Research (Taiwan), 2009).

The first-level task is informal and involves constructing a demonstration of the interior angle sum; the second-level task requires some reasoning about angles, applying previous knowledge that there are 360° in a full rotation; the third task requires a different kind of reasoning, involving formal proof. It is the responsibility of the teacher to make the links between the three levels so that it does not have the appearance of circular reasoning. In cultures where the emphasis is on efficient actions of teachers and learners, it is hard to introduce the messier aspects of problem-solving in which solutions may not be arrived at through the optimal use of time, effort, and method. In the given example, there is a further difficulty, namely, that the practical, spatial reasoning expected for the first two tasks gives way to formal proof for the third task, a shift which is recognized as a major learning obstacle and pedagogical challenge (e.g., Bell, 1976).

In the texts available in our discussions at the ICMI Study Conference, we identified different emphases on mathematical behavior. Learners were expected to develop efficiency (e.g., Dindyal et al., 2013), abstraction (e.g., Chang et al., 2013), and applications (e.g., Maaβ et al., 2013) or investigate social problems (e.g., Movshovitz-Hadar and Edri, 2013) in various cultures. There are variations within cultures as well. No text can do any of this on its own; in most cases, the effects of the texts are mediated by the actions of the teacher. Ensuring that teachers understand the pedagogic purpose of a task is another design issue that has cultural variations.

In some systems (e.g., China, see Ma (1999)), the teacher guide is understood to be the authority for pedagogic knowledge and the national curriculum includes very detailed information about curriculum and pedagogic purpose and design. That is, the guide provides information to teachers about the *curriculum vision* so that teachers understand the overall goals and how materials fit together; such understanding is important in building *curriculum trust* so that any adaptations made by the teachers are consistent with the overall pedagogical and epistemological vision of the materials (Drake and Sherin, 2009).

In other systems, teacher guides exist but might be ignored or used in different ways, particularly if the pedagogic purpose for tasks or their sequence is not clearly laid out in the guide. For example, it might be assumed, incorrectly, that merely using a textbook in its given order assures coverage of the curriculum and coherence with the designers' intent (Thompson and Senk, 2010). Another reason teachers might ignore teacher guides is when they expect to interpret and structure the curriculum for themselves, using textbooks as one of several resources and determining their own pedagogic purpose for tasks for use with their students.

Differences in pedagogic purpose also play out in how various cultures design tasks to address learner diversity. In Japan, one task might be offered to an entire class with the emphasis on collaboration, recognizing that different learners will learn from this process in different ways. In the UK, it has been common to have different but parallel textbook series aimed at learners whose level of attainment differs, so that those with lower prior attainment have textbooks in which the conceptual and cognitive material is less challenging. In Sweden, the law requires equal access for all, so it is seen to be against the law to differentiate between learners by placing them in different curriculum tracks. As we write, there is debate about the interpretation of these laws (Lundberg, personal communication, 6 January 2014), but at present it means that there is no differentiation of textbooks, except for those with varied communication abilities.

One implication of non-differentiated materials is that tasks need to be designed to enable learners with previously low attainment to gain higher-level understandings and also for those with high understanding to extend their knowledge. This means that mathematics cannot be presented as a linear accumulation of ideas with assumptions about prior learning, but instead task design needs to develop concept images and dispositions that will be sustainable across a range of mathematical activity and enable learning at several levels. That is, tasks need to be designed so there are multiple entry points, with options for extensions and adaptations.

To illustrate what we mean by zooming in to learners' experience, consider the following task:

- *Find $9 + 7 =$ ___?___ .*

As written, this is a fairly closed task and students generally know the fact or they have to work it out. Now consider the following adaptation:

- *Write as many number sentences as you can for 16.*

This version of the task addresses a similar ultimate goal but has many points of entry. Some learners may start by writing $15+1=16$, $14+2=16$, and so on. Depending on judgments of learners' potential and their past achievements, the teacher can ask students to use more than two addends, more than one operation, etc.

The point is that all learners in a classroom could experience success with the task, some with simple number sentences and others with more complex ones. In the process, students are investigating number relationships and completing many more computations than would have occurred from the original task. When elementary learners have been given such a task, it is not uncommon to have pairs of learners write upward of 20 number sentences in a relatively short time span. If pairs of learners check the work of other pairs, learners have opportunities to consider other potential ways to combine numbers in appropriate number sentences.

Similar adaptations of closed tasks might occur in many contexts. Rather than consider a task with a single answer and one way to obtain that answer, teachers might adapt tasks to encourage multiple answers or multiple pathways to an answer. The intent of such tasks is to provide access to diverse learners, and many teacher guides provide information about the pedagogic purpose of such adaptations through examples of adapting tasks for students who need remediation or extension. Such features of text-based tasks, especially the expectations of organization of work, are most likely to be identified through textbook comparison studies in which assumptions and expectations about ways of working can become more clear (e.g., Pepin and Haggarty, 2001; Stylianides, 2009, 2014).

5.3.2 How Purpose Is Presented to Learners

In the task at the end of the previous section, all learners are investigating number relationships with differing levels of difficulty. We now zoom in again to examine more possible purposes of tasks and how these might be expressed to learners.

In their classic study of children reading mathematics, Shuard and Rothery (1984) present five main purposes for mathematical texts:

1. Teach concepts, principles, skills, and problem-solving strategies.
2. Give practice in the use of concepts, principles, skills, and problem-solving strategies.
3. Provide revision of 1 and 2 above.
4. Test the acquisition of concepts, principles, skills, and problem-solving strategies.
5. Develop mathematical language, for instance, by broadening the pupils' mathematical vocabulary and their skill in the presentation of mathematics in a written form (pp. 5–6).

Shuard and Rothery's five purposes apply to texts in their entirety. Applying these five purposes to individual tasks would imply that such tasks might address individual purposes, such as the various desirable goals and outcomes presented in Chap. 2. A more helpful approach would be to use these purposes as parameters for task sequence intentions, so a task might incorporate some revision content, some new concept content, some relevant language, and so on, and a task sequence might present all these purposes in a developmental order. To some extent these purposes would be teacher and learner specific, or even topic specific, so that tasks that support learning to resolve right-angled triangles would look very different to tasks that

Table 5.1 MATH taxonomy categories (From Smith et al., 1996)

Group A	Factual knowledge (A1)
	Comprehension (A2)
	Routine use of procedures (A3)
Group B	Information transfer (B1)
	Applications in new situations (B2)
Group C	Justifying and interpreting (C1)
	Implications, conjectures, and comparisons (C2)
	Evaluation (C3)

Table 5.2 Categorizations to analyze assessment tasks for mathematical processes (From Thompson et al., 2013)

Reasoning and Proof
The item directs students to provide or show a justification or argument for ***why they gave that response***
Opportunity for Mathematical Communication
The item directs students to communicate ***what they are thinking*** through symbols, graphics/pictures, or words
Connections
The item is set in a real-world context outside of mathematics
The item is *not* set in a real-world context, but explicitly interconnects two or more mathematical concepts (e.g., multiplication and repeated addition, perimeter and area)
Representation: Role of Graphics
A graphic is given and must be interpreted to answer the question
The item directs students to make a graphic or add to an existing graphic
Representation: Translation of Representational Forms
Students are expected to record a translation from a verbal representation to a symbolic representation or vice versa
Students are expected to record a translation from a symbolic representation to a graphical (graphs, tables, or pictures) representation or vice versa
Students are expected to record a translation from a verbal representation to a graphical representation or vice versa
Students are expected to record a translation from one graphical representation to another graphical representation

support learning to prove properties of triangles. Another approach is that of the MATH taxonomy derived by Smith et al. (1996) from examination questions but which could be applied to opportunities afforded by individual text-based tasks for learning. Their categories are outlined in Table 5.1.

A more detailed categorization designed to analyze assessment tasks is that of Thompson, Hunsader, and Zorin (2013). In Table 5.2, we give a summarized form to show the range of foci that can be present in a task.

Although these categorizations apply to assessment tasks, they can also be related to the purpose of learning tasks. For example, if some assessment tasks focus on translation between representations, this kind of mathematical action needs to be met throughout lessons and also needs to be accompanied with a coherent theory that connects translation with some desirable learning outcomes. While using any of these categorizations to ensure that a learning management system addresses the associated desirable learning outcomes, it is not the case that merely setting a task that requires a particular mathematical approach ensures learning. The three-part task sequence given previously about triangles gives no help to learners who cannot see how to proceed. By contrast, the text-based tasks in Burn's approach (e.g., *A Pathway into Number Theory*, 1982) promote guided learning by anticipatory dialogue. Burn's opening four questions are (p.18):

1. Look at table 1.1 [below]. If the same pattern was extended downwards, would it eventually incorporate any positive integer $\{1, 2, 3, ..., n, n+1, ...\}$ that we might care to name?
2. What is the relation between each number in table 1.1 and the number below it?
3. Give a succinct description of the full set of numbers in the column below 0.
4. If you choose two numbers from the column below 0 and add them together, where in the table must their sum lie?

0	1	2	3
4	5	6	7
8	9	10	11
12	13	14	15
etc.			

Burn's "answers" for these tasks are unusual as they often set up new ideas for learners and encourage dialogue with the text. As an answer to question 1, he gives formal definitions and notations for natural numbers, integers, rational numbers, real numbers, and complex numbers. For question 2, he writes *four less and four more* and for question 3, *three multiples of 4*. Question 4 he answers with $4n + 4m = 4(n + m)$. So the first answer situates what is meant by *integers* in the context of different classes of number. The second and third confirm the learner's reasoning. The fourth indicates that it is time to shift to symbolic representations and shows how such representations can be a tool to express reasoning. In this fashion, Burns leads the reader through a number theory course in which the learner's activity initiates the closest thing to a dialogue that one can get from a static text. Note also the fact that only two of these *questions* are imperatives; even then, the first instructs the learner to *look* before posing a question. In terms of the previous categorizations of Thompson et al., Burns' questions provide opportunities for students to communicate, make connections, and interpret a graphic.

It would be possible to dismiss Burns' approach as a teaching style relevant only for adult learners or those who are studying mathematics through choice. However, The School Mathematics Project (SMP) in England experimented widely with dialogic teaching for 11-/12-year-old learners of average attainment from the early 1960s. We do not have good measures of readability that take into account the need to interpret mathematical ideas and multilingual classrooms, but it is likely that learners with restricted literacy would find a text-based dialogic approach hard to understand and successive versions of SMP materials reduced the reading requirements and hence the dialogic opportunities significantly (there were multiple editions, now all out of print).

You have been given some meter strips of paper and a meter stick and a table to enter your findings. Fold the strip into 2 equal lengths; measure the length of one piece in centimeters and write the measurement in the table. Fill in any other cells that you can in the same row of the table. Look at the column heading to decide what to write. Now fold strips into other numbers of equal lengths and continue to complete the table.

Equal pieces (n)	Fraction of a metre: $\frac{1}{n}$	Measurement in cm	Percentage of a metre	Decimal fraction of a metre	Calculate $1 \div n$
2					
4					
8					
5					
10					
3					
6					

Fig. 5.7 An example of a task to format new knowledge

Anticipation of learners' responses is a key idea in designing tasks to promote mathematical dialogues with the text, and published examples are often trialed before publication. Burns anticipates responses in his tasks previously described (1982). A Portuguese textbook (Gregório, Valente, and Calafate, 2010) provides examples of a new method intended to be used in that page, but the first example $(1 + 2 = 1 + 1 + 1 = 3)$ is ambiguous and could be taken by learners to mean individual counting, rather than an instance of $n + (n + 1) = n + n + 1$. Here, anticipation has not led authors to imagine alternative interpretations. Some texts have specific examples in which there is a thinking frame to help learners think about a solution. Then an actual solution is written in a different font to model the type of response that the learner would be expected to provide; such a method is another way to engage in dialogue between a learner and a static text.

The task in Fig. 5.7 was given to a class of 11-year-olds with diverse prior knowledge. The task shaped the mathematical activity in such a way that the teacher was able to identify particular problems of understanding. It also provided a structure within which learners could work together to show each other how to measure accurately, how to use the algebraic information, how to express a fraction of 100 as a percentage, and so on. Because the class had been enculturated into making conjectures, connections within rows were identified by learners. However, if this had not happened, the teacher could have used a digital version of the table, bringing appropriate columns adjacent to each other so that conjectures could be made. While all learners were working on connections between decimal fractions, vulgar fractions, and percentages, new learning varied from learner to learner depending on what they already knew and could do.

The task in Fig. 5.7 demonstrates a particular strength of text-based tasks, namely, that they offer formats that bring particular features together so that comparisons and connections can be made to show relationships, equivalence, and so on. The provision of tables, grids, sequences, columns, and so on to organize mathematical data can draw attention to connections between different representations or different instances. Comparisons and connections could be engineered to ensure that a critical feature of a concept is foregrounded and that data are structured so that patterns and relations can be sought and conjectures made. Formatting the outcomes of activities is one way that text-based prepared tasks can provide scaffolding for new insights and relational ways of thinking. Texts developed by the University of Chicago School Mathematics Project contain guided examples, with blanks to help students get started on a solution. The well-known use of ratio tables in RME (Corcoran and Moffett, 2011; Van den Heuvel-Panhuizen, 2003) and bars in Singapore (Hoven and Garelick, 2007) shows how consistent use of images in text-based tasks can scaffold understanding.

One issue, however, in all these task examples is that the purpose for the task is not always made clear to students. Particularly when tasks have engaged students in inquiry and discovery, there is a need to bring closure or summary to ensure that students take from their engagement with the task the expected learning objectives. So, opportunities to summarize learning are an essential feature of task design and associated pedagogy.

5.3.3 Coherent Purpose in Collections of Tasks

The role of the teacher with regard to text-based tasks is to mediate between the text and the learner. If that role is passive, the teacher is neither augmenting nor limiting what is offered by the text, whether compliant or dialogic. Of course, there is no guarantee that a learner will use a static text interactively, even if there are interactive prompts such as those in Fig. 5.2. By contrast, teachers who assume responsibility to provide conceptual and pedagogic coherence through their teaching inevitably mediate tasks through the construction of classroom cultures in which tasks direct and shape existing forms of mathematical activity. Between these two extremes, published collections of tasks can themselves provide conceptual and pedagogic coherence through the consistent application of design principles. Firstly, we look at a coherent approach to epistemology and pedagogy.

Herbart (1904a, 1904b) suggested that teachers should raise learners' interest before formal teaching. His approach contrasts with classical texts in which a formal definition is provided first. In Herbart's model, the learner's first task is to think imaginatively about the phenomenon; in the classical model, the learner's first task is to decode the definition and possibly imagine some examples of it. The role of *direct apprehension*, i.e., being provided with a situation or image that embodies a concept, is more than merely motivational; it suggests that mathematics is a process of abstraction of structures, properties, and relationships from specific contexts,

2a. *Find a number in Z_{10} that you can add to 6 to get 0 mod 10. Such a number is the additive inverse of 6 in Z_{10}.*

2b. *Find a number in Z_{10} that you can add to 2 to get 0 mod 10. Such a number is the additive inverse of 2 in Z_{10}.*

2c. *What is the additive inverse of 3 in Z_{10}? Explain.*

2d. *What is the additive inverse of 3 in Z_8? Explain why this answer is different than the answer you got for the additive inverse of 3 in Part c.*

Fig. 5.8 Task sequence to scaffold conceptualization of a new idea leading to a definition (Hart, 2013, p.340)

whereas the definitional approach suggests that mathematics consists of instantiations and use of abstract ideas. An example of direct apprehension is the use of a domino rally to introduce proof by induction. Love and Pimm (1996, p. 375) note that starting with an exploratory task involving inquiry also has implications for authority because the work starts with learners' activity or with the learner's mental model, whereas starting with a definition implies an external authority. They contrast texts which consist of exposition of given ideas with texts which develop meaning through learners' construction.

As Hart says (personal communication, 24 July 2013), "A good definition encapsulates a core idea … But, in terms of learning and task design, they seem to be more effective if they come after instruction, not before. After wrestling with an idea, figuring it out, seeing how it naturally comes up when trying to solve interesting problems, then you say, well, let's call this idea ___, and then define it." Applying this perspective to task design, he provides carefully structured sequences of examples that, on reflection, can be treated as data for conjecture and conceptualization of a new idea (Fig. 5.8). As Hart notes, the sequence of numbers used is critical as part of the design process. Numbers that are special cases and do not assist in developing a generalization are not appropriate as examples in the development of the concepts.

To establish in learners the habit of automatic reflection on collections of examples, this kind of task would need to be used regularly (see also Watson and Mason, 2006). It is more usual, however, for a sequence of procedural questions to be treated by learners and teachers as a sequence of isolated cases. When cases are used to encourage looking for patterns, the specific instances must be chosen with care to ensure they lead to the desired generalization and do not generate a misconception. For example, consider attempts at conceptual understanding of division using the examples in Fig. 5.9.

Note how the task embeds practice of division and encourages a comparison between division and subtraction that may connect them via a *repeated subtraction* procedure. As with some other tasks in this chapter, practice is not only associated with fluent use of procedures but also with insights into relationships. There is a potential problem, however. In the two cases, the divisor and the quotient have the same value. So, learners may correctly write $16 \div 4 = 4$ and $25 \div 5 = 5$, respectively,

5 Design Issues Related to Text-Based Tasks 165

Fig. 5.9 Attempts to connect conceptual and procedural views on division (Adapted from Zorin, Hunsader, and Thompson, 2013)

16 − 4 = 12	25 − 5 = 20
12 − 4 = 8	20 − 5 = 15
8 − 4 = 4	15 − 5 = 10
4 − 4 = 0	10 − 5 = 5
	5 − 5 = 0

Rewrite each set of subtractions as a division.

for the two examples but reverse the meanings of the divisor and the quotient. Modifying the example to be 20–5 = 15, 15–5 = 10, 10–5 = 5, and 5–5 = 0 avoids the potential confusion because the appropriate division sentence is 20 ÷ 5 = 4, and the divisor and quotient cannot be switched.

In the previous two tasks, there is an imperative approach to what needs to be done, but not how to do it. In the Chinese textbook and *Helice* series mentioned earlier in this chapter, an important feature is multiple approaches to each mathematical situation. In a German textbook described by Barzel, Leuders, Prediger, and Hußmann (2013), a consistent set of characters regularly display preferred approaches to solving problems throughout the series, for example, "Till likes to try numbers and to begin a table. Pia likes to explore patterns and to describe a situation algebraically". This embeds the idea that there are alternative approaches to problems that might be valid.

Zooming out to the overall context, another way in which published task collections can establish cultures of mathematical activity is through the inclusion of assessment tasks whose design aligns with the curriculum and pedagogic aims (Thompson et al., 2013). In the cases just considered, method is less important—so long as it is correct—than reflection on the outcomes. If both the formative and summative assessment tasks provide the same coherence and consistency as the intended curriculum, then even if the teacher and learners approach mathematics with a test-focused lens, broad aims might be achieved. Such consistency can be achieved through aligning curriculum, pedagogy, and assessment so that the expected forms of reasoning, the expected communication methods, the connections between and within topics, the representations used, and the connections between them are evident. Consistency also requires that similar things are prioritized and foregrounded in assessment tasks as in the curriculum, not only in conceptual and procedural aspects but also in the nature of mathematical activity.

For example, consider the development of an assessment task as in Fig. 5.10 showing how apparently similar tasks can make different demands on learners, emphasizing first counting and representation, second the action of sharing and a representation, and third the action of sharing and producing two related representations. In the final version, the connection between pictorial and symbolic representation has to be explicit, thus providing insight into conceptual understanding by having to change register, as described by Duval (2006). This task involves more than simple translation, because in each representation some process has to take place, and the processes are different. In words, some imaginary modeling needs to

Adaptation 1. Five friends have 20 pieces of candy to share equally. How many pieces of candy will each friend get? Write a number sentence to show how many pieces of candy each friend will get.

Adaptation 2. Five friends have 20 pieces of candy to share equally. How many pieces of candy will each friend get? Write a number sentence to show how many pieces of candy each friend will get. Use the picture to explain your thinking about the problem.

Adaptation 3. Five friends have 20 pieces of candy to share equally. Draw a picture to show how many pieces of candy each friend will get. Write a number sentence to represent your picture.

Fig. 5.10 Three adaptations to a basic division task to engage learners in different mathematical processes (From Thompson et al., 2013, p. 406)

be done; in the arrays of sweets, some methods of enumeration have to take place. The varied forms of the assessment can align with the nature of the tasks incorporated as part of instruction. Mathematics is presented as being about expressing relations between quantities in alternative representations. In addition, mathematical activity consists of following instructions to draw these different representations, with the implication that comparison will take place to support cognition. These versions show how assessment tasks within a learning management system can express overall aims of the system and also give learners direct insight in the associated values of that system. Appropriate assessment tasks are essential to ensure that collections of tasks have a coherent purpose overall, both instructionally and in evaluative aspects.

Of course, high-stakes assessment tasks structure purpose and pedagogy to some extent, but a recurrent problem is how such tasks can recognize and even measure the development of mathematical behavior. Dindyal et al. (2013) have attempted to make problem-solving activity assessable by using a practical worksheet, similar to worksheets or data recording sheets used in laboratories. By providing a format in which learners can record the stages and processes of problem-solving (see Fig. 5.11), teachers not only enculturate learners into the habits of exploratory group work but also are able to monitor competence and progress in relevant behavior. This is another situation where textual formats can scaffold mathematical enquiry and insights.

Collections of tasks need to have a consistent approach to the conceptual development of the content. We have already compared how different ideas about learning addition can be enacted throughout a text, presenting either the additive relationship or addition techniques. We also gave an example in Fig. 5.5 of how geometric reasoning can be differently enacted. In the respective textbook series as a whole, these differences are sustained; the one based on similarity assumes

> **Devise a plan**
> Write down the key concepts that might be involved in solving the equation.
>
> Plan 1
> 1. Define variables (x and y)
> 2. Set up equations

Fig. 5.11 Outline of part of the practical worksheet with student's response (From Dindyal et al., 2013, p. 319)

that this idea has been understood earlier and the relevant notation already adopted. In the use of mathematical notation, there is little room for variation throughout a task collection as notation tends to be standardized, but in the use of images, development of mental images, inner language, and promoted action, there is more room for variety, sometimes based on cultural expectations.

At times, clashes of images need to be anticipated in collections of tasks within a textbook or between textbooks in a series. For instance, difficulty can occur when learners have depended on a balance model for solving linear equations but are then introduced to a "find roots" approach for quadratics, thus fragmenting their knowledge of solving equations. A similar problem for younger students is having an array understanding of multiplication and then trying to use multiplication for scaling quantities. Text can introduce particular images, but these have to be used in a coherent manner.

5.3.4 Embedding New Knowledge with Existing Knowledge

So far we have discussed tasks for learners to introduce them to new ideas, problems and procedures, and assessment tasks that might follow and convey a particular view of what is valued in mathematics. An intermediate range of tasks can be designed to connect new to existing knowledge, help learners recognize the value of that knowledge, and make it available for future use. Such tasks are particularly important when the main teaching/learning mode is exploratory and divergent because the tasks help relate the exploration to conventional knowledge. There are various ways to address this range of tasks, as described in more detail in Chap. 2: in Realistic Mathematics Education, *vertical mathematization* describes the necessary process of transforming methods used for individual problems into tools for future use (Treffers, 1987); Brousseau refers to *institutionalization* as a process of legitimizing the work done in conventional mathematical terms (1997). In Japan a process of *neriage* (kneading) takes place to bring students' into a coherent whole (Takahashi, 2011). In these approaches, the teaching is vital. In the KOSIMA project (Hußmann, Leuders, Prediger, and Barzel, 2011b), this exploratory process has

Ratios and Fractions!
Consider this statement: The ratio of boys to girls in a class is 1 : 2. $\frac{1}{3}$ of the class are boys, $\frac{2}{3}$ are girls. The table below is based on statements like these. Can you arrange the given numbers into the right place to make similar statements correct? You must use all the numbers and once only. For example, for the statement I have just made the given numbers would be: 1, 1, 2, 2, 3, 3.

Numbers	Ratio of boys to girls	Fraction of class that are boys	Fraction of the class that are girls
3, 3, 5, 5, 8, 8			
3, 3, 4, 4, 7, 7			
4, 4, 5, 5, 9, 9			
1, 2, 3, 4, 4, 6			
1, 2, 3, 3, 3, 6			

Fig. 5.12 Formatted task to help students connect ratios and fractions (From Jim Noble, personal communication)

been a specific focus for research in determining how to coordinate the individual efforts of learners with the intended conventional knowledge and how to support this through the text. Note that all these approaches suggest that the teacher's intellectual input needs to relate to students' activity.

Barzel et al. (2013) offer three components of knowledge organization necessary for learners to incorporate their experience of exploratory tasks into a repertoire of conventional knowledge: *systematization* (structuring results and connecting them to other knowledge), *regularization* (transforming into the conventional repertoire), and *preserving* (writing in an accessible form). These processes do more than *institutionalization* by also focusing on personal conceptual development and recording. Tasks can align divergent experiences toward shared understanding of concepts and procedures by embedding technical language and conventional symbolism, relating definitions to recognition, and providing individuals with opportunities to express in words and symbols.

A good teacher can provide knowledge organization experiences by orchestrating students' ideas as seen in the task designed by teacher Jim Noble (personal communication, 8 April 2014) when he found that some of his students were confused by information in a textbook that ratios could be expressed as fractions, e.g., 1:2 could be expressed as 1/2. He created a formatted task to help them compare meanings of ratios and fractions (Fig. 5.12). Note that his use of fractions differs from that in the textbook they were using.

Published tasks can also provide these reflective perspectives. KOSIMA includes many strategies for knowledge organization as specific tasks, for example, after studying parallel and perpendicular pairs of lines, students are offered several statements from which to choose *the best* descriptions of parallelism and perpendicularity (Barzel et al., 2013).

5.3.5 Summary

In this section, we have talked mainly about the purposes a designer might have for tasks to address learning or assessment and have given examples of how these purposes are turned into design parameters for collections and sequences of tasks. In the next section, we focus more systematically on desirable forms of mathematical activity and whether these can be shaped by text-based tasks.

5.4 Intended/Implemented Mathematical Activity

The final node in the task design intention triangle relates to the mathematical activity of the task. Design issues related to mathematical activity relate to principles about learning and mathematical aims of particular tasks. We discuss both of these issues in turn.

5.4.1 Principles About Learning

We cannot talk about connections between tasks and learning without some theories about how pedagogy shapes mathematical learning and hence design frameworks as described in Chap. 2. Here, we add to the arguments in Chap. 2 by providing examples of these principles in action in individual test-based tasks and what they might look like on a page or screen.

The idea behind a cognitive conflict approach is that learners are presented with situations that conflict with ideas they already have, so that their ideas have to be modified to incorporate new experiences. Tasks have to bring these conflicts to learners' attention by leading them to become stuck if they continue with the old idea. Learning in this theoretical frame means to adapt, alter, or extend a previous notion, so tasks have to present opportunities to use previous notions and then find contradictions or puzzles that need to be resolved. Many other examples of tasks evincing the resolution of cognitive conflicts created through paradoxes that can be integrated in school curriculum can be found in Movshovitz-Hadar and Webb (2013) and also in Swan (2006).

As an example of how cognitive conflict can be used to extend learning, Barabash (personal communication, March 2014) points to potential conflicts between learners' early conceptions of the tangent concept that do not hold in more advanced settings. In the Israeli curriculum, the concept of *tangent* is introduced in geometry as a tangent to a circle and then in precalculus as a tangent to a parabola. For both the circle and the parabola, the tangent is intuitively understood or actually defined as "if a line has common points with a curve, then it **either** intersects it (in two points!) **or** is tangent to it." However, this result is only true provided the curve is seen as convex and smooth; developing the concept of tangency on this basis is not

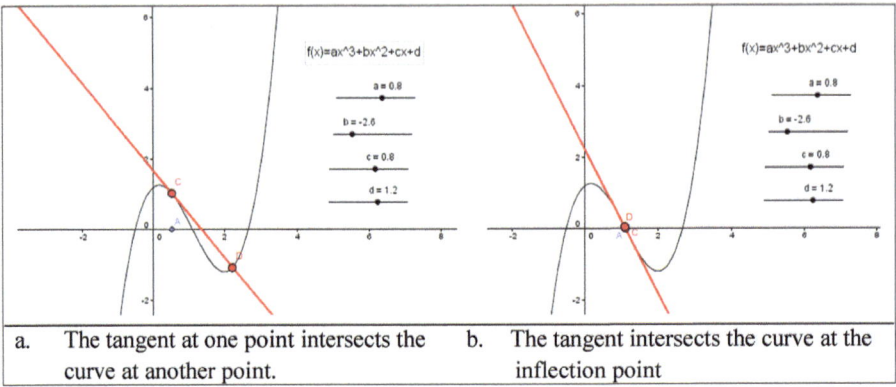

Fig. 5.13 Illustrations in which the concept of tangent conflicts with formal definition as a line that intersects a *curve* at only one point (diagrams from Barabash, personal communication). (**a**) The tangent at one point intersects the *curve* at another point. (**b**) The tangent intersects the *curve* at the inflection point

always valid as illustrated in Fig. 5.13a, b when the curve has power higher than 2. Thus, the means for introducing tangent in early instances and with the given definition are in conflict with later, more advanced perspectives on the concept. The task is to redefine tangency given these two juxtaposed examples. In presentational terms, the graphs can be compared side by side to identify what is the same and what is different.

In Variation Theory (VT), the idea is that learners will notice what is varying against an invariant background (Marton, 2014). Mathematics tasks should be designed so that the desired key idea (known as the *critical aspect* in the theory) is varied and learners can see this and the effects of such variation in successive examples. A full application of VT requires the initial identification of a critical aspect to be learned, and it is this that will then be varied. Such identification takes place through a phenomenographic analysis of learners' work in a particular context, so it could be argued that no static text can fully use VT. Nevertheless, the theory does draw attention to the importance of organizing variation in learners' experience, the *space of learning*. Dynamic digital environments are very useful for this type of variation as variations of a parameter of an object and variation in a representation of the object can be seen at the same time or soon after each other. For instance, imagine learners using a graphing tool to graph $y=x$, $y=2x$, $y=3x$, $y=-x$, $y=-2x$, and so on. Learners should quickly be able to determine that all lines of the form $y=mx$ go through the origin and that the value of m determines the slant and steepness of the line. In static environments, near-simultaneous or adjacent presentation can have a similar effect. The presentation of the sequence of examples needs to make clear what features of a concept are varying in order to show relationships between different aspects of the mathematical idea. The learner is to recognize and generalize the relationship between variables.

ATD (Anthropological Theory of Didactics) and RME are both ways to engineer a need for a formal mathematical idea. Fujii gave an example of such a task (2015,

Chap. 9, this volume) in which learners were given various distance-time relationships and asked to identify which relationship represented the *fastest*. It is impossible to give a typical example of an RME task in text, because the nature of the resources varies widely. One type of task is to present a picture, such as a stack of oranges, and invite questions to be posed and generalizations made about piles of oranges of various heights and widths (Dickinson and Hough, 2012). Another type is to offer a realistic problem and a format for the findings of the problem-solving process, such as a bar diagram, which can then become a model for future reasoning (Van den Heuvel-Panhuizen, 2003).

In all these theoretical frameworks, the main pedagogic purpose is to learn new mathematical concepts and methods. In the first two (cognitive conflict and variation theory), this is generally achieved by working on given examples and then comparing and reflecting on the outcomes. In the second two (RME and ATD), learners have often to be exploratory and exert some mental effort to suggest solution paths. Most of the tasks presented so far can be viewed as examples of one or more of these approaches, but Fig. 5.3 offers nothing but the opportunity to mathematize a situation. The teacher could use this figure to develop the need for the idea of proportionality or as a context for applying proportional reasoning. By contrast, Fig. 5.4 is a relatively closed task that can be treated merely as practice in completing linear sequences, but a teacher could then encourage reflection and comparison of outcomes to develop algebraic understanding of linearity, possibly using learners' conjectures to do so.

Tasks on their own are unlikely to address all the complex aims of the mathematics curriculum, particularly those that are about developing mathematical behavior, and the authors' intentions depend on associated pedagogy. Note that the author might be the teacher (as in Jim Noble's task in Fig. 5.12) and may have produced the text-based task to support complex pedagogic aims. The extent to which the teacher understands and supports the pedagogic aims of the text influences the manner in which adaptations are conducted in order to maintain those aims—the issue of curriculum vision and trust (Drake and Sherin, 2009) discussed previously.

5.4.2 Aims Enacted Through Individual Tasks

In earlier sections, we have sometimes described the mathematical activity prompted by a task. We now systematize this from the perspective of desirable mathematical activity. In Chap. 2, there is a call for more focus on the grain size of frameworks as research in task design would benefit from further clarity about different levels of activity. Because the purpose of tasks is to promote mathematical action, we look at grain size from the point of view of actions:

Grain Sizes of Mathematical Actions

i. Basic actions
ii. Transformative actions

iii. Concept-building actions
iv. Problem-solving, proving, and applying
v. Interdisciplinary activity

Compliant and passive learners expect to undertake basic actions of type (i) and can begin to get stuck with transformative actions of type (ii). A common pedagogic approach to overcoming these difficulties is to routinize type (ii) actions by providing rules for transformation, such as *change the side, change the sign* as a routine when solving linear equations, or *FOIL* (first, outside, inside, last) as a routine for multiplying two binomials. Type (iii) actions have been available to learners in many of the task examples presented so far, sometimes explicitly and sometimes implicitly. Where these are implicit, explicit pedagogy and the development of appropriate cultures of classroom mathematical work can ensure they take place. Actions of type (iv) are expected in most national curricula and statements of educational aims, but an overreliance on routinizing type (i) and type (ii) actions and a lack of explicit focus on type (iii) actions can make type (iv) actions, and hence type (v) actions, hard to achieve. We also note that some would say the outcomes of type (iv) activity are essential components of concept building, and we would agree. However, here we are focusing on how a designer needs to imagine what the learner is actually going to DO in response to the task, rather than only imagining what MIGHT be possible with supportive teaching.

It is helpful to think in terms of desirable mathematical behavior that needs to be promoted by tasks. Cuoco, Goldenberg, and Mark (1996) describe *habits of mind*, or what several people call the *verbs of mathematics*, as starting points for tasks (i), (i), and (iii). Similarly, Schoenfeld (1985) and Mason, Burton, and Stacey (2010) provide ways to think about tasks of types (iv) and (v). For this chapter, our main goal is to indicate behavior which can be triggered by text-based tasks, how this can happen, and what remains the domain of pedagogy, particularly the creation of certain classroom cultures.

In Table 5.3, we elaborate on actions at different grain sizes but do not claim that these are mutually exclusive or that the table is complete.

This approach to thinking about what learners need to do omits some aspects of mathematical experience, such as:

- The need to talk, write, and listen to mathematics
- Reasoning for different purposes, e.g., to conjecture, persuade, and prove
- The need to use mathematical feedback, such as self-correcting, monitoring overall sense, understanding comments from others, and appreciating a need for consistency
- Seeing mathematics as part of citizenship—information for understanding the world
- Relating mathematical work to other human values

We see these aspects as pertaining to all grain sizes of mathematical actions. None of these can be embedded in learners' experience solely by indicating them in text or including an opportunity in a task; there has to be the associated pedagogy to ensure they happen. For example, Simon's design of a questioning sequence used

5 Design Issues Related to Text-Based Tasks

Table 5.3 Actions for different grain sizes of mathematical work

Grain size	General focus	Examples of specific actions
i	Basic actions	Calculating, doing procedures, stating facts
ii	Transformative	Organizing, rearranging, systematizing, visualizing, representing
iii	Concept building	Sorting, comparing, classifying, generalizing, structuring, varying, extending, restricting, defining, specializing, relating to familiar and intuitive ideas
iv	Problem-solving, proving, applying	Conjecturing, assuming, symbolizing, modeling, predicting, explaining, verifying, justifying, refuting, testing special cases
v	Interdisciplinary activity	Incorporating other epistemologies, identifying variables and structures, recognizing similarities, comparing familiar and unfamiliar knowledge

to move Erin toward understanding the traditional fraction division method depends on being responsive to her (described in Chap. 2). He tries to identify, through observing patterns in talk, when she has developed a new schema, so the next question he poses must provide reflection on this new schema toward abstraction. In this way, Simon structures a task that sounds like type (i) but has the effect of moving through to type (iii) as she extends the domain of application and then generalizes her ad hoc and visual reasoning into an algorithm. Although it would be possible to write the questions as a sequence in text, it would not be possible to hold them until the right pedagogic moment arises for them to be effective in changing understanding. Human dialogue is necessary, even though the outcome will be a calculation procedure.

5.4.3 *Complex Aims Enacted Through Large-Grain-Size Tasks*

It is possible to shape experiences related to more complex activity through providing tasks in textual form. Movshovitz-Hadar and Edri (2013) developed an approach to help teachers bring values into the mathematics classroom (Fig. 5.14).

By reverse engineering the outcomes of the project (as is also applied to educational tasks by Amit and Movshovitz-Hadar, 2011, p. 176), the authors present four issues related to design that make this approach manageable for teachers and learners:

1. Tasks are based in the mathematics curriculum and designed to last one class or homework session.
2. They include a clearly phrased introduction followed by two kinds of short questions: (i) mathematical exploration or thinking and (ii) dialogue to clarify values using mathematical and other perspectives.
3. Editing to avoid obstacles.
4. Advice about the social mode of working: group, individual, discussion, and so on.

> In 2007, minorities (Arabs, Druze, and Circassians) were one fifth of the population in Israel. Despite this, only 6.2% of all civil service employees in this year were minorities. Over the years, the Israel government has made decisions (in 2004, 2006, and 2007) to promote suitable representation of minorities in the civil service, setting 10% as a target for the percentage of employment of minorities in the civil service.
> 1. What do you think about the goal that was set by the government?
> 2. The Ministry of Housing and Construction had 741 employees in 2007. Had the target set by the government been achieved, how many members of minorities would have worked in the Ministry of Housing and Construction?
> 3. Twelve employees in the Ministry of Housing and Construction were minorities in 2007. What percentage of all employees in the ministry were minorities?
> 4. What do you have to say about the two results you obtained?

Fig. 5.14 Values task for Israeli classrooms (From Movshovitz-Hadar and Edri, 2013, p. 382)

Consideration of values-focused tasks raises the question of the starting points for task design and for presentation of the task in text. Is priority given to the context and the mathematical perspectives that emerge from it? Or is priority given to a mathematical idea and the context is then built around it? Design might have problem orientation, concept orientation, or context orientation (Nikitina, 2006). For example, within variation theory the critical aspects of the mathematical concept have priority; however, when interdisciplinary work is an explicit aim, priority might be given to context. In the COMPASS project, Maaβ et al. (2013) report on their development of design issues for interdisciplinary tasks. This was an international project, so thought had to be given to different prevailing pedagogic attitudes and ICT use in the participating countries. The aim was to produce digital and paper-based resources for dissemination beyond the development process, and these had to communicate clearly the key mathematical and scientific ideas so that teachers and learners could see how these emerged from their work. Each task had to make reference to the appropriate national curriculum to encourage teachers to use it. The most appropriate pedagogy for interdisciplinary tasks of this type would have been extended exploratory project-type work, but the designers also provided more structured versions to support teachers who were not confident enough to undertake long tasks. Task designers gave considerable thought to how complex materials could be made teacher friendly and easy to use. Further issues, many of which emerged during the design research process with teachers, included:

- The need for an overview of the lesson sequence, with tasks and subtasks.
- Tasks presented so they could be used directly in lessons without transformation.
- Questions needed for guiding learners.
- Clear links between subtasks and the big contextual questions.
- Possible solutions.
- Information about adapting tasks for different learners.
- Different materials (e.g., task sheets, solutions, background information) had to be easily distinguished at first glance.

5 Design Issues Related to Text-Based Tasks   175

What are the main sources that cause severe water shortage in Europe? How do different harmful elements pollute freshwater? What methods / strategies can be used to save water?	**Task 4: Desalination World!** Students learn about potential future sources of drinking water. Students, through a hands-on activity, learn about the desalination process.
What methods can be used to provide people with freshwater? How much freshwater can be conserved if strategies and attitudes at a personal level are adopted?	**Task 5: Building a new Desalination Plant!** Investigate which is the optimal place to build a new desalination plant in order to serve in a fair way the needs of four cities. Students explore different quadrilaterals and their properties (diagonals, collinear points, angles)

Fig. 5.15 Examples from cross-curriculum task in COMPASS (2013© 2010–2011 COMPASS project reuse under Creative Commons Licence)

One such project is about water shortage. The subtask materials are too complex to give in full here. The guiding questions and two of the subtasks are shown in Fig. 5.15.

The issues of design for COMPASS are useful guidelines for the production of any multimedia curriculum package. They address key questions in interdisciplinary work about the balance between context for mathematical learning and the use of mathematics to support learning in other disciplines.

Staats and Johnson (2013) tackled the problem of algebraic competence at college level through a novel interdisciplinary approach. They adopted the typical social science pedagogic method and provided tasks coauthored by a mathematics teacher, a disciplinary specialist, and a creative writer. The creative writer prepares an engaging presentation of a scenario. This is followed by explicit learning goals for both disciplines for class discussion, with scaffolding questions and a short supportive bibliography. We do not have room here to present a full narrative but will summarize the content of one module (for more details see z.umn.edu/icmi22). The core of the module is a short story titled *Indebted,* in which a young man wrestles with the question of how to pay for his college education. The young man visits his grandfather, who suffers from Alzheimer's disease. The grandfather hoped to contribute to his grandson's education but instead had to use his savings for his own care. The young man considers mathematical scenarios associated with indebtedness, such as rapidly rising college tuition and the per capita value of the national debt. Finally, he signs his college loan papers. The learner recognizes the social and emotional dimensions of the problem as well as the mathematical issues.

So far in this section, we have talked about tasks that address the largest of our grain sizes but include aspects of smaller grain size, mainly type (i) and type (ii) with other types to obtain models for prediction purposes. This is not a surprise as most of the research in task design addresses complex aspects of mathematical work. However, there is much to be understood about the design of tasks that scaffold basic, transformative, or concept-building activity. Each of the actions in Table 5.3 could be triggered by an imperative, such as *put in order*; *classify according to …*; *give three contrasting examples of …*; and *prove that ….* As the professional development procedures of lesson study and learning study have shown,

Fig. 5.16 Marks left by *solids* from static contact with a surface or by rolling; the task is to determine which solid could have left the given mark (Barabash and Guberman, 2013, p. 297)

the actual examples used and their availability to learners make a difference in opportunities to learn. We give some insight here, and there are others throughout this chapter.

The example comes from Barabash and Guberman (2013). In writing textbook tasks about shapes, they had to choose between (a) allocating a lesson to each solid in turn and then asking for comparison or (b) looking at all relevant solids together and finding their common and special properties. They chose (b), an approach which brings type (i), (ii), and (iii) actions together, rather than (a) which climbs up from (i) to (iii). The authors embedded the approach in a problem situation in which an intruder has left traces of certain solids in the form of stains made from static contact or traces left by rolling solids (see Fig. 5.16), thus working with type (iv) actions. Their overall aim is to develop *mathematical insight*, which is akin to the organization of knowledge described by Barzel et al. (2013). Their approach is shaped at the start of the work by the expectation that mathematics is going to provide the analytical tools to identify the solids. Students explore with the solids, make conjectures, and then refine their conjectures.

Note that we have addressed all grain sizes from the zoomed-in view of learners' experience, rather than from a zoomed-out view of curriculum aims. How learners can be "ramped" from simpler tasks to more complex tasks, where their previous experience has not prepared them for complexity, is a related problem. One useful approach is to use a gradient of "novice", "apprentice", and "expert" tasks (MARS, 2014). These descriptions were developed for designing assessment tasks but could be used equally well for the development of complex mathematical habits of mind.

5.5 Visual Features of Text-Based Tasks

Text-based tasks are planned, prepared, and presented to learners visually; they are not the tasks that arise within teacher-learner dynamics. Most of our discussion is about the work that goes into planning and preparation, but the experience for the learner is firstly visual. For this reason, research about learners' experiences with text-based tasks needs to take into account many of the same perceptual impacts as might be considered by graphic designers. However, we found little research which related perceptual impact to mathematical cognition. In the graphic design literature, we read that the learners' attention can be directed in sequence to particular features, as in statistical representations, but making mathematical sense of those features and coordinating with other features are not widely discussed across mathematics. In a few textbook comparisons, researchers draw attention to the use of color, pictures, text boxes, and so on and whether these are used for cognitively specific purposes or whether they are merely to make a page appear attractive to young learners. Thompson et al. (2013) classify the use of graphics as was previously shown, but we repeat a shortened form here: no graphic; graphic does not illustrate inherent mathematics; graphic explicitly illustrates inherent mathematics; graphic has to be interpreted to answer a question; and make a graphic.

When we talk of visual features as part of a task, we are not interested in the use of color, pictures, or position merely for visual attraction but at how those features contribute to learners' mathematical activity by enabling coordination of the eye and brain. In the static environments that are our focus in this chapter, we are also interested in how pictures and diagrams can suggest action. For example, in Fig. 5.6 the pictures suggest both collation and separation using place value. In Fig. 5.10, the sweets suggest systematic enumeration.

In Fig. 5.6, the pictures have a deliberate cognitive purpose in that they illustrate actions which can contribute to an understanding of the symbolic representations that follow immediately. In Chap. 2, Fig. 2.7 demonstrates a similar set of pictures and diagrams. Symbolic statements are placed next to each other when they relate to each other in particular structural ways or follow each other in a deliberately varied fashion. Without the need for mediation through speech, a learner who deciphers the page from top to bottom and left to right has the information they need to complete the suggested statements. In Fig. 5.4, the layout confirms for the learner that they need to fill in some blank spaces, and when this is done there are some relationships to be found. In other words, the layout encourages comparison, conjecture, and generalization between sequences rather than merely completing them separately. However, visual similarity does not always imply mathematical equivalence and learners have to sort out when it does and when it does not.

Learners have to make a distinction between pictures and diagrams. For example, Curcio (1987), among others, describes learners being over-influenced by the shape of graphs when matching them to situations that might be generating the relevant data and points to confusion between words such as higher, faster, and lower and the shape of associated graphs. In several studies, learners are seen to react visually to diagrams that need to be interpreted symbolically (Dörfler 2005; Radford, 2008). In geometrical reasoning, a diagram has to be understood not as an accurate case but as a representation of a system of relations and properties. Moreover, both diagrams and pictures can introduce simplifications or elaborations which could mislead novice learners. For example, a vertex could look as if it is a right angle when it is supposed to be general; a learner might assume that the base of a triangle has to be parallel to the page edge. Love and Pimm (1996, p. 380) point out that dynamic digital technology will help learners to understand that a single example is an instance of a class of figures and that a specific example can be manipulated to address these potential misconceptions. Puphaiboon, Woodcock, and Scrivener assume a dynamic environment, saying "The graphic representation must portray the relationship between the graphical parts in time and space to reinforce cause and effect relationships" (2005, p. 3). Static images have to embed this dynamic relationship and be understood as instances of a variable class so that they direct the learner toward the salient features of the class.

Tufte shows a variety of ways in which quantitative and dynamic data can be represented through static diagrams, such as through labeling, encoding, relating data to familiar scales, etc. (1997, p. 13). He claims that good design enables attention without clutter so that "clear and precise seeing becomes as one with clear and precise thinking" (p. 53). Shuard and Rothery (1984, p. 61) draw attention to the use of arrows in text to indicate some movement and action, the static equivalent of mouse clicks and dragging; the use of arrows is a convention that learners know from outside the classroom but that might have a special meaning and use inside mathematics. Diagrams make a difference to learning; so long as the diagram and its associated text are near each other so the eye can move back and forth, relative position does not matter (Shuard and Rothery, 1984, p. 53). Some recent research using eye tracking to determine if experts and novices *read* tabular data differently has shown no differences, but each participant in the study seemed to have a personal pattern of engagement with such data (Crisp, Inglis, Mason, and Watson, 2011). More work is needed in this area to find out how learners acquire and coordinate information from a mathematical text.

The use of color for specific mathematical purposes became established in nineteenth-century geometry teaching, for example, Byrne's edition of Euclid (1847) color to draw attention to different objects and quantities whose relations could then be understood spatially. Fig. 5.17 gives a sense of the use of color to compare objects.

Color is widely used in the teaching of algebra to draw attention to like terms or to provide spatial patterns to be generalized (e.g., showing $(x+y)^2 = x^2 + 2xy + y^2$ with appropriate shading), as a way to draw attention to area as the space inside a closed 2-d shape and so on. Indeed color is used for a mixture of mathematical and

1.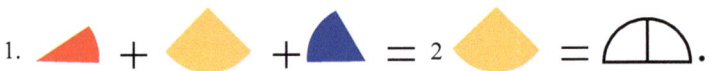
 That is, red angle added to the yellow angle added to the blue angel, equal twice the yellow angle, equal two right angles.

2.
 Or in words, the red angle added to the blue angle, equal the yellow angle.

Fig. 5.17 Use of *color* to relate objects (Byrne, 1847, p. x)

attentional reasons to support the distinctions learners make when understanding a mathematical idea, but it is also used purely for visual variety. We wonder how learners learn when to use color cognitively and when it has no mathematical purpose. The use of specific color words disadvantages color-blind learners (up to 10 % of boys and 0.5 % of girls can be color blind). In Fig. 5.17, the words *red*, *yellow*, and *blue* will confuse a significant number of people, so although different shades may be perceived, it is better to refer to them in some other way, as is done in Japan (Ohtani, personal communication, 22 May 2014). The key idea is how the text draws learners' attention to examples and how they relate to each other. As well as the use of emphasis through color and layout, the control of variability and the near-simultaneous presentation of variation are key factors in the kind of attention that is necessary for type (ii) and (iii) activities. Tasks in Figs. 5.5 and 5.9, among others, demonstrate the use of juxtaposition to elicit conjecturing and generalization.

Thinking about the page as a whole, Kress and Van Leeuwen (1996) use the metaphor of an art gallery or museum (see Yerushalmy 2015, Chap. 7, this volume) to think about where to position items to direct the learner in a logical or developmental order while making ancillary information and elaborations available through hyperlinks. Although difficult to replicate on the printed page, such links could easily be provided in digital text. Cognitive load theory, which is concerned with finding the optimal number of ideas that can be handled to understand a concept while not oversimplifying it, also has a role to play in the preparation of a page. The learner should be able to follow a pathway through the text that allows access to core ideas, possibly through various representations and instances, without becoming too distracted by irrelevant details; in essence, the learner needs to distinguish the core idea from other material and cannot learn to do that if it is always presented in isolation (Love and Pimm, 1996). In cognitive load theory, researchers are concerned with whether the content is intrinsically necessary for the object of learning or germane to it or extraneous (e.g., Paas, Renkl, and Sweller, 2003). We would argue that the grain size of the pedagogic intentions determines whether these loads are desirable or not. For type (i) activity, only intrinsic content is necessary; for the other types of activity, mixtures of intrinsic and germane and even extraneous content are desirable. In variation theory, it is suggested that background variables (i.e., those that are not the critical aspect for learning) should be kept invariant but need to be present to enable variation in the critical aspect to be observed in the foreground. An alternative is to have variation of many features but invariance of a key feature. Examples could be a collection of contextual problems that all have the same underlying structure when the structure is the intended object of learning or a collection of quadratics that all have the same roots when roots are the intended object of learning.

An associated factor is whether collections of exercise questions in grid form are done horizontally or vertically and whether it matters. The example from an old algebra text in Fig. 5.18 shows that it does matter whether the learner follows rows or columns. The authors of the exercise seem to be aware of the value of organizing

Fig. 5.18 From *Elementary Algebra*, Part 1 (Godfrey and Siddons, 1915, p. 43, Cambridge University Press)

1. $\frac{3}{11} \times 5$ 4. $\frac{5}{12} \times 3$

2. $\frac{5}{11} \times 5$ 5. $\frac{7}{12} \times 3$

3. $\frac{x}{11} \times 5$ 6. $\frac{x}{12} \times 3$

Find the sum of:
1. $a + b$ and $a - b$ 2. $2x - a$ and $3x + a$
3. $-x + a$ and $x + a$ 4. $2x + a$ and $3x + a$
5. $a - 3b$ and $a + 2b$ 6. $2a - b$ and $3a - b$

Fig. 5.19 From W. Baker and A. Bourne (1937, p. 19)

variation in examples, and the numbering order encourages comparisons, so that the role of the numerator can be reflected upon.

In Fig. 5.19, the authors appear to be aware of the importance of variation, but the layout and order provide several variations between successive questions, so that little reflective awareness is available. If the first two questions had been: $a + b$ and $a–b$; $−b + a$ and $b + a$, there would have been something to notice and justify which could have supported conceptual learning. A type (i) task could have become a type (i), (ii), (iii), and even (iv) task in this way.

The amount of writing may be an issue for some learners. Shuard and Rothery (1984) studied 400 learners' reactions to versions of a task with more or less writing required and found that reactions varied (p. 130). Learners' reactions may be due to past experience, causing Shuard and Rothery to conclude that people need help to learn how to read mathematical text. Reactions to an exposition presented in comic strip style also varied inconclusively, but there is now a range of popular resources of this style on the internet.

The use of background effects such as grids, frames, fills, and so on can enhance attention and avoid the *flatness* of appearance of the page (Cuoco, 2001), in which every part of the text appears to have a similar status. However, care has to be taken that such effects do not mislead readers. For example, presenting rectilinear shapes on squared grids can mislead learners into counting boundary squares to find the perimeter.

In addition, the choice of font, or variation of fonts, and length of lines of print can influence learners' attention and reading capability. Such features can even influence the ways in which teachers and their learners engage with the text, an interesting example of this being the use of handwriting in the RLDU materials (e.g., Fig. 5.2) establishing a sense of mathematics as a human and exploratory endeavor. However, in Shuard and Rothery's study of 400 learners aged 11–12 using handwritten text, some found this font friendly and helpful, but others found "funny writing" harder to read (1984). Literacy in the language of instruction is an omnipresent issue as are broader issues about language and mathematics.

In the discussion so far, we have recognized the difficulties designers have in enacting their intentions through their design. In every case, we have focused not on what WILL be learned but the opportunities made available to learn by each task. We now turn to how designers need to anticipate teachers' use of tasks (see also Chap. 3).

5.6 Teachers' Use of Tasks

Design and use are like two sides of a coin and both are influenced by the educational system, assessment system, culture, and other contextual features. The job of design is to communicate to teachers and learners through text the mathematical intentions; the teacher's role includes modification to enable learners to connect to the core ideas and learning goals (Love and Pimm, 1996; Rezat, 2006; Tzur, Zaslavsky, and Sullivan, 2008). Assuming that the designer is not the teacher, we ask whether the design assumes that teachers can use the task as written, or whether teachers will need to adapt the task for their context. The latter assumes that teachers have the motivation, time, and knowledge to make adaptations, whereas the former masks the fact that teachers are likely to make adaptations anyway, deliberately or not. Teachers are integral actors in the whole process of design when they use published tasks in their classrooms, with the tools, cultural expectations, and norms of classroom life. An important aspect of professional learning is to become critical users of an externally designed task. There has always been debate about this. Wittmann (1995) wanted to preserve task design as a specialist process undertaken by those who have time and experience to develop tasks that are to some extent *teacher proof*. Stein, Grover, and Henningsen (1996) point out that there will always be adaptation of tasks in use, at least because of classroom dynamics and at most because of teachers who alter the goals and demands of tasks. These issues are explored further in Chap. 3 of this volume.

Prestage and Perks (2007) offer a collection of task-adapting tools, with which busy teachers can develop complex tasks from textbook resources: change a given, add a mathematical constraint, change representations, and so on. Whereas designers have more time, experience, and access to research than busy teachers, teachers have more local knowledge but need design adaptation tools of this kind. Swan (2006) also provides design heuristics that could be used by teachers to create and adapt tasks:

- Is a statement always, sometimes, or never true?
- Interpret, match, and classify different representations of similar objects.
- Diagnose and correct examples of common mistakes.
- Resolve cognitive conflicts.
- Create new problems by reversing given problems or varying givens.

The final type can be extended by turning givens into variables, a technique also suggested by Prestage and Perks (*ibid.*). In an elementary example, given that $8=3+5$, tasks could be created to find x where $x=3+5$; $8=x+5$; $8=3+x$ or x and y where $x=y+5$; $8=x+y$ and so on. This sequence shifts learners from a number fact to dealing with an unknown, to dealing with variables and, finally, $z=x+y$ could be an exploration of structure. Watson and Mason (1998) collected generic task design heuristics by providing a range of actions that can be applied to mathematical objects: classifying, ordering, defining, constructing, varying, reversing, exemplifying, and so on. Using these, teacher task design can be a repertoire of in-the-moment strategies rather than a time-consuming process.

Lee et al. (2013) observed how teachers modified tasks and noticed they would typically change the givens or the context or the demands of the question. One example they shared is as follows: a rectangle of paper is folded in half and then cut along the diagonal of the new shape; the textbook asks for the name of the resulting shape and where equal angles might be found. One teacher made the task more open ended by asking students to find the properties of the resulting shape. In their research, the mathematical knowledge of the teachers played an important part in their decisions to improve text-based tasks. In contrast, Lundberg and Kilhamn (2013) show how problems inherent in a published task derailed a teacher who relied on the textbook to prepare learners to solve some ratio problems. The problem involved mixing lemon squash using juice, water, and sugar and asked for measures in liters. They report widespread confusion about whether the sugar contributes any volume to the drink, confusing everyday knowledge and mathematical assumptions. They also report that teachers resorted to ad hoc additive methods rather than setting up a multiplicative equation as the authors expected.

In several papers referred to in this chapter, teachers have been involved throughout the design process (e.g., Hußmann et al., 2011a; Movshovitz-Hadar and Edri, 2013). In many countries, teachers are involved in collaborations that produce banks of tasks, shared among teachers (e.g., Sesamath, Wikitext, SMILE). There is a growing use of digital sharing which enables individual teachers to adapt the text, the examples, and the language for their own learners. This massive growth of resources places an increasing burden on teachers who design or selectively choose tasks rather than rely on the authority of an unknown author from the web. It is safe to assume that the reason for proliferation of such resources is teachers' dissatisfaction with commercially published materials and inability to find published tasks which address precisely their teaching goals. There is also software which supports teachers' creation of worksheets, sometimes from banks of individual questions, and video resources. This could be seen as rejection of the authority of textbook authors, publishing houses, or outside designers.

Designers, by contrast, are often concerned with how to make their intentions explicit to teachers and what support to provide to enable tasks to be used as intended. It is likely there needs to be a professional development component in the textual presentation of the task, and teachers need time to prepare themselves to use a task fully. Even though there is a need for research about whether, how, and why teachers pursue the explicit intentions of task designers, such research will always

5   Design Issues Related to Text-Based Tasks   183

be contextualized within the normal pedagogic practices of the research sites. So, such research might be seen as local, specific evaluation.

Other designers might want tasks to provide something that can be used directly by learners without teacher mediation. When these are published as collections, such as in the textbook or as a package of worksheets, there may be consistent, strong messages about how to study. For example, readers might be asked to predict answers or reflect on key ideas that arise in a task. In a package of tasks that is developed over time with classroom trialing, the question of how, whether, and why learners take up these messages would be a key aspect of evaluation.

One particular aspect of teacher adaptation that may be of general concern is when tasks are adapted so that learners of different capabilities can work with them. Such adaptations can simplify or extend the learning goals or simplify access or both. Designers might indicate ways in which tasks can be adapted that maintain the core learning afforded by the task.

Moving from the use of individual tasks to the sequencing of tasks, again much depends on the prevailing culture. For example in the UK, the roles of good mathematics teachers are to provide the curriculum and cognitive mathematical coherence and to adapt text not only to engage learners but also to help them organize their knowledge (in the sense offered by Barzel et al., 2013). Ancillary materials and teacher guides might be available but are not necessarily used consistently. The use of teacher guides and textbooks in some cultures is seen as central to professional practice; for example, the *kyozaikenkyu* phase of Japanese lesson study uses these, while Chinese teachers claim to learn most from the teacher guides (Fan, 2013), and *concept study* is a growing practice in Canada (Davis, 2008). Some schemes provide no learner textbook but offer teacher-friendly guides and resources such as photocopiable masters, apps, and online resources. Teachers who engage in these schemes have to engage with the guidance, possibly collaboratively, and develop their own teaching from the scheme. By contrast, there are schemes which provide detailed lesson scripts or videos to copy. There is, therefore, a spectrum of practice and expectations, ranging from using the task sequence provided so as to adhere to the implied theories and goals of learning and development behind such sequencing, to teachers developing their own sequencing and populating it with tasks from a variety of sources and media. The relations between tasks and teaching are pluralistic and situated.

5.7   Conclusion: What Text-Based Tasks Can and Cannot Do

To introduce this section, we present a thought experiment as an extension to an example given earlier from Taiwan about plausible pedagogic approaches to the interior angle sum of any triangle, assuming that learners understand what angles are.

- Teachers could state the property and then give learners various triangles with two angle measures so that learners use the property to find the measure of the third angle.

- Teachers could have learners draw various triangles, measure all the angles, and then put the sum of the measures of the angles at the front of the room, presumably having most sums close to 180°.
- Teachers could construct a triangle using geometric software, have learners measure the angles, place the measures in a table, and then drag one vertex of the triangle and record the angle measures as they update, again finding that all sums are 180°.
- Teachers could have learners draw various triangles, tear the triangles into three pieces without tearing through the angles, and then place the angles adjacent to each other to demonstrate that the three angles form a straight line.
- Learners could be told that angles round a point add up to 360°, given a tessellation of the plane by congruent triangles, and asked to use logical reasoning to deduce the angle sum of any triangle.

The underlying mathematics concept is the same in all five tasks, but the nature of the mathematical activity embedded within each instantiation influences the degree to which learners develop sensemaking, reasoning, and a justification that such a relationship is true for all possible triangles. The version of the task used by the teacher depends to some extent on a curricular and pedagogical vision of learning. It is possible to imagine all of these presented as written text to learners, especially the prepared triangles in the first suggestion. However, could learners follow the instructions (especially in the fourth version), and, if so, would they come to the conclusion that the angle sum is 180° without a further lesson phase of regularization, systematization, and verbalization as described by Barzel et al. (2013)? In this thought experiment, nothing needs to be prepared on paper apart from a bank of examples in version 1 or the materials for version 5.

So why do we need text-based tasks at all? Most teachers cannot initiate all mathematical activity from their own creativity and resources, due to a range of workplace limitations. As we have shown, tasks can offer engagement in mathematical processes and opportunities to demonstrate, practice, and apply knowledge. They can offer suggestions for action, in an order, with some intentions for learning, with a range of visual and verbal stimuli in planned positional relation to each other (possibly using hypertext). Text-based tasks can offer models of structuring questions and prompts at all grain sizes of mathematical activity, planned sequences of tasks, conceptual focus and development, representations, pedagogic assumptions, and triggers to organize knowledge and can also provide simultaneous or sequential representational variety, possibly using hyperlinks. At the level of complex tasks, text can provide realistic resources which would be hard for individual teachers to find or construct and can bring together documentary resources for enquiry tasks. Text-based tasks can introduce teachers and learners to new ways of engaging in mathematics even if these are not taken up. We have also shown that text can provide formats which structure mathematical information and make comparisons and connections available for learners and teachers. Text can offer visual repetition of useful images and layouts. Text can also provide frames and methods of self-evaluation.

Many of these features can be provided digitally, but it is important to note that text can offer learning management systems in which tasks are presented in an order based on mathematical and educational principles; static text can offer immutable structure, data, images, and layouts; static text can be used in a variety of on-screen and off-screen modes of working. Static text cannot provide direct haptic experience of mathematical change nor instant feedback from all possible learners' actions nor models of continuous variability in mathematical and other phenomena. Furthermore, text cannot provide the important phases of learning that take place through interaction and mathematical reflections on what has been done by a particular set of learners. In other words, to echo what was said earlier in Chap. 2, tasks are only one element of a complex interactive learning ecology.

5.7.1 A Potential Research Agenda

The design principles described throughout this chapter and illustrated with varied examples from a range of international sources suggest areas where future research might be conducted. Research into textbook design and use is being undertaken widely and addresses concerns about the relationships between curriculum authorities, publishers, author teams, teachers, pedagogy, and learners from many perspectives and has led to international conferences (International Conference on Mathematics Textbook Research 2014) and several publications (e.g., Thompson and Usiskin, 2014).

Where individual tasks are concerned, teachers' use (Chap. 3) and learners' perspectives (Chap. 4) make critical contributions to the act of design. In thinking about the actual words, diagrams, and appearance of text-based tasks, we can ask: *how do differences in authority and voice in text-based tasks influence learning* and *how do visual aspects of text-based tasks influence attention and learning*? There is little robust research about how these aspects of text-based tasks influence learning. More attention to these, such as is undertaken in Learning Study, might help answer the question: *what different conceptual experiences arise from different task treatments of the same concept*? Working on such comparisons will generate more knowledge about the relationships between task and learning.

We can also ask: *what are the relationships between grain size of tasks, types of mathematical activity, and learners' mathematical development*? We are not convinced that these are always fully matched in practice. In structuring this chapter, we offered a triangular, interdependent relationship between the nature and structure of a task, its purpose, and the resulting mathematical activity for consideration. This structure has given a way to think about design and selection of tasks that places the task at the heart of the connection between teaching and learning.

The questions above should all be seen in the context of more general research about design principles, implementation, teacher knowledge, learners' perspectives, and digital affordances as described in other chapters.

References

Amit, B., & Movshovitz-Hadar, N. (2011). Design and high-school implementation of mathematical-news-snapshots - An action-research into 'Today's news is tomorrow's history'. In E. Barbin, M. Krongellner, & C. Tzanakis (Eds.), *History and epistemology in mathematics education: Proceedings of the Sixth European Summer University* (pp. 171–184). Austria: Verlag Holzhausen GmbH/Holzhausen Publishing Ltd. Based upon a workshop given at ESU 6, Vienna, July, 2010.

Anderson, J. R., & Schunn, C. D. (2000). Implications of the ACT-R learning theory: No magic bullets. In R. Glaser (Ed.), *Advances in instructional psychology: Educational design and cognitive science* (Vol. 5, pp. 1–34). Mahwah, NJ: Lawrence Erlbaum Associates.

Baker, W., & Bourne, A. (1937). *Elementary algebra*. London: G. Bell and Sons.

Barabash, M., & Guberman, R. (2013). Developing young students' geometric insight based on multiple informal classifications as a central principle in the task design. In C. Margolinas (Ed.), *Task design in mathematics education: Proceedings of ICMI Study 22* (pp. 293–302). Oxford, UK. Available from http://hal.archives-ouvertes.fr/hal-00834054

Barzel, B., Leuders, T., Prediger, S., & Huβmann, S. (2013). Designing tasks for engaging students in active knowledge organization. In C. Margolinas (Ed.), *Task design in mathematics education: Proceedings of ICMI Study 22* (pp. 283–292). Oxford, UK. Available from http://hal.archives-ouvertes.fr/hal-00834054

Bell, A. W. (1976). A study of pupils' proof-explanations in mathematical situations. *Educational Studies in Mathematics, 7*(1), 23–40.

Brousseau, G. (1997). *Theory of didactical situations in mathematics 1970–1990*. Translation from French: M. Cooper, N. Balacheff, R. Sutherland, & V. Warfield. Dordrecht, The Netherlands: Kluwer Academic (1998, French version: *Théorie des situations didactiques*. Grenoble, France: La Pensée Sauvage).

Burns, R. P. (1982). *A pathway into number theory*. Cambridge, England: Cambridge University Press.

Byrne, O. (1847). *The first six books of Euclid*. London: William Pickering. Available from http://www.math.ubc.ca/~cass/Euclid/byrne.html.

Chang, Y., Lin, F., & Reiss, K. (2013). How do students learn mathematical proof? A comparison of geometry designs in German and Taiwanese textbooks. In C. Margolinas (Ed.), *Task design in mathematics education: Proceedings of ICMI Study 22* (pp. 303–312). Oxford, UK. Available from http://hal.archives-ouvertes.fr/hal-00834054

Chesné, J.-F., Le Yaouanq, M.-H., Coulange, L., & Grapin, N. (2009). *Hélice 6e*. Paris: Didier.

Common Problem Solving Strategies as links between Mathematics and Science (COMPASS). (2013). Retrieved from: http://www.compass-project.eu

Corcoran, D., & Moffett, P. (2011). Fractions in context: The use of ratio tables to develop understanding of fractions in two different school systems. In C. Smith (Ed.), *Proceedings of the British Society for Research into Learning Mathematics, 31*(3), 23–28. Available from http://www.bsrlm.org.uk/IPs/ip31-3/BSRLM-IP-31-3-05.pdf

Crisp, R., Inglis, M., Mason, J., & Watson, A. (2011). Individual differences in generalization strategies. In C. Smith (Ed.), *Proceedings of the British Society for Research into Learning Mathematics, 31*(3), 35–40.

Cuoco, A. (2001). Mathematics for teaching. *Notices of the AMS, 48*(2), 168–174.

Cuoco, A., Goldenberg, E. P., & Mark, J. (1996). Habits of mind: An organizing principle for mathematics curricula. *Journal of Mathematical Behavior, 15*, 375–402.

Curcio, F. R. (1987). Comprehension of mathematical relationships expressed in graphs. *Journal for Research in Mathematics Education, 18*, 382–393.

Davis, B. (2008). Is 1 a prime number? Developing teacher knowledge through *concept study*. *Mathematics Teaching in the Middle School, 14*(2), 86–91.

Dickinson, P., & Hough, S. (2012). *Using realistic mathematics education in UK classrooms*. MEI. http://www.mei.org.uk/files/pdf/rme_impact_booklet.pdf. Retrieved November 20, 2014.

Dindyal, J., Tay, E. G., Quek, K. S., Leong, Y. H., Toh, T. L., Toh, P. C., et al. (2013). Designing the practical worksheet for problem solving tasks. In C. Margolinas (Ed.), *Task design in mathematics education: Proceedings of ICMI Study 22* (pp. 313–324). Oxford, UK. Available from http://hal.archives-ouvertes.fr/hal-00834054

Dole, S., & Shield, M. (2008). The capacity of two Australian eighth-grade textbooks for promoting proportional reasoning. *Research in Mathematics Education, 10*(1), 19–35.

Dörfler, W. (2005). Diagrammatic thinking: Affordances and constraints. In M. H. Hoffman, J. Lenhard, & F. Seeger (Eds.), *Activity and sign: Grounding mathematics education* (pp. 57–66). Berlin/New York: Springer.

Drake, C., & Sherin, M. G. (2009). Developing curriculum vision and trust: Changes in teachers' curriculum strategies. In J. T. Remillard, B. A. Herbel-Eisenmann, & G. M. Lloyd (Eds.), *Mathematics teachers at work: Connecting curriculum materials and classroom instruction* (pp. 321–337). New York: Routledge.

Duval, R. (2006). A cognitive analysis of problems of comprehension in learning of mathematics. *Educational Studies in Mathematics, 61*(1–2), 103–131.

Even, R., & Olsher, S. (2012). *The integrated mathematics wiki-book project*. Available from http://www.openu.ac.il/innovation/chais2012/downloads/e-Even-Olsher-61_eng.pdf

Fan, L. (2013). *A study on the development of teachers' pedagogical knowledge* (2nd ed.). Shanghai: East China Normal University Press.

Fan, L., Zhu, Y., & Miao, Z. (2013). Textbook research in mathematics education, development status and directions. *ZDM: The International Journal on Mathematics Education, 45*, 633–646.

Fujii, T. (2015). The critical role of task design in lesson study. In A. Watson & M. Ohtani (Eds.), *Task design in mathematics education: An ICMI Study 22*. New York: Springer.

Godfrey, C., & Siddons, A. (1915). *Elementary algebra, Part 1C*. Cambridge, England: Cambridge University Press.

Gregório, M., Valente, N. M., & Calafate, R. (2010). *Segredos dos Números 1 - Manual -Matemática/1.° ano do Ensino Básico*. Lisboa: Lisboa Editora.

Gueudet, G., Pepin, B., & Trouche, L. (2013). Textbooks' design and digital resources. In C. Margolinas (Ed.), *Task design in mathematics education: Proceedings of ICMI Study 22* (pp. 325–336). Oxford, UK. Available from http://hal.archives-ouvertes.fr/hal-00834054

Harel, G. (2009). *A review of four high school mathematics programs*. Retrieved from http://www.math.jhu.edu/~wsw/ED/harelhsreview2.pdf

Hart, E. W. (2013). Pedagogical content analysis of mathematics as a framework for task design. In C. Margolinas (Ed.), *Task design in mathematics education: Proceedings of ICMI Study 22* (pp. 337–346). Oxford, UK. Available from http://hal.archives-ouvertes.fr/hal-00834054

Herbart, J. F. (1904a). *Outlines of educational doctrine*. New York: Macmillan.

Herbart, J. F. (1904b). *The science of education*. London: Sonnenschein.

Herbel-Eisenmann, B. A. (2009). Negotiating the "presence *of* the text": How might teachers' language choices influence the positioning of the textbook? In J. T. Remillard, B. A. Herbel-Eisenmann, & G. M. Lloyd (Eds.), *Mathematics teachers at work: Connecting curriculum materials and classroom instruction* (pp. 134–151). New York: Routledge.

Hirsch, C. R. (Ed.). (2007). *Perspectives on the design and development of school mathematics curricula*. Reston, VA: National Council of Teachers of Mathematics.

Hoven, J., & Garelick, B. (2007). Singapore math: Simple or complex? *Educational Leadership, 65*(3), 28–31.

Hußmann, S., Leuders, T., Prediger, S., & Barzel, B. (2011a) *Mathewerkstatt 5*. Cornelsen.

Hußmann, S., Leuders, T., Prediger, S., & Barzel, B. (2011b). Kontexte für sinnstiftendes Mathematiklernen (KOSIMA) – ein fachdidaktisches Forschungs-und Entwicklungsprojekt. *Beiträge zum Mathematikunterricht*, pp. 419–422.

Huntley, M. A., & Terrell, M. S. (2014). One-step and multi-step linear equations: A content analysis of five textbook series. *ZDM: The International Journal on Mathematics Education, 46*(5), 751–766.

Kress, G. R., & Van Leeuwen, T. (1996). *Reading images: The grammar of visual design*. New York: Psychology Press.

Lee, K., Lee, E., & Park, M. (2013). Task modification and knowledge utilization by Korean prospective mathematics teachers. In C. Margolinas (Ed.), *Task design in mathematics education: Proceedings of ICMI Study 22* (pp. 347–356). Oxford, UK. Available from http://hal.archives-ouvertes.fr/hal-00834054

Li, Y., Zhang, J., & Ma, T. (2009). Approaches and practices in developing school mathematics textbooks in China. *ZDM: The International Journal on Mathematics Education, 41*(6), 733–748.

Llinares, S., Krainer, K., & Brown, L. (2014). Mathematics, teachers and curricula. In S. Lerman (Ed.), *Encyclopaedia of mathematics education* (pp. 438–441). New York: Springer.

Love, E., & Pimm, D. (1996). "This is so": A text on texts. In A. J. Bishop, K. Clements, C. Keitel, J. Kilpatrick, & C. Laborde (Eds.), *International handbook of mathematics education* (pp. 371–409). Dordrecht: Kluwer Academic.

Lundberg, A. L. V., & Kilhamn, C. (2013). The lemon squash task. In C. Margolinas (Ed.), *Task design in mathematics education: Proceedings of ICMI Study 22* (pp. 357–366). Oxford, UK. Available from http://hal.archives-ouvertes.fr/hal-00834054

Ma, L. (1999). *Knowing and teaching elementary mathematics: Teachers' understanding of fundamental mathematics in China and the United States*. Mahwah, NJ: Lawrence Erlbaum Associates.

Maaβ, K., Garcia, F. J., Mousouides, N., & Wake, G. (2013). Designing interdisciplinary tasks in an international design community. In C. Margolinas (Ed.), *Task design in mathematics education: Proceedings of ICMI Study 22* (pp. 367–376). Oxford, UK. Available from http://hal.archives-ouvertes.fr/hal-00834054

MARS (Mathematics Assessment Resource Service) (2014). *Summative assessment*. Available from: http://map.mathshell.org.uk/materials/background.php?subpage=summative

Marton, F. (2014). *Necessary conditions of learning*. London: Routledge.

Mason, J., Burton, L., & Stacey, K. (2010). *Thinking mathematically* (2nd ed.). Harlow, England: Prentice Hall.

Mathematics Textbook Developer Group for Elementary School. (2005). *Mathematics*. [in Chinese]. Beijing: People's Education Press.

Mayer, R. E., Sims, V., & Tajika, H. (1995). A comparison of how textbooks teach mathematical problem solving in Japan and the United States. *American Educational Research Journal, 32*, 443–460.

Movshovitz-Hadar, N., & Edri, Y. (2013). Enabling education for values with mathematics teaching. In C. Margolinas (Ed.), *Task design in mathematics education: Proceedings of ICMI Study 22* (pp. 377–388). Oxford, UK. Available from http://hal.archives-ouvertes.fr/hal-00834054

Movshovitz-Hadar, N., & Webb, J. (2013). *One equals zero and other mathematical surprises*. Reston, VA: National Council of Teachers of Mathematics. http://www.nctm.org/catalog/product.aspx?id=14553

National Academy for Educational Research. (2009). *Mathematics, grade 8* (Vol. 4). Retrieved from http://www.naer.edu.tw/bookelem/u_booklist_v1.asp?id=267&bekid=2&bemkind=1. Accessed on February 07, 2010.

National Council of Teachers of Mathematics. (1989). *Curriculum and evaluation standards for school mathematics*. Reston, VA: Author.

Nicely, R. F., Jr. (1985). Higher order thinking in mathematics textbooks. *Educational Leadership, 42*, 26–30.

Nicely, R., Jr., Fiber, H., & Bobango, J. (1986). The cognitive content of elementary school mathematics textbooks. *Arithmetic Teacher, 34*(2), 60–61.

Nikitina, S. (2006). Three strategies for interdisciplinary teaching: Contextualising, conceptualizing, and problem-centring. *Journal of Curriculum Studies, 38*(3), 251–271.

Paas, F., Renkl, A., & Sweller, J. (2003). Cognitive load theory and instructional design: Recent developments. *Educational Psychologist, 38*(1), 1–4.

Pepin, B., & Haggarty, L. (2001). Mathematics textbooks and their use in English, French, and German classrooms: A way to understand teaching and learning cultures. *ZDM: The International Journal on Mathematics Education, 33*(5), 158–175.

Prestage, S., & Perks, P. (2007). Developing teacher knowledge using a tool for creating tasks for the classroom. *Journal of Mathematics Teacher Education, 10*, 381–390.

Puphaiboon, K., Woodcock, A., & Scrivener, S. (2005, March 25). Design method for developing mathematical diagrams. In P. D. Bust & P. T. McCabe (Eds.), *Contemporary ergonomics: 2005 Proceedings of the International Conference on Contemporary Ergonomics (CE2005)*. New York: Taylor & Francis.

Radford, L. (2008). Diagrammatic thinking: Notes on Peirce's semiotics and epistemology. *PNA, 3*(1), 1–18.

Remillard, J. T., Herbel-Eisenmann, B. A., & Lloyd, G. M. (Eds.). (2009). *Mathematics teachers at work: Connecting curriculum materials and classroom instruction*. New York: Routledge.

Rezat, S. (2006). A model of textbook use. In J. Novotná, H. Moraová, M. Krátká, & N. Stehliková (Eds.), *Proceedings of the 30th Conference of the International Group for the Psychology of Mathematics Education* (Vol. 4, pp. 409–416). Prague: Psychology of Mathematics Education.

RLDU, Resources for Learning and Development Unit (n.d.). Available from www.nationalstemcentre.org.uk/elibrary/maths/resource/6910/an-addendum-to-cockcroft

Rotman, B. (1988). Toward a semiotics of mathematics. *Semiotica, 72*(1/2), 1–35.

Schoenfeld, A. H. (1985). *Mathematical problem solving*. New York: Academic.

Sesamath. (2009). *Le manuel Sésamath 6e*. Chambéry: Génération 5.

Shuard, H., & Rothery, A. (1984). *Children reading mathematics*. London: John Murray.

Small, M., Connelly, R., Hamilton, D., Sterenberg, G. & Wagner, D. (2008). *Understanding mathematics: Textbook for Class VII*. Thimpu, Bhutan Curriculum and Professional Support Division, Department of School Education.

SMILE. (n.d.). Available from http://www.nationalstemcentre.org.uk/elibrary/resource/675/smile-wealth-of-worksheets

Smith, G., Wood, L., Coupland, M., Stephenson, B., Crawford, K., & Ball, G. (1996). Constructing mathematical examinations to assess a range of knowledge and skills. *International Journal of Mathematical Education in Science and Technology, 27*(1), 65–77.

Staats, S., & Johnson, J. (2013). Designing interdisciplinary curriculum for college algebra. In C. Margolinas (Ed.), *Task design in mathematics education: Proceedings of ICMI Study 22* (pp. 389–400). Oxford, UK. Available from http://hal.archives-ouvertes.fr/hal-00834054

Stacey, K., & Vincent, J. (2009). Modes of reasoning in explanations in Australian eighth-grade mathematics textbooks. *Educational Studies in Mathematics, 72*(3), 271–288.

Stein, M. K., Grover, B. W., & Henningsen, M. (1996). Building student capacity for mathematical thinking and reasoning: An analysis of mathematical tasks used in reform classrooms. *American Educational Research Journal, 33*(2), 455–488.

Stigler, J. W., Fuson, K. C., Ham, M., & Kim, M. S. (1986). An analysis of addition and subtraction word problems in American and Soviet elementary mathematics textbooks. *Cognition and Instruction, 3*, 153–171.

Stylianides, G. J. (2009). Reasoning-and-proving in school mathematics textbooks. *Mathematical Thinking and Learning, 11*, 258–288.

Stylianides, G. J. (2014). Textbook analyses on reasoning-and-proving: Significance and methodological challenges. *International Journal of Educational Research, 64*, 63–70.

Sun, X., Neto, T., & Ordóñez, L. E. (2013). Different features of task design associated with goals and pedagogies in Chinese and Portuguese textbooks: The case of addition and subtraction. In C. Margolinas (Ed.), *Task design in mathematics education: Proceedings of ICMI Study 22* (pp. 409–418). Oxford, UK. Available from http://hal.archives-ouvertes.fr/hal-00834054

Sun, X. (2011). "Variation problems" and their roles in the topic of fraction division in Chinese mathematics textbook examples. *Educational Studies in Mathematics, 76*(1), 65–85.

Swan, M. (2006). *Collaborative learning in mathematics: A challenge to our beliefs and practices*. London: National Institute for Advanced and Continuing Education.

Takahashi, A. (2011). The Japanese approach to developing expertise in using the textbook to teach mathematics. In Y. Li & G. Kaiser (Eds.), *Expertise in mathematics instruction: An international perspective* (pp. 197–220). Dordrecht: Springer.

Thompson, D. R., Hunsader, P. D., & Zorin, B. (2013). Assessments accompanying published curriculum materials: Issues for curriculum designers, researchers, and classroom teachers. In C. Margolinas (Ed.), *Task design in mathematics education: Proceedings of ICMI Study 22* (pp. 401–408). Oxford, UK. Available from http://hal.archives-ouvertes.fr/hal-00834054

Thompson, D. R., & Senk, S. L. (2010). Myths about curriculum implementation. In B. J. Reys, R. E. Reys, & R. Rubenstein (Eds.), *Mathematics curriculum: Issues, trends, and future directions* (pp. 249–263). Reston, VA: National Council of Teachers of Mathematics.

Thompson, D. R., Senk, S. L., & Johnson, G. J. (2012). Opportunities to learn reasoning and proof in high school mathematics textbooks. *Journal for Research in Mathematics Education, 43*(3), 253–295.

Thompson, D. R., & Usiskin, Z. (2014). *Enacted mathematics curriculum: A conceptual model and research needs*. Charlotte, NC: Information Age.

Treffers, A. (1987). *Three dimensions: A model of goal and theory description in mathematics education: The Wiskobas project*. Dordrecht: Kluwer Academic.

Tufte, E. (1997). *Visual explanations: Images and quantities, evidence and narrative*. Cheshire, CT: Graphics Press.

Tzur, R., Zaslavsky, O., & Sullivan, P. (2008). Examining teachers' use of (non-routine) mathematical tasks in classrooms from three complementary perspectives: Teacher, teacher educator, researcher. In *Proceedings of the Joint Meeting of the 32nd Conference of the International Group for the Psychology of Mathematics Education, and the 30th North American Chapter* (Vol. 1, pp. 133–137).

Van den Heuvel-Panhuizen, M. (2003). The didactical use of models in realistic mathematics education: An example from a longitudinal trajectory on percentage. *Educational Studies in Mathematics, 54*(1), 9–35.

Wagner, D. (2012). Opening mathematics texts: Resisting the seduction. *Educational Studies in Mathematics, 80*(1–2), 153–169.

Watson, A., & Mason, J. (1998). *Questions and prompts for mathematical thinking*. Derby, UK: Association of Teachers of Mathematics.

Watson, A., & Mason, J. (2006). Seeing an exercise as a single mathematical object: Using variation to structure sense-making. *Mathematical Thinking and Learning, 8*(2), 91–111.

Wittmann, E. C. (1995). Mathematics education as a 'design science'. *Educational Studies in Mathematics, 29*(4), 355–374.

Xin, Y. P. (2008). The effects of schema-based instruction in solving mathematics word problems: An emphasis on prealgebraic conceptualization of multiplicative relations. *Journal for Research in Mathematics Education, 39*, 526–551.

Yerushalmy, M. (2015). E-textbooks for mathematical guided inquiry: Design of tasks and task sequences. In A. Watson & M. Ohtani (Eds.), *Task design in mathematics education: An ICMI Study 22*. New York: Springer.

Zorin, B., Hunsader, P. D., & Thompson, D. R. (2013). Assessments: Numbers, context, graphics, and assumptions. *Teaching Children Mathematics, 19*(8), 480–488.

Chapter 6
Designing Mathematics Tasks: The Role of Tools

Allen Leung and Janete Bolite-Frant

with *Ferdinando Arzarello, Christian Bokhove, Peter Boon, Orly Buchbinder, Yip-cheung Chan, Alison Clark-Wilson, Paul Drijvers, Vince Geiger, Lulu Healy, Marie Joubert, Kate Mackrell, Ami Mamolo, Anthony Or, Elisabetta Robotti, Sophie Soury-Lavergne, Mike Thomas, Floriane Wozniak, Michal Yerushalmy*

and additional contributions from *Michiel Doorman, Solange Hassan Ahmad Ali Fernandes, Catherine Lin, Michela Maschietto, Trevor Redmond, Sietske Tacoma, Jaiwant Timotheus, Walter Whiteley, Orit Zaslavsky*

6.1 Introduction

A mark of human intelligence is our ability to create and use tools to extend our abilities to achieve tasks that are otherwise only unimaginable and to create and use representations to afford interpretation and communication in the objective and personal worlds.

There are different perspectives to interpret tools and representations. Tools can be regarded as mediators between the phenomenological world and the conceptual world, or, according to Radford, "artefacts do much more than mediate: they are a constitutive part of thinking and sensing" (2013, p. 8). Hence, our interaction with tools, artifacts, and culture material should be considered as more than auxiliary elements. Tools influence cognition, and for the purpose of this chapter, they impact on mathematical knowledge. Development of mathematical ideas and concepts has been closely associated with development of technology that, according to Abramovich (2001), can be interpreted as cultural tools in contemporary educational practices. Mathematics teaching that involves tools, among them technology and representations, enables teachers to guide students to (re)invent and visit mathematics.

> Children should repeat the learning process of mankind, not as it factually took place but rather as it would have done if people in the past had known a bit more of what we know now. (Freudenthal, 1991, p. 48)

A. Leung (✉)
Hong Kong Baptist University, Hong Kong, China
e-mail: aylleung@hkbu.edu.hk

J. Bolite-Frant
LOVEME Lab, UNIBAN, São Paulo, Brazil

This chapter has been made open access under a CC BY-NC-ND 4.0 license. For details on rights and licenses please read the Correction https://doi.org/10.1007/978-3-319-09629-2_13

© The Author(s) 2021
A. Watson, M. Ohtani (eds.), *Task Design In Mathematics Education*, New ICMI Study Series, https://doi.org/10.1007/978-3-319-09629-2_6

This chapter concerns designing teaching and learning tool-based tasks in school mathematics. Tools are broadly interpreted as physical or virtual artifacts that have potential to enhance mathematical understanding. A *tool-based task* is seen as a teacher/researcher design aiming to be a thing to do or act on in order for students to activate an interactive tool-based environment where teacher, students, and resources mutually enrich each other in producing mathematical experiences. In this connection, this type of task design rests heavily on a complex relationship between tool mediation, teaching and learning, and mathematical knowledge (Fig. 6.1).

Different epistemological approaches to mathematical learning will have different implications on designing tool-based tasks. For instance, using Sfard's (2008) two metaphors for learning, a *participationist* orientation would favor design with potential for students to participate in the construction of mathematical knowledge/experiences, whereby a more *acquisitionist* orientation would favor design that explores and discovers established mathematical knowledge. Tools can act as a mediator for mathematical discourse; hence, it is important to have views on mathematical discourse in tool-based task design. Two such views are (1) by observing student's discourse, a teacher may find what is missing from the mathematical discourse legitimated by the mathematics community and (2) mathematics is a discourse, so learning mathematics is participating in discourse (Sfard, 2008). These two orientations on the nature of mathematical discourse are relevant considerations for task design because both consider mathematics to be discursive: in (1), students acquire knowledge by learning a specific mathematics language/content; in (2), because mathematics is a discourse, learning mathematics is participating in this discourse, and learning is conceived as changing participation (Rogoff, 1998).

The choice of tool for pedagogical purpose can be considered as a function of how mathematical knowledge is perceived epistemologically by the teacher. The same tool can be used in two task designs that are at opposite epistemological poles. For example, using compasses and ruler may be seen in different ways: for students to construct their own geometrical models in order to explain a certain mathematical phenomenon that they experienced or for students to follow a teacher-given construction procedure to check the validity of a given theorem. An exemplar illustration of the former can be found in the work of DiSessa, Hammer, and Sherin

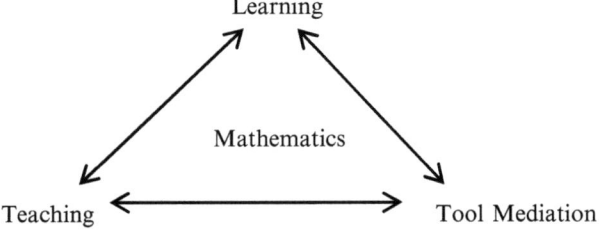

Fig. 6.1 A relational *triangle* highlighting the interrelationships among different components in a tool-based pedagogical environment

(1991) where a cooperative activity of a sixth-grade class focused on inventing adequate static representations of motion (graphing) using a computational medium, such as Boxer (the Berkeley Boxer Project), was described. The findings indicated students possessed strong meta-representational competence. Tools have potential to empower students to hurdle over didactical and epistemological obstacles (Brousseau, 1998). A tool-based task design could harvest this power to shorten any *distance* between students' prior mathematical experiences and the intended mathematical knowledge to be learned. Rabardel's (1995) theory of *instrumental genesis* was an explication of how usage of a tool can be turned into a cognitive instrumentation process for knowledge acquisition or construction. This approach sees artifacts as psychological tools in a utilization process. It focuses on how a learner develops a scheme of usage for a particular tool while she/he is using it to solve a problem. Pedagogically, this scheme can be attached to the tool to make it into an instrument for teaching and learning. A Vygotskian approach would see artifacts as psychological tools in the context of social and cultural interaction developed through the zone of proximal development and internalization processes (Vygotsky, 1978, 1981). A concrete tool can be transformed into a mind tool through a carefully designed pedagogical environment. In particular, internalization is a socio-semiotic process.

With these constructs as backdrop, Bartolini Bussi and Mariotti (2008) proposed the notion of a *tool of semiotic mediation* for the mathematics classroom, taking on multiple pedagogical functions. On one hand, personal meanings are related to the use of the tool while students accomplish a task; on the other hand, mathematical meanings are related to the tool and its use. This dual relationship constitutes the semiotic potential of a tool. An artifact is regarded as a tool of semiotic mediation if the teacher, intentionally using it, mediates mathematical content through designed didactical intervention. Thus, teachers and students play critical roles in the process of tool mediation. In fact, the teacher, the students, and tools are parts of the mediation process (c.f. Fig. 6.1). Thus, the semiotic potential of tools can be a principle consideration in tool-based task design.

TELMA (see Bottino & Kynigos, 2009) and ReMath projects (see Artigue & Mariotti, 2014) are tool-based task design examples, and we shall describe their work in more detail later. The Technology Enhanced Learning in Mathematics (TELMA) project focused on representation systems and learning contexts. The ReMath Project was a follow-up which produced an academic description of cross-case studies of classroom activity emerging from the implementation of pedagogical plans using digital media.

Other important tool-based task design considerations are how a piece of mathematics can be embodied in a tool and how feedback in tool usage can be used pedagogically. These considerations relate to actual tool manipulation skills and practices; these evolve into amplified learner abilities to discern critical features in a mathematical situation and finally foster the development of mathematical discourses that may or may not be tool dependent.

Confrey et al. proposed that

> we need engaging environments, in which the mathematics is actually needed for students to achieve goals that they find compelling, and made visible to students and expressed in a language with which they can connect. (2010, p. 20)

In order to accomplish these concerns, they presented four major thematic approaches for software design: *modeling, project-based instruction, learning progression,* and *microworlds* (as proposed by Seymour Papert). They present two design examples, Graphs 'n Glyphs animation and Lunar Land Game, that followed those recommendations. Knowing the rationale behind the design of a tool is an important consideration for tool-based task design as these rationales may form pedagogical bases to harvest the mathematical potential of the tool. A powerful illustration of these connections is presented by Yerushalmy (2015, Chap. 7, this volume).

This introduction sets a stage to delve into a comprehensive exploration of the questions and matters raised. This chapter will present theories, issues, and cases concerning the use of tools in mathematics task design presented and discussed in the ICMI Study 22 Conference Tools and Representation Theme Group (Margolinas, 2013) and other state-of-the-art research. In doing so, we indicate and exemplify heuristics or principles (theoretical or pragmatic) for tool-based task design that are conducive to teaching and learning of mathematics.

In this chapter, we discuss the epistemological, mathematical, representational, and pedagogical considerations of tool-based task design. We then present various theoretical frames and heuristics and give examples of tasks throughout.

6.2 Considerations in Designing Tasks that Make Use of Tools

6.2.1 *Epistemological and Mathematical Considerations*

Epistemological and mathematical considerations play central roles in task design. Different epistemological approaches to mathematical knowledge have different implications on task design. Sfard's participationist epistemological orientation would favor a tool-based design with the potential for students to participate in the construction of shared mathematical experiences or discourses, whereby the acquisitionist epistemological orientation would use tools to explore and consequently construct personal mathematical knowledge (c.f. Sfard, 2008). The same tool can be used in task designs with different epistemological stances. Thus, tool-based design should couple with how a piece of mathematics content can be learned under a preferred epistemological disposition.

A challenge to tool-based task design is to determine the possible range of epistemological orientation and the type of mathematical knowledge that a tool can afford and to choose them appropriately for pedagogical situations. Dynamic digital tools, like dynamic geometry software, can be used in task design to cover a large epistemic spectrum from drawing precise robust geometrical figures to exploration of new geometric theorems and development of argumentation discourse. Non-digital tools, such as a set of transparent grid papers with different grid sizes, can be used to design tasks for learning the concept of area, arithmetic operation, etc. depending on how the teacher who designs the tasks understands how the mathematics involved can be manifested and taught using grid systems. Epistemological and mathematical considerations should be closely linked with the tool's potential to represent or manifest mathematical knowledge. Ironically, tools themselves have different degrees of epistemological obstacle through the constraints and limitations of their use, but these that can be used to create cognitive conflict to stimulate the learning process (see Chap. 2 for discussion about cognitive conflict). When thinking of obstacles to learning, there are two possibilities: one is a teaching gap in classroom practices, and the other is a distance between learner's prior mathematical knowledge and the intended mathematical knowledge to be learned. Brousseau (1998) called these, respectively, a *didactical obstacle* and an *epistemological obstacle*. Epistemological obstacles are those knowledge gaps which students need to overcome by construction of new knowledge (Joubert, 2013). Using the notion of epistemological obstacle as a design consideration for a tool-based mathematics classroom should enhance the knowledge construction process because tools can be used as a concrete bridge between students' prior mathematical experiences/knowledge and the intended mathematical knowledge to be learned. For example, the use of a dynamic graphing tool could help students hurdle over the hidden relationships among the a, b, and c parameters in a quadratic equation. A visualization tool can be powerful to overcome many epistemological obstacles in mathematics learning. There is a strong connection between the ways in which students will talk about their mathematical activity and the tools they are using.

6.2.2 Tool-Representational Considerations

Acquiring or constructing mathematical knowledge can be described in terms of creation of representations. The question of the way a chosen tool represents mathematical knowledge is at the heart of tool-based task design. School mathematics is basically symbolic in nature, which is a linguistic type of representation. There are at least two tool-representational considerations that are of design interest for a mathematical topic. The first one is how far away from the expected symbolic representation is in the tool's potential to represent the mathematical concept? Consideration of this distance could become the main concern in the task design approach. This is of particular interest in the use of simple and basic tools that leave ample room for a teacher-designer to exercise his/her pedagogy.

A second consideration is more philosophical and open-ended in nature. Is the tool "capable enough" of representing the targeted mathematical knowledge parallel to the corresponding symbolic representation, and if so, in what sense? That is, can a tool embed or represent a particular mathematical concept? In this case, the task design is not just to shorten an epistemological distance, but to propose a possible alternative tool-based representation that is compatible with conventional symbolic representation. This second consideration becomes more relevant when emerging powerful ICT pedagogical environments are extending human cognitive abilities and becoming more in alignment with human intellect. For example, blind students can learn geometry through a digital sensory tool via haptic representation of shape. In the work of Healy, Fernandes, and Bolite-Frant (2013), a special tool was designed and implemented with appropriate tasks for blind students to learn about functions (Fig. 6.2). The tool was a digitally controlled board made up of a rectangular matrix of pins, each of which represented a point on the plane. When a particular point is requested or a graph of a given function plotted, the relevant pins are raised up (sequentially as the value of the independent variable increases in the case of the graph of a function), allowing the student to feel, to sense, the image as it is produced (Fig. 6.2). For blind students, this tool-dependent "sensory feeling of the pins" may become their understanding of the concept of function. The tool influences the way the concept is initially understood.

6.2.3 Pedagogical Considerations

Tool-based task design must be supported by a suitable pedagogical environment. Different types of tools afford different mathematical task activities and discourses and have constraints that can be capitalized in constructive ways. Designers and teachers need to take into account possible epistemic interactions among different

Fig. 6.2 Digitally controlled board made up of a *rectangular matrix of pins*, each of which represents a point on the plane (Healy et al., 2013, p. 67)

types of tools and resources in a pedagogical situation. In Bruner's theory of instruction on the cognitive development of children (Bruner, 1966), three modes of representation were proposed: *enactive representation* (action based), *iconic representation* (image based), and *symbolic representation* (language based). These three instructional modes and the possible cognitive sequences connecting them can serve as a model to guide the design of tool-based tasks that involve different types of tools and classroom resources. With action-based tools, students gain mathematics experiences like number sense or spatial sense through physical sensory means (e.g., Dienes' blocks). Image-based tools lead students to interpret mathematics through visual (dynamic or static) reasoning (e.g., graphics calculator), and linguistic-based tools are conducive to development of mathematical discourse (e.g., commands in CAS tools). Different combinations of Bruner's instructional modes can serve as pedagogical guidelines for tool-based task design. Recent development in ICT environments, like GeoGebra, combined these three modes into one single dynamic multi-representational tool. The enactive and iconic modes connect through the relationship between the mouse, or other driving apparatus, and what appears on the screen. The language used might refer to moving an on-screen object, rather than moving the mouse. Recent developments with touchpad technology connect actions, imagery, and language even more closely to each other and to the intended mathematical structure (Sinclair, 2013; Sinclair & Pimm, 2014). This opens up a rich pedagogical space for task design to explore.

Familiarity with a tool and how to use it effectively to teach and learn are important pedagogical considerations for tool-based task design. Rabardel's (1995) tool-ergonomic theory of *instrumental genesis* focuses on how a learner turns the tool into an instrument by developing a *utilization scheme* that can be mentally attached to the tool and frames how it is used. Furthermore, a tool can progressively be instrumentalized for specific uses by loading it with different potentials through a range of tasks (c.f. Artigue, 2002). Through well-designed utilization tasks, a concrete tool can be internalized to become a psychological tool. A tool can be placed at the center of a pedagogical situation as a means to create a *didactical situation* (Brousseau, 1998) or *didactical intervention* (Bartolini Bussi & Mariotti, 2008); it need not be an adjunct to the didactical event, but can motivate and shape the event itself. Task design should consider the mediation potentials of the tool used in terms of bridging any gaps between the phenomenal world and the conceptual world, the teacher's and students' perceptions of mathematical knowledge, and other pedagogic gaps. Morgan, Mariotti, and Maffei (2009) discuss epistemological and social distances between representations in computational environments:

> Distance between representations in different media may be epistemological, affecting the nature of the mathematical concepts available to students, or may be social, affecting pedagogic relationships in the classroom and the ease with which the technology may be adopted in particular classroom .... (p. 241)

This distance concept can be extended to any tool-based environment and should be an important pedagogical consideration for task design.

6.2.4 Discursive Considerations

Sfard's (2008) commognitive view sees mathematics discourse as a type of communication with special features: keywords, visual mediators, routines, and endorsed narratives. Tool usage can be a means to facilitate the development of these features through task design. Leung (2011) proposed a nested sequence of three epistemic modes in tool-based task design consideration to foster the expansion of a learner's cognitive space: skills and practices, critical discernment, and situation discourse. Practicing to use a tool to accomplish a task involves formation of appropriate tool-based vocabularies in the development of utilization routines. The tool then becomes a mediator for discernment of critical features of the mathematical ideas being sought in the task. Afterward, a tool-based narrative can be constructed to explain the mathematics. This progressive task design epistemic sequence can be seen as a tool-based mathematical discourse development in the commognitive sense. This echoes Radford's idea of a tool as a "constitutive part of thinking and sensing". An overarching question to address is *how to design tool-based tasks that can bring about (situated) discourses for mathematical knowledge mediated by tools in the mathematics classroom, and furthermore, how do these discourses relate to mathematics knowledge?*

6.3 Theoretical Frames for Designing Tool-Based Tasks

6.3.1 Didactical Theories

How can using a tool open up a space of learning where a learner makes epistemic strategies or choices from feedback while using the tool, and what is the role of interaction in this usage? In the Theory of Didactical Situations (Brousseau, 1998), knowledge "is a property of a system constituted by a subject and a 'milieu' in interaction" (Mackrell, Maschietto, & Soury-Lavergne, 2013, p. 80). Balacheff and Sutherland (1994) described *milieu* as "the specific part of the environment of the learner which is accessible and relevant to his or her actions" (p. 140). Interaction can be understood as "a 'dialogue' between the student and the feedback from the milieu" (Joubert, 2013, p. 72). *Milieu* can be considered as a space of learning constituted by didactical variables. A *didactical variable* is a pedagogical parameter that opens a dimension in which students engage in epistemic interaction and which takes on different values depending on student responses to feedback. Thus, the usage of tools in a pedagogical situation can be designed within a milieu in such a way as to create different didactical parameters that are conducive to learning mathematics. In particular, tool-based tasks can be designed to create adidactical situations in which students need to construct mathematical knowledge themselves to resolve problems posted by the task.

In Chevallard's Anthropological Theory of the Didactic (described more fully in Chap. 2), mathematics is seen as a product of human activities into which he introduced the notion of an *ostensive* (Chevallard, 1994). An ostensive is "any object which can be handled concretely by the body, the voice, the vision" (Wozniak, 2013, p. 142). Non-ostensives are "objects" which can only be evoked through the handling of associated ostensives. This implies that mathematical conceptualizations are consequences of acting on and doing using artifacts or tools. Thus, didactically, techniques of using a tool create a dialectic between skills and understanding. Artigue (2002) called these two faces of tool techniques *pragmatic* values and *epistemic* values:

> A technique is a manner of solving a task and, as soon as one goes beyond the body of routine tasks for a given institution, each technique is a complex assembly of reasoning and routine work. I would like to stress that techniques are most often perceived and evaluated in terms of pragmatic value, that is to say, by focusing on their productive potential (efficiency, cost, field of validity). But they have also an epistemic value, as they contribute to the understanding of the objects they involve, and thus techniques are a source of questions about mathematical knowledge. (p. 248)

These two values are framed in terms of praxeology: praxis for pragmatic value and logos for epistemic value. "The praxis component contains techniques to achieve a kind of tasks. The logos component includes the theoretical discourses… that describe, explain, justify or develop the techniques used…" (Wozniak, 2013, p. 143). Therefore, a tool can be designed as a didactic variable in a task that forms part of a learner's space of learning; this variable takes evolving pragmatics and epistemic values which have the potential to transform a learner's mathematical experiences from experimental to theoretical.

Because tools are potent with pragmatic and epistemic values, designing a tool-based task needs to pay attention to the interplay between techniques and discernment of mathematical concepts and to those features of the tool that have epistemic value (Thomas & Lin, 2013). Kieran et al. (2006) reported a study on the dialectical relation between theoretical thinking and technique, as they co-emerge in a combined computer algebra system (CAS) and paper-and-pencil environment. In their study, the anthropological view was seen as a Task-Technique-Theory triad. A three-part task activity sequence (seeing a pattern, refining a generalization, and proving) was designed based on a CAS and paper-and-pencil environment for students to explore the factoring of $x^n - 1$. The conclusion was that CAS techniques, together with paper-and-pencil techniques, were found to be significant in the deepening of students' theoretical thinking. Making sense of the CAS outputs and coordinating these with both theoretical notions and paper-and-pencil techniques were fundamental processes in the students' theoretical and paper-and-pencil progress. This finding supports the existence of a complex pedagogical relationship between the pragmatic and epistemic values of tool use.

6.3.2 Instrumentation Approach to Tool Use

Verillon and Rabardel's (1995) seminal paper on instrumental approaches to relate cognition and artifacts opened up a vast research arena in the context of using technology in the mathematics classroom (see, e.g., Gueudet & Trouche, 2009;

Trouche, 2004, 2005). An instrument is a psychological tool resulting from instrumental genesis through a learner using the tool to accomplish tool-based tasks. Instrumental genesis is a combination of two processes. Instrumentation is the techno-centric process by which the tool "prints its mark on the subject, i.e., allows him/her to develop an activity within some boundaries" (Trouche, 2004, p. 290). That is, the user is mastered by the tool rather than the other way round. Instrumentalization is the process through which a user develops his/her own ideas of what the tool is designed for and how it should be used. The psychological component of the instrument is the utilization scheme developed by the user (see examples involving TI-92™ in Trouche, 2004; Artigue, 2002). It is usually not easy to distinguish between instrumentation and instrumentalization due to the complexity of the user-tool interactions. To further the idea of instrumental genesis, instrumental orchestration and documentational genesis have been proposed to describe how a tool or a system of tools can be integrated into classroom tasks or purposeful mathematical activities (see, e.g., Gueudet & Trouche, 2009; Trouche, 2004, 2005). Gueudet and Trouche (2009) extended the idea of tool (artifact) to "resources" which are tools (artifacts) with the potential to promote semiotic mediation in the process of learning. Geiger and Redmond (2013) explained:

> Resources include entities such as computer applications, student worksheets or discussions with a colleague. A resource is appropriated and reshaped by a teacher, in a way that reflects their professional experience in relation to the use of resources, to form a schema of utilisation—a process parallel to the creation of a schema of instrumented action within instrumental genesis. The combination of the resource and the schema of utilisation is called a document. (p. 121)

Instrumental theories concern the development of a utilization scheme, synergetic interactions among tools, and systematic collation of tool usages in a multiple-tool environment.

6.3.3 Cultural Semiotic Frame

Based on a Vygotskian perspective on signs and psychological tools (Vygotsky, 1978), the question of how far tool usage in a pedagogical situation carries mathematical meaning is studied using a tool of semiotic mediation framework proposed by Bartolini Bussi and Mariotti (2008). The framework concerns the relationship between accomplishing a task through the use of an artifact (tool) and learning. It considers the crucial role of human mediation under semiotic and educational perspectives in the teaching and learning process through the artifact. Students produce personal signs while using a tool, and the production and transformation of these signs link to their construction of mathematical knowledge. The link between a tool, a task, and mathematical knowledge is called the *semiotic potential* of the tool in the task context (Arzarello, Bartolini-Bussi, Leung, Mariotti, & Stevenson, 2012, p. 107; Bartolini Bussi & Mariotti, 2008, p. 754). Figure 6.3 is a diagrammatic

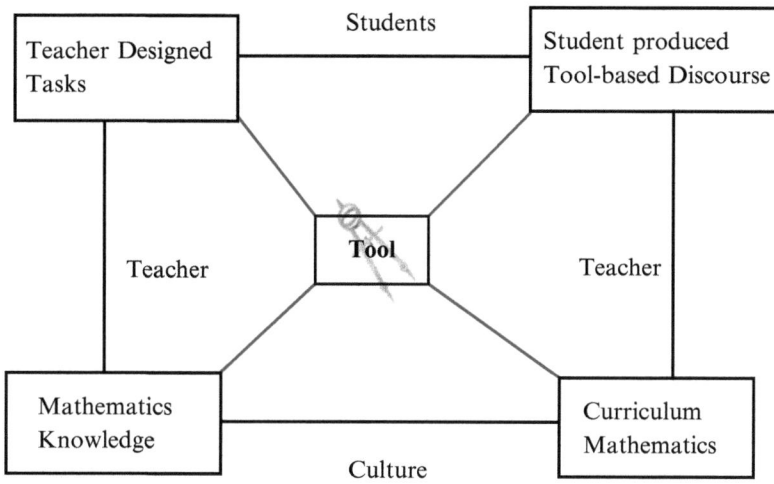

Fig. 6.3 A schematic of the semiotic mediation framework

summary of the semiotic mediation framework, highlighting the interrelationships among the different components of mediation.

Teachers play an important mediator role creating the circumstances in which the mediation process occurs through task design and didactic intervention. Didactic intervention takes the form of didactic cycle "where different categories of activities take place, each of them contributing differently but complementarily to develop the complex process of semiotic mediation" (Mariotti, 2012, p. 29).

Healy, Fernandes, and Bolite-Frant (2013) reported a Brazilian study on how researchers, high school teachers, and deaf/blind students worked together to develop suitable tools to teach and learn matrices. The teacher, who was fluent in Libras (the Brazilian sign language), suggested "that a lack of specific signs for the vocabulary associated with matrices served as a complicating factor in teaching the topic" (*ibid.*, p. 65). Thus, a mediating sign was needed for meaningful feedback to occur. They developed the tool MATRIZMAT, which were boxes that can be joined together by magnets representing rows and columns of matrices. The set for deaf students had numbers put inside the boxes (Fig. 6.4a), while the set for blind students had lids with Braille numbers on them (Fig. 6.4b).

MATRIZMAT became a sensory sign tool (thus "vocabulary" for the deaf and the blind students) that teachers and students used to develop their mathematical language. For the deaf students, new hand gestures were developed to talk about matrices; for the blind students, new spatial positions were explored to experience the meaning of matrices. Thus, feedback generated from a collective (teachers and students) designed tool can be used to develop meaningful classroom mathematical discourse. The semiotic mediation framework gives us a cultural semiotic frame with which to structure tool-based activities that favor production of different categories of signs as mediators for mathematical knowledge.

Fig. 6.4 MATRIZMAT in two forms for (**a**) deaf students with written numerals and (**b**) blind students with numbers in Braille (Healy, Fernandes & Bolite-Frant, 2013; p. 65)

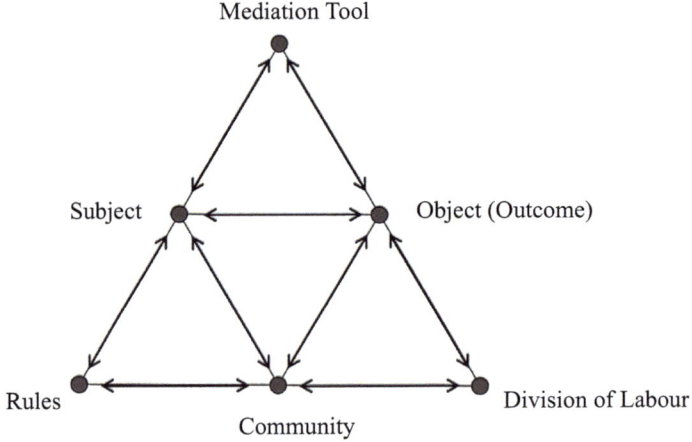

Fig. 6.5 A version of Engeström's activity system (e.g., 1987)

6.3.4 Activity Theory

Another way to look at the relationship between tool use and the pedagogic environment is to look at the interrelationships offered by Engeström (1987) in his structural representation of activity theory. The activity system is seen as an interacting network composed of subject, mediating artifacts or tools, object, division of labor, community, and rules. Figure 6.5 depicts the main components of Engeström's activity system depicted as a mediational triangle.

A tool and its interaction with other components in the activity system play a major role; the arrows in the mediational triangle in Fig. 6.5 serve as guides to design tool-based tasks. For example, the subject is a class of students, the object of activity is to solve an algebraic problem, the tool (mediating artifact) is a CAS, the rules are how to use the CAS's functionalities, the community is the teacher and students, and the division of labor is organized by pedagogical approaches and arrangements. In this scenario, task design then focuses on how the CAS can create a process of learning comprising of possible connecting activity routes that would lead to possible outcomes.

6.3.5 TELMA and ReMath

Attempts have been made by European mathematics education communities to try to connect different theoretical frames previously mentioned in the context of teaching and learning mathematics using digital tools. From these research projects, important guidelines and concerns have been established relevant to tool-based task design.

TELMA (Technology Enhanced Learning in Mathematics) was a joint collaboration project consisting of six European research teams engaged in cross-experimentation in classrooms using ICT as a tool aiming to develop a common language to analyze the intertwined influence played by different contextual characteristics and theoretical frames (Bottino & Kynigos, 2009). Instead of looking for a unified compromising theory, the TELMA team developed an operational methodology to connect the theoretical frames used by the different teams to study the design and use of ICT tools in the mathematics classroom. In this methodology, the notion of *didactical functionality* (Cerulli, Pedemonte, & Robotti, 2007) provided a common perspective on the use of interactive learning environments in mathematics education, and the Concern Methodology Tool (Artigue, Cerulli, Haspekian, & Maracci, 2009) was used to express key concerns for the didactical functionalities. Three key elements in the definition of didactical functionalities of an ICT tool are (1) tool features/characteristics, (2) educational goals, and (3) modalities of employing the tool in a teaching/learning process. To each element, the Concern Methodology Tool associated a list of concerns: to (1) concerns regarding the ergonomic of the tool, semiotic representation, interaction between student and mathematical knowledge, etc.; to (2) epistemological concerns focusing on specific mathematical contents or specific mathematical practices, cognitive concerns focusing on specific cognitive processes, specific cognitive difficulties, etc.; to (3) concerns regarding the functions given to the tool, instrumental issues, instrumental genesis, etc. (*ibid.*, pp. 221–222). These three functionalities and most of the concerns are relevant tool-based task design considerations that could go beyond the use of the ICT tool. Another key idea developed by the TELMA team was the notion of *distance* (Morgan et al., 2009), in particular, epistemological distance and didactic distance are significant to tool-based task design. *Epistemological distance* refers to distance between different representations (tools) focusing on the difference between the affordances for meaning offered by the representations (tools). *Didactic distance* refers to the nature of and quantity of feedback by the tool asking questions like: *What forms of feedback are provided? How are solutions validated and by whom, e.g., by the tool itself, by a teacher, and by peer- or self-validation?* (*ibid.*, p. 251). Tool feedback is a key issue in tool-based task design which will be explored further in the later part of this chapter.

The European Research and Development project ReMath (Representing Mathematics with Digital Media) was a continuation of TELMA to "experientially develop and apply strategies for integrating theoretical frameworks and constructs-in-use to enhance knowledge in the uses of digital media for mathematics education" (Kynigos & Lagrange, 2014) based on the work in TELMA. An agreed assumption on a shared idea of (mathematical) representation was the Minimal Theoretical Framework used as a skeleton of implicitly shared theories concerning representation:

1. No direct access to mathematical objects is possible; rather, mathematical meanings are represented through language, formal mathematical notations, and informal idiosyncratic representations.
2. Representations play a fundamental role in the "generation" of mathematical meanings, and this role is assumed to be crucial in the teaching/learning of mathematics.
3. Digital artifacts can provide representations of mathematical objects with a clear potential of generating mathematical meanings (Artigue & Mariotti, 2014, p. 338).

These highlight further the critical role that tools play in learning mathematics and hence the importance of developing tool-based task design.

6.3.6 Other Theories Relevant to Tool-Based Task Design

Mathematics tasks involving the use of tools can be compared with experiments using scientific instruments or construction with tools in the spirit of exploration and invention. The basic tenet of the teaching and learning theory, Realistic Mathematics Education (RME) (Freudenthal, 1973), is described in detail in Chap. 2.

One manifestation of RME is the Freudenthal Institute's Digital Mathematics Environment (DME), an ICT tool in the form of Java applets designed by teachers, textbook authors, or educators for adapting existing online modules or for designing new ones. It consists of systems to manage and distribute mathematical content and to monitor student progress. Figure 6.6 depicts the DME Authoring Tool interface (Drijvers et al., 2013, p. 55).

Drijvers et al. (2013) illustrated with DME examples how this Authoring Tool with its multiple pedagogical functions supports the three Realistic Mathematics Education (RME) principles that inform task design: guided reinvention, didactical phenomenology, and emergent modeling (*ibid.*, p. 54). These principles point to opening up variant situations and opportunities for students to explore and to investigate and creation of micro-digital worlds (mathematical or not) that are

6 Designing Mathematics Tasks: The Role of Tools

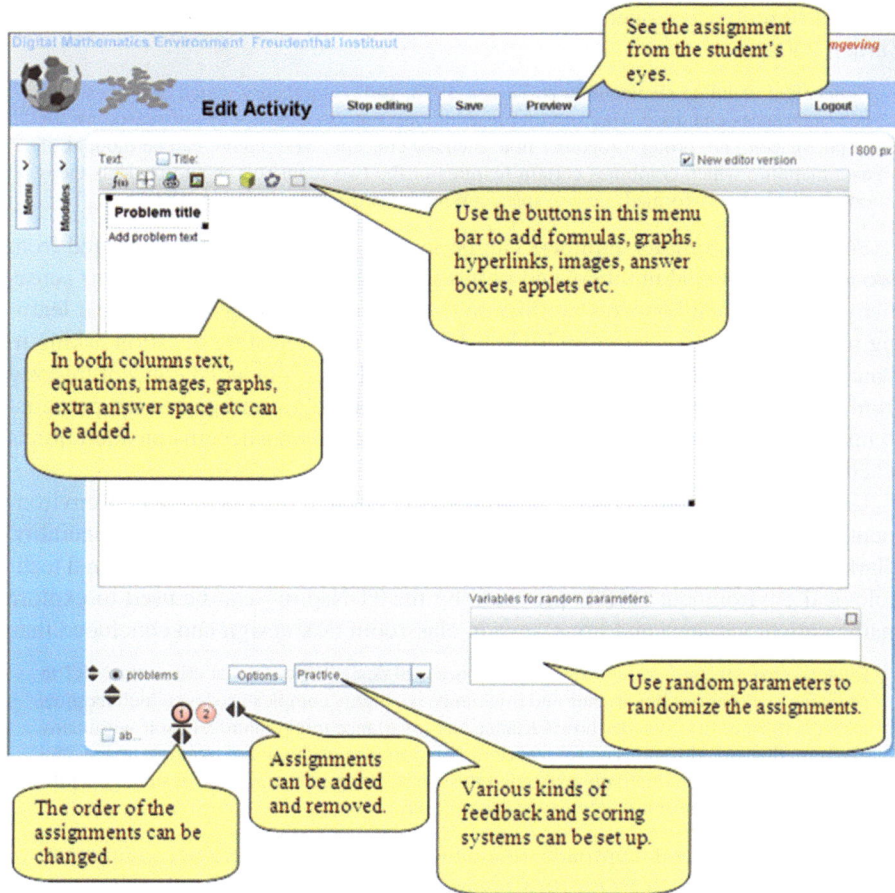

Fig. 6.6 The digital mathematics environment authoring tool (Drijvers, Boon, Doorman, Bokhove, & Tacoma, 2013, p. 55)

real to students. They also point to the evolution of a model developed through different phases of activity by

> an increasing repertoire of representations and techniques in the digital environment, or by increasing options to dynamically use these representations, connect them, and switch between them. (Drijvers et al., 2013, p. 5)

When different tools possessing different epistemic values are chosen to put together into a task design, these tools afford different possibilities for learners in ways of "providing them with different types of experiences and activities, as well as different ways to represent ideas and concepts" (Whiteley & Mamolo, 2013, p. 130).

These different ways open up a space for conceptual blending (Fauconnier & Turner, 2002):

> Conceptual blending (Fauconnier & Turner, 1998, 2002) is a theory which describes how new inferences can arise when two representations and associated ways of reasoning (or 'input spaces') are brought together in a 'blended concept'. The 'blend' can be thought of as a mapping which combines certain features of the two input spaces and projects them onto a third (newly formed) mental space. (Whiteley & Mamolo, 2013, p. 133)

For example, complex numbers can be seen as a blended concept initiated from two input spaces: real numbers and points in planar geometric space. In this sense, many mathematical concepts are blended concepts. Designing teaching and learning mathematics tasks using multiple tools raises an interesting question asking to what extent a mathematical concept can be regarded as a blended concept inferred from mental spaces generated by the tools. If multiple tool usage has the potential to infer a blended concept in a mathematical activity, then a design consideration is whether or not to design this inference into the task.

Using different tools can create a multi-representational pedagogical environment which provides teacher and student with opportunities to express generality. Clark-Wilson and Timotheus (2013) studied how the multiple representational technological environment (MRT) provided by the TI-Nspire™ can be used to explore mathematical variance and invariance in classroom task design and concluded that

> Evidence from the study suggests that the process of designing tasks that utilise the MRT to privilege explorations of variance and invariance is a highly complex process which requires teachers to carefully consider how variance and invariance might manifest itself within any given mathematical topic. The relevance and importance of the initial example space, and how this might be productively expanded to support learners towards the desired generalisation is a crucial aspect of activity design. (p. 51)

An epistemological approach to mathematics is to experience mathematical knowledge through discernment of invariant under variation (see also the description of Variation Theory in Chap. 2). Task design in a tool-based environment opens up pedagogical spaces for exploration of variation and invariant. Leung and Lee (2013) designed a task-based dynamic geometry tool providing a platform to explore variation in students' geometrical perception. Collecting student responses from dynamic geometry drag-based tasks, the platform generated task perceptual landscapes which were used to form and explore students' collective example spaces (see Sect. 6.4.4) and personal example spaces. Variation in the drag-based task design produces different example spaces and hence results in variation in students' geometrical perception. Arzarello, Bairral, Danie, and Yasuyuki (2013) initiated task designs using touch screen multi-figure dragging in the Geometric Constructer (GC) software to study students' perceptions on geometrical concept formation (Fig. 6.7). This new type of dynamic geometry tool will become a new niche to study using variation in task design. Students drag and manipulate dynamic figures with more than one finger on a touch screen.

Special tools can be designed to enhance other abilities in the case of students with a physical disability like blindness (Healy et al., 2013). This tool substitution

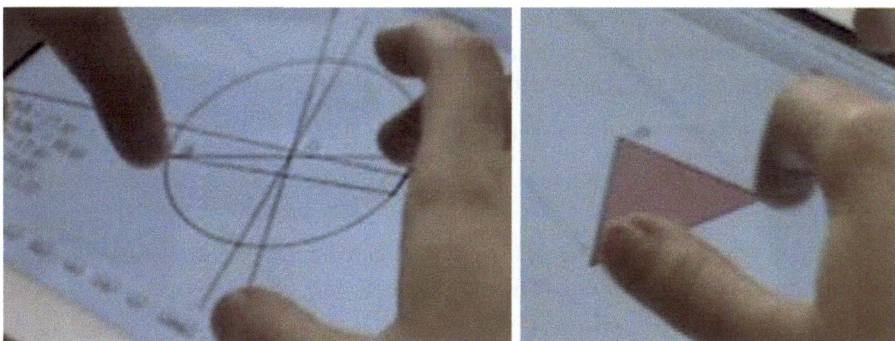

Fig. 6.7 Geometric constructor: a multi-figure dragging touch screen dynamic geometry environment (Arzarello et al., 2013, p. 62)

theory may produce different forms of activity and interaction that could generate different types of mathematical insight and reasoning. "We need to pay attention to—and when necessary create—a multitude of (substitute) semiotic systems to mediate mathematics learning" (Healy et al., 2013, p. 64). Research in tool-based design for inclusion should provide interesting perspectives in mathematics epistemology.

6.4 Further Design Considerations and Heuristics

6.4.1 *Dialectic Between Pragmatic and Epistemic Values*

A gap exists between how to use a tool and how the tool usage leads to mathematics concept development. This gap can be widened if tools are used without meaningful design in the teaching and learning process, especially when skills are involved in using the tools. Without careful consideration on how a skill can be transformed into a cognitive tool to shape a mathematical concept, the use of tools may result in epistemological and didactical obstacles for students, depriving them from reaching the desired or intended mathematical knowledge. The feedback produced by the tool could have negative impact that goes in a direction opposite to the teacher's intention. Thus, tool-based task design must take into account the pragmatic/empirical-mathematical/systematic gap (Noss, Healy, & Hoyles, 1997) that exists in pedagogical usage of a tool. Joubert (2013) discussed this area of concern in the context of a milieu supported by the computer:

> When computers are used, very often the task students are given requires them to construct an artifact on the screen, such as, for example, a graph or a geometric transformation and they begin by working in the pragmatic/empirical field. It seems that frequently the task does not require movement between this and mathematical/systematic fields and mathematical learning is limited. (Joubert, 2013, p. 75)

One perspective to discuss the shortening of the gap is to pay attention to the interaction between the pragmatic and epistemic values of a tool to develop dialectic between mathematical practices and mathematical discourses—*praxeology*:

> The teacher must consider the techniques he wants to make alive in the classroom to perform a specific task; he must consider the technological discourses he want the pupils to develop for describing, explaining, justifying and developing the techniques used. (Wozniak, 2013, p. 148)

As the pragmatic value of a tool increases via formation of utilization schemes, its epistemic value may also increase:

> The instrumental genesis (Rabardel, 1995) does not facilitate only the use, and thus the instrumentality, of an ostensive. It develops also its semiotic value: an increase of the pragmatic value allows also an increase of the semiotic value. (Wozniak, 2013, p. 148)

Robotti (2013) discussed visuospatial tasks for linear equations using the dynamic algebra artifact AlNuSet to make it possible for students to interpret algebra by a kind of drag-to-fit strategy. For example, Fig. 6.8 depicts snapshots from AlNuSet exploring the meaning of $x + 2 = 2x + 3$. The expressions $x + 2$ and $2x + 3$ are positions on an "algebra line" that depends on the position (value) of x (left figure). Drag x along the algebra line; when it reaches -1, $x + 2$ and $2x + 3$ automatically line up with each other vertically at 1, which has the same value for the two expressions when $x = -1$ (right figure). This is a dynamic visualization of equality of two linear expressions (Robotti, 2013, p. 104).

In this example, the task was designed to bridge the mathematical conceptual notion "equality" and a discursive notion "drag until they line up". This discursive notion depends on drag-based visual spatial feedback and forms a dialectal language

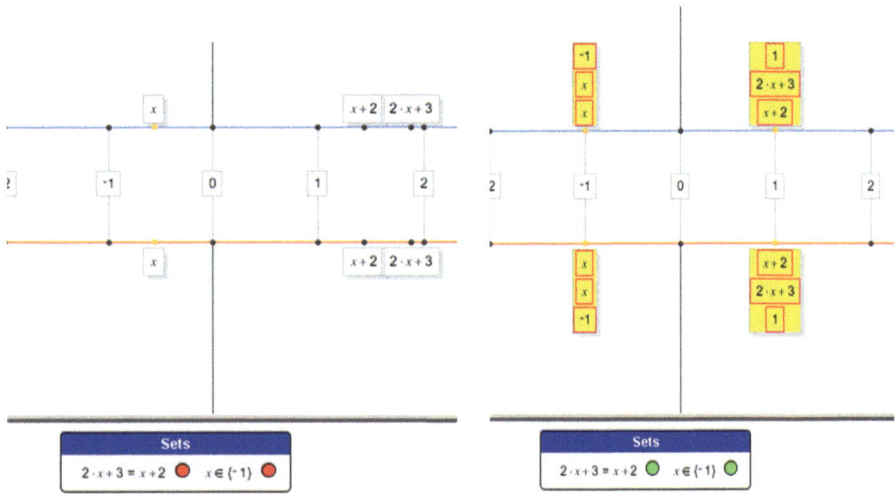

Fig. 6.8 An AlNuSet task exploring the meaning of $x + 2 = 2x + 3$ (Robotti, 2013, p. 104)

in which the pragmatic value of the dragging technique can be transformed into a mathematical concept. The discourse depends on the designing skills of the teacher who is using the software. The AlNuSet also has a Manipulator Component to allow students to solve equations using different symbolic approaches (*ibid*). Combining the algebra line and the manipulator, teachers can design tasks to make a connection between student-produced situated algebra discourse, resulting from the pragmatic value of the tool and the conventional algebraic concepts.

There are further design considerations arising from the dialectic between pragmatic and epistemic values:

- Design cognitive aspects into didactical situations to appropriate the instrumental genesis process.
- Design tool-based activities to explore optimally the epistemic value of techniques.
- Distinguish between, and connect, non-ostensive and ostensive aspects of thinking and doing.

6.4.2 Mediation and Feedback

There is another kind of gap that exists in the mathematics classroom, especially in tasks that promote student discourse. Student-produced discourse often differs from what is expected in the mathematics curriculum. One reason may lie in the disparity between mathematical concepts defined by teachers and mathematics experienced by students in the classroom (see Chap. 4). Students usually don't use teacher-defined mathematical concepts in learning activities, but rather use their own taken-for-granted mathematical ideas and discourses in these activities. For example, a teacher defines the notion "perpendicularity" in the classroom while students use a corner of a book to talk about it. To shorten this "discourse gap", a well-designed tool with associated tasks may help, such as a 2-D representation of a 3-D figure in which the corner of a book does not indicate perpendicularity. Tools can become a mediator between different mathematical discourses.

Different types of feedback from tool use can play a major mediation role. In the dynamic geometry environment Cabri Elem, teachers can design their own milieu in which dynamic objects are regarded as didactical variables. Figure 6.9 depicts a snapshot of the opening page in a teacher-designed Cabri Elem activity book (Mackrell et al., 2013). The dots are counters that can be dragged into the different regions of the target scoreboard. When a counter is dragged into the outermost region, the Score will show 1, into the next region will show 10, and into the central region will show 100. Thus, the Score shows the numbers formed by the student. Students can form different 3-digit numbers by dragging counters into the scoreboard in different ways to try to score the preset target number 955. This particular predesigned activity page starts the process of instrumental genesis by dragging counters.

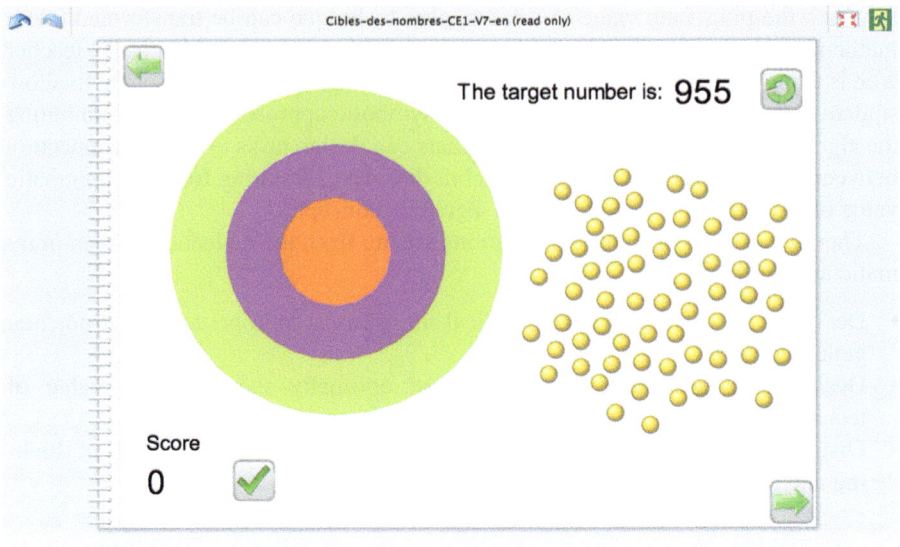

Fig. 6.9 Title page and a task page from the "Target" activity book (Mackrell et al., 2013, p. 83)

The visibility of the score, the movability of a counter, the number of counters, and the target number are all didactical variables that teachers can control. Subsequent activity pages can be designed with different control of the didactical variables in which

> an evolution in student strategies may be provoked through changes in the value of the various didactical variables. The ability to choose the way in which pages are linked also enables the provision of optional help in tool use, differentiated tasks, notes for teachers, etc. (Mackrell et al., 2013, p. 83)

This Cabri Elem activity book enables strategic feedback and an interplay between evaluative and direct manipulation feedback (*ibid.*). This evolution can be seen as a kind of *feedback trajectory* that mediates between students' tool-based discourse and (curriculum) mathematics.

Dragging in a dynamic geometry environment (DGE) affords another important mediation feedback. It is known as soft construction. A soft construction in dynamic geometry is a construction

> in which one of the chosen properties is purposely constructed by eye, allowing the locus of permissible figures to be built up in an empirical manner under the control of the student. (Healy, 2000, p. 107)

Constructed by eye means drag a base point of a DGE figure to position(s) that would satisfy a certain condition or property. In the context of task design, Or (2013) used soft construction as a tool to design a GeoGebra task exploring the circumcircle of a triangle using the drag-to-fit strategy. Figure 6.10 depicts snapshots of a soft dragging sequence [(a) to (f)] starting with a predesigned configuration [(a)].

6 Designing Mathematics Tasks: The Role of Tools

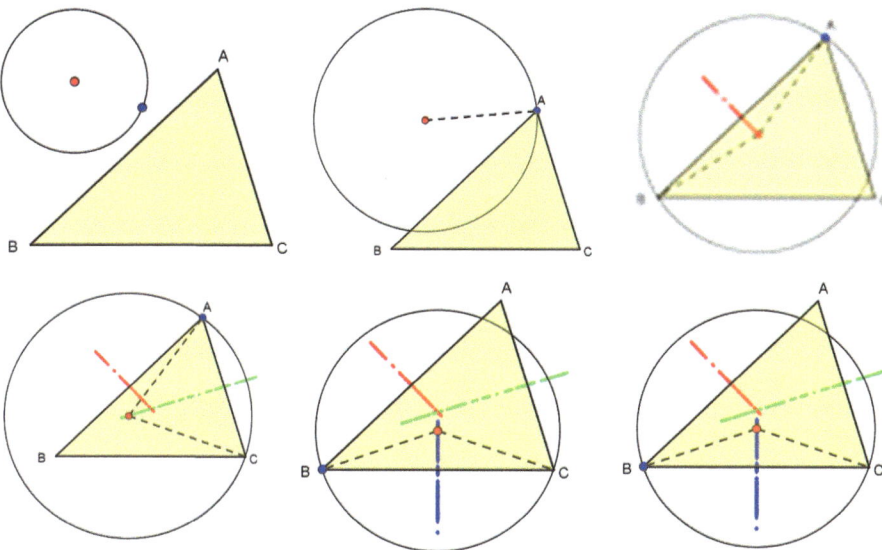

Fig. 6.10 (**a–f**) Snapshots in a soft dragging exploration sequence of a predesigned DGE task exploring *circumcircle* of *triangle* (Or, 2013, p. 94)

When the point on the circle is dragged to one of the vertices of triangle *ABC*, say, *A*, a dotted radius is shown [(b)]. Soft dragging the center of the circle to positions where the circle passes through another vertex, say, *B*, another radius is shown and the trace of these positions of the center is marked automatically, resulting in a visual trace [(c)]. At this stage of the task, the teacher can discuss with students the insights gained by what they see on the screen while dragging the center around. Students then continue the discourse by performing the same dragging strategy to the other pairs of vertices [(d) to (f)] and eventually reach the final intended learning outcome [(f)]. The drag-to-fit strategy and the Trace functionality in DGE can be regarded as a kind of recursive feedback that instigates a mediation loop between visualization and reasoning. This soft construction task in DGE is an example of task design making use of this type of feedback.

A specific feedback trajectory can be designed into a task sequence. Bokhove (2013) used a model with elements of *crises*, *feedback*, and *fading* to design a sequence of online near-similar tasks to foster procedural fluency and conceptual understanding in solving algebraic equations. Crisis was intentionally designed into the task sequence at the point where students were beginning to get acquainted with some routine problem-solving strategies. The crisis took the form of modified problems that were similar to what students were familiar with, but may need different problem-solving strategies. Feedback in terms of hints and suggestions was given by the online system to the students in a fading manner as the problem sequence progresses. This model anticipates that students' cognitive load will go through a piecewise decreasing trend in real time with a sharp discontinuous jump at the crisis

Fig. 6.11 A crisis and fading feedback model (Bokhove, 2013, p. 18)

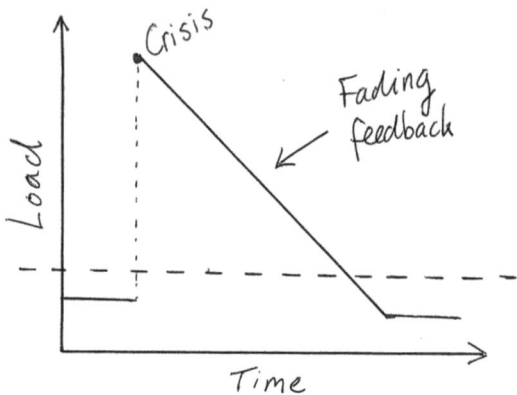

moment (Fig. 6.11). A lowering of cognitive load may imply internalization leading to conceptual development; thus, an online designed feedback trajectory like this one can mediate between a designed discourse and the targeted object of learning, in this case, solving algebraic equations.

The previously discussed cases illustrated different types of tool-dependent feedback that serve to mediate between different aspects in mathematics teaching and learning. Feedback can be regarded as designed or unanticipated didactical intervention. Thus, a tool with designed feedback (actual or potential) can be regarded as a tool of semiotic mediation when used by the teacher to mediate a mathematical content. When teachers design feedback in a tool-based task, they need to anticipate the affordances and constraints of the task as a frame for mathematical activity, types of interaction with the tools, ease of use, uncertainty, available choices, didactic variables, and multi-representation. In terms of mathematics content, feedback acts at the boundary between mathematical and pedagogical fidelity in the sense that the tool-learner interaction creates a bridge between pedagogical design and mathematical meaning. This fidelity boundary can be seen as *discrepancy potential*, which will be the discussed in the next section.

6.4.3 Discrepancy Potential

In this section, the idea of "distance" that has been manifested throughout this chapter continues with the introduction of the idea *discrepancy potential*. The discrepancy potential of a tool is a pedagogical space generated by (i) feedback due to the nature of the tool or design of the task that possibly deviates from the intended mathematical concept or (ii) uncertainty created due to the nature of the tool or design of the task that requires the tool users to make decisions. For a tool used to teach and learn mathematics, there is an instrumental distance between what the tool can represent and the intended mathematics to be taught. Within that distance,

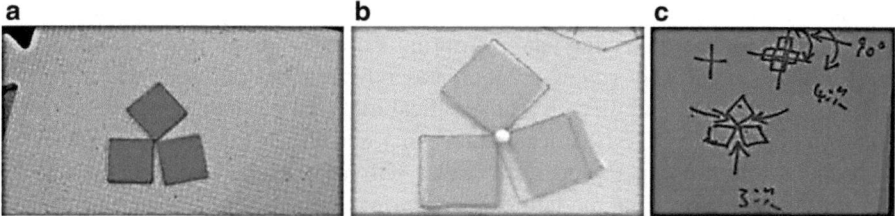

Fig. 6.12 A transparency toolkit used to teach rotational symmetry (Chan & Leung, 2013)

a teacher has the space to introduce disturbances and perturbations, but these might also occur unintentionally.

To illustrate (i), Chan and Leung (2013) reported a study on using a transparency toolkit to teach rotational symmetry in a Primary Five classroom. Students were asked to arrange a rotational symmetric figure using plastic-shaped pieces (Fig. 6.12a). An overhead transparency sheet was placed on top of the figure with a pushpin to pivot it at the center of rotation. Students traced an outline of the figure on the pinned transparency sheet and rotated the transparency sheet to see if the figure satisfied being rotationally symmetrical by paying attention to the overlapping (see Fig. 6.12b).

This was a crude tool and not very accurate. However, because of its crudeness, a pedagogical space was opened for teacher-student discourse that led to development of the mathematical concept. For the episode depicted in Fig. 6.12, the teacher asked students why the overlapping was not exact under the rotation and how to modify the figure to correct this discrepancy. This discussion led to students discovering that for rotational symmetry, the gap angles in the figure must be the same; the more number of times of overlapping in one rotation, the smaller the gap angles become. Afterward, the teacher led the class to explore the angle of rotation for rotation symmetry (Fig. 6.12c). This episode was an unexpected disturbance due to a student-produced figure using the transparency toolkit. The teacher made use of this opportunity to develop a mathematical concept that was not intended in the lesson plan. Thus, a task "which provides an opportunity to make use of the discrepancy embedded in the tool may initiate meaningful mathematics discussion which could lead to deeper conceptual understanding" (Chan & Leung, 2013, p. 41). The notion of discrepancy potential does not have an inherent good/bad value.

To illustrate (ii), uncertainty can be intentionally designed into the task to create cognitive conflict that could bring about discernment:

> the potential *to evoke uncertainty and doubt*, which are widely recognized as powerful diagnostic tools as well as vehicles for creating situations in which the need to prove arises intrinsically ... the uncertainty evoked by the tasks trigger students' discussion and attempts to convince each other through argumentation, which would allow for various aspects of their understanding to be revealed. (Buchbinder & Zaslavsky, 2013, p. 28)

The potential for using discrepancy to foster meaningful mathematical activities can be considered to be a design heuristic. DGE soft construction was discussed in the last section with respect to mediation and feedback. The epistemic potential of soft construction lies in asking learners during a dragging exploration: *What needs to be in order to get what you want?* It creates a discrepancy potential within which learners can develop and discern mathematical ideas. In particular, relaxing conditions is a heuristic in DGE task design that can probe into different ways to generate conjectures and construct proofs. Arzarello, Olivero, Paola, and Robutti (2002) identified seven dragging modalities (wandering, guided, bound, dummy locus, line, linked, drag test) while trying to analyze conjecture-making episodes by students working on a geometrical problem. The following are two different DGE soft construction task design approaches exploring the cyclic quadrilateral making use of dragging modalities.

Design One

1. Construct a general quadrilateral *ABCD*.
2. Measure two opposite interior angles, say, ∠*ABC* and ∠*CDA*.
3. Calculate ∠*ABC* + ∠*CDA*.
4. Turn the Trace function on for point *C*.
5. Drag point *C* continuously to keep ∠*ABC* + ∠*CDA* as close to 180° as possible.
6. Observe the shape of the path that point *C* traces out.
7. Make a conjecture on the shape of the path.
8. Explain why the conjecture is true.

Design Two

1. Construct a general quadrilateral *ABCD*.
2. Construct a circle that passes through *A*, *B*, and *C*.
3. Construct a circle that passes through *D*, *B*, and *C*.
4. Measure two opposite interior angles, say, ∠*ABC* and ∠*CDA*.
5. Calculate ∠*ABC* + ∠*CDA*.
6. Drag the quadrilateral to keep ∠*ABC* + ∠*CDA* as close to 180° as possible, and observe what happens to the two circles *or* drag the quadrilateral to make the two circles overlap and observe what happens to ∠*ABC* + ∠*CDA*.
7. Make a conjecture and explain why the conjecture is true.

In Design One, tracing the point to satisfy the required condition is not obvious. In fact, the trace can be wriggling and hard to discern. However, this discrepancy demands that the learner refines his/her dragging strategies and heighten his/her critical discernment, consequently turning the DGE dragging tool into a mental cognitive tool. Thus, the DGE Trace function can be used (intentionally) to create disturbances that the learner needs to resolve (Fig. 6.13).

In Design Two, an uncertainty is designed in Step 6 where the learner can choose to take one of the two dimensions of variation (the varying values of ∠*ABC* + ∠*CDA* and the "overlappedness" of two circles) as the drag-to-fit target. This intentional designed uncertainty creates a situation where the learner needs to interpret dragging feedback with logical inference sequences and, hence, develops geometrical reasoning (Fig. 6.14).

Fig. 6.13 Design One

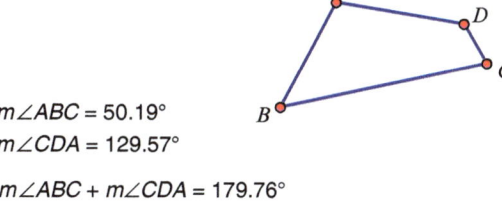

$m\angle ABC = 50.19°$
$m\angle CDA = 129.57°$

$m\angle ABC + m\angle CDA = 179.76°$

Fig. 6.14 Design Two

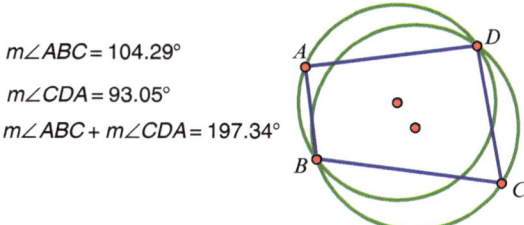

$m\angle ABC = 104.29°$
$m\angle CDA = 93.05°$
$m\angle ABC + m\angle CDA = 197.34°$

Possible DGE conjectures resulting from these two tasks are:

Design One

For a quadrilateral to satisfy the condition "a pair of interior opposite angles adds up to 180°", the vertices of the quadrilateral should lie on a circular path.

Design Two

Given a quadrilateral *ABCD*, its vertices lie on the same circle *if and only if* a pair of interior opposite angles adds up to 180°.

The idea of discrepancy potential in tool-based task design highlights the importance of paying attention to how the nature or design of a tool can bring about mathematical learning. Different tools can be used to design tasks that foster the teaching and learning of the same mathematical topic in different ways. Their different discrepancy potentials could interact in meaningful ways to bring about rich mathematical experiences for students, as suggested by the overlapping shaded areas in Fig. 6.15. This leads to the issue of multi-representations, which is the discussion in the next section.

6.4.4 Conceptual Blending and Multiplicity

Historically, mathematical ideas have been developed through tools and cultural artifacts. Thus, in retrospect, a mathematical concept should be able to be represented by different tools and artifacts. Furthermore, new and developing technologies play revolutionary roles in representing mathematical concepts and even creating new ones. Multiplicity is a key idea in instrumental orchestration and documentation.

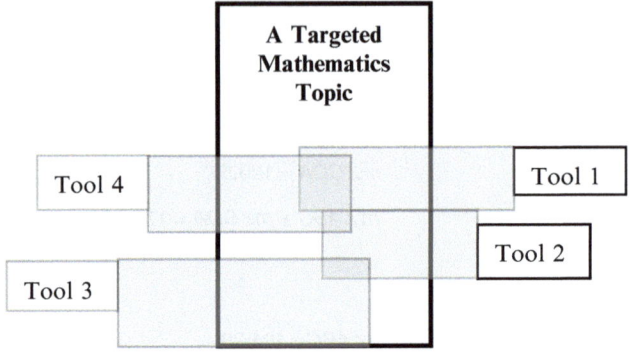

Fig. 6.15 The *shaded regions* are the tool discrepancy potentials for different tools

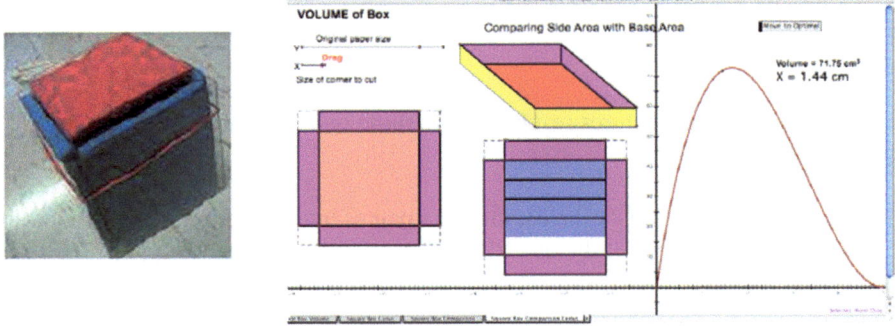

Fig. 6.16 A multi-tool environment consisting of concrete models and interactive 2-D representations of 3-D *boxes* and function graphs in Geometer's Sketchpad (GSP) (Whiteley & Mamolo, 2013, pp. 137–138)

Often, abstract generalizations come about when critical aspects from multiple mathematical representations and discourses fuse and blend together. Therefore, in a multi-tool pedagogical environment, interactions among the discrepancy potentials of different tools, as discussed in the previous section, and the bridging and switching between tools and representations are key task design considerations.

Whiteley and Mamolo (2013) used a framework of conceptual blending (Turner & Gilles Fauconnier, 2002) to design a multi-tool task that deals with a box optimization problem. The multi-tool environment consisted of concrete models and interactive 2-D representations of 3-D boxes and function graphs in Geometer's Sketchpad (GSP) (Fig. 6.16).

It was found that teachers and students had multiple ways of reasoning about the task and created different conceptual blends for these representations. Nevertheless, the selected problem encouraged all participants to anticipate that there was an optimal shape and guess a candidate shape (always incorrectly). Even simple activities could indicate that better reasoning tools were needed to resolve the conflict and find the optimum. For teachers, the task was about allowing an opportunity to unpack the basic processes of calculus and optimization and to realize a novel spatial sense in optimizing geometrical problems. For students, the task afforded them with accessible tools to solve a problem in different ways, which supported or laid a foundation for a symbolic/algebraic approach (Whiteley & Mamolo, 2013, p. 134). Using a conceptual blending framework to analyze the multi-tool environment and the activity emerging from the task gives information about the discrepancy potentials of the individual tools and also the conceptual development of the students (*ibid.*, p. 136). The concrete model became a thinking tool and the GSP applet invited rapid (or fluent) switching among multiple ways of reasoning which were supported by the multiple representations. Task design should consider the mental processes involved in rapid (fluent) switching back and forth between representations and students' alternative ways of reasoning. With the lens of conceptual blending, individual student difficulties can be tracked to gaps in how particular representations are manipulated to support reasoning. With practice, learners can apply the reasoning elicited by one tool within the context of a second tool, for example, taking the focus on "sign of the change" in the model to the slope of a secant in the graph.

Addressing the issue of students' multiple feedback responses to a tool, Leung and Lee (2013) designed predesigned dragging tasks in a DGE-based platform that was able to record multiple student responses in a collective fashion. The platform was capable of generating a collective image map of student geometrical perceptions for a predesigned dragging task. This map, called a *task perceptual landscape*, is visually interpreted as students' qualitatively different ways of perceiving a geometrical phenomenon under the drag mode, ways which are quantified and categorized in a collective way (see, e.g., Fig. 6.17). The scatterplot on the left consists of students' responses superimposed on the dynamic task template. The task perceptual

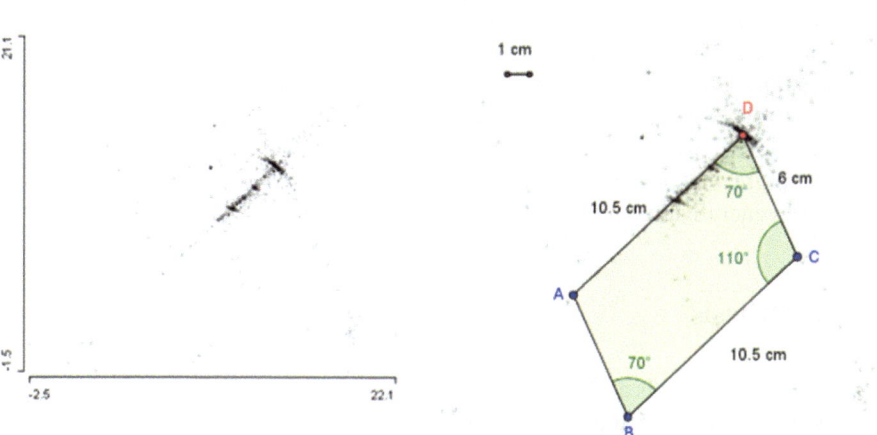

Fig. 6.17 Dynamic task template (Leung & Lee, 2013)

landscape is a *collective example space* of students' perceptions of a task. This collective student example space can be used to design teaching and learning tasks to explore students' perceptions on geometrical understanding. In this case, multiplicity is seen in terms of amalgamation of multiple student responses in a dynamic interactive tool environment. This task-based dynamic geometry platform opened up the idea of an *instrumental documentation environment* that should have potential pedagogical implications which can be explored.

Kaput (1986) postulated that a multiple representational environment supported by technology might enhance high-level engagement with mathematics. In this connection, Clark-Wilson and Timotheus (2013) studied the design rubric of a multi-representational technological environment (MRT) supported by the TI-Nspire. In particular, the study focused on how MRT helped to explore mathematical invariance and variance. A key task design consideration was how to switch and flow different representations (e.g., graphical, tabular, and linguistic) so their interactions would bring about mathematical generalization and how the situated discourses that developed under these representations communicate with each other with respect to the generalization. The findings gave rise to seven questions that are relevant to task design in a multi-tool environment (Clark-Wilson & Timotheus, 2013, p. 51). They take seriously the assertion that "a lesson without the opportunity for learners to express a generality is not, in fact, a mathematics lesson" (Mason, Graham, & Johnston-Wilder, 2005, p. 297). In that sense, these questions apply equally well to all tool-based task design contexts:

1. What is the generalizable property within the mathematics topic under investigation?
2. How might this property manifest itself within the multi-representational technological environment—and which of these manifestations is at an accessible level for the students concerned?
3. What forms of interaction with the MRT will reveal the desired manifestation?
4. What labeling and referencing notations will support the articulation and communication of the generalization that is being sought?
5. What might the "flow" of mathematical representations (with and without technology) look like as a means to illuminate and make sense of the generalization?
6. What forms of interaction between the students and teacher will support the generalization to be more widely communicated?
7. How might the original example space be expanded to incorporate broader-related generalizations?

6.5 Synthesis

For some theories, tools and representations act as extensions to our sensory/mental cognition to experience and perceive mathematics. Tools may start as external artifacts, but they have the potential to turn into internal cognitive tools or cognitive extensions. For embodied theory, in which there is no external/internal dichotomy

6 Designing Mathematics Tasks: The Role of Tools   219

Choices of Tool	Visible crises
Open-endedness	Formative assessment
Connectivity	Didactic phenomena
Accessibility	Task sequence design
Freedom to choose	Variation of didactic variables
Immersion in the technology	Perturbation and disturbance
Possibility of student as a designer	Feedbacks: Controlled, uncontrolled,
Constraining or respecting diversity	fading, student-generated, teacher-designed

Fig. 6.18 Issues around pragmatic and epistemic dialectic

and cognition is dependent on the activity, tools are regarded as part of cognition. In either case, whether tools are conducive or disruptive in the process of teaching and learning depends on how they are designed to be used in pedagogical situations. As discussed in the previous sections, choices of tool and design of tool usage open up pedagogical spaces where the pragmatic and epistemic dialectic is at play and where students and teachers are invited to develop mathematical discourses. Figure 6.18 displays some pedagogical issues surrounding the tool-based pragmatic and epistemic dialectic that task design needs to pay attention to.

To focus further on tool-based task design, considerations, and heuristics, there are at least four broad areas of concerns:

1. Use implicit, explicit, and strategic feedback from the tool-based environment to create crises and/or fading that are conducive to student learning.
2. Design pragmatic/instrumental and epistemic/semiotic activities that consider the boundary of mathematical and pedagogical fidelity, tactile and meaningful interactions with the tools, and substitution or enhancement of sensory abilities.
3. Make use of the affordances and constraints created by the tool. For example, unintentional disturbances caused by the tool quality, different design given by relaxing conditions/functionalities, (epistemological) obstacles created by the tool, etc.
4. Fluent switching between representations/tools.

These heuristics should not be seen as separate and distinct design considerations; rather they are various foci of attention in the design process, and they inform and reinforce each other. A few key heuristics discussed in this chapter are further highlighted in the following.

6.5.1 Strategic Feedback and Mediation

Feedback is a key to promote skill and understanding in any type of tool usage. Types of feedback can be regarded as didactical variables that entail possibly different pragmatic and/or epistemic values conducive to mathematical concept development.

Thus, designing appropriate strategic feedback into tool-based tasks should be a major design attention. Feedback always has potential to open up a space for unanticipated actions and/or discourse (especially coming from didactical and epistemological obstacles) with potential pragmatic and/or epistemic values. Task design should be flexible enough to allow development of this contingent pedagogical space. Tool usages can be developed as semiotic signs that can mediate mathematical knowledge. Whether this process is natural, designed, or a combination of both depends on the choice of tool, the mathematical content, and the choice of pedagogical approach. Feedback and mediation are complementary processes leading internalization and/or transformation of knowledge.

6.5.2 The Instrumental/Pragmatic-Semiotic/Epistemic Continuum

Evolution of a tool's instrumented action scheme (Trouche, 2004) transforms the instrumental/pragmatic-semiotic/epistemic dialectic. The instrumental genesis and tool of semiotic mediation frameworks (Sect. 6.3) both deal with this transformation from physical tool to psychological tool. Task design needs to sequence tool-based activities to progress through the *epistemological continuum* where mathematical skills/techniques are gradually transformed into abstract concepts and understanding. Tools can be assigned or designed to play different epistemic roles at different points of the continuum. For example, tools can enable: action, formulation, and validation (Brousseau, 1998). Action involves familiarity with the tool and how to implement a procedure to use the tool. Formulation concerns a task situation when tool feedback produces difficult unfamiliar problems for students to tackle that need to be resolved mathematically. It is in this mode that a tool begins to turn into a psychological tool for discernment of critical mathematical ideas. Validation dialectics involve explaining, justifying, theorizing, and proving. This echoes with many aspects in Leung's nested epistemic modes task design framework mentioned in Sect. 6.2.4.

6.5.3 Boundary Between Mathematical and Pedagogical Fidelity

Task design can situate a tool at the boundary between mathematics and pedagogy. This is in the sense that a tool can be designed to embody a pedagogical dimension of mathematical knowledge. Thus, the mathematical and pedagogical usage of a tool becomes a task design issue. A pedagogical dimension may be, for example, a didactical situation that is relevant to students' mathematical experiences but may not immediately access to the rigorous mathematics dimension, or vice versa. Task design should be sensitive to this boundary and aim to create a symbiosis between mathematics and pedagogy.

6.5.4 Discrepancy Potential

Tools arouse uncertainty and doubt because feedback from tool usage may deviate from the intended mathematical concept to be learned or may create cognitive conflicts for students. This opens up a potential pedagogical space where teacher and students have opportunities to use such discrepancies to develop mathematics discourse that could lead to formation of mathematical concepts. Task design can intentionally make use of a tool's discrepancy potential to create uncertainties and cognitive conflicts which are conducive to student learning.

6.5.5 Multiplicity

Different tools extend or amplify different abilities. A multi-tool teaching and learning environment provides learners a milieu where they can interact with different tools and representations. In such a pedagogical situation, task design should pay attention to design connection, switching, and transition between tools and representations to bring about contrasting experiences for learners which may lead to critical discernment of invariant mathematical concepts. Mathematical activities are mostly concerned with discerning invariants in variation or seeing how invariants appear in variation. Thus, designing variation tasks in a multi-tool environment should enrich the mathematics knowledge acquisition/construction process.

As said, the above heuristics are not separate considerations for tool-based task design. They are all closely related to each other and are critical aspects in the whole design process. At present, no overarching or unified theory for tool-based task design exists. The TELMA and ReMath projects experimented with a cross-experimentation and cross-case analysis methodology aiming to create a dialogue platform where different theoretical frameworks were used for task design in an ICT environment; some of these frameworks are discussed in this chapter (Kynigos & Lagrange, 2014). Their findings should serve as a solid reference for tool-based task design. A direction for tool-based task design research is to see how far the findings from these two projects extend to non-ICT tools. The discussion in this chapter is a preamble to this course of research. Technology is developing at an exponential rate, and the nature of reality is becoming less easy to define. Integrating advanced technological tools into the mathematics classroom challenges the nature of mathematical knowledge and, in particular, the question of what represents mathematics. In this direction, tool-based task design principles and heuristics evolve with emerging technology. Therefore, teachers who design tasks should be aware of his/her disposition toward the nature of mathematics as this will influence how the designed tool and its use will be seen to represent mathematics.

References

Abramovich, S. (2001). Cultural tools and mathematics teacher education. In F. R. Curcio (Ed.), *Proceedings of the Third U.S.-Russia Joint Conference on Mathematics Education and the Mathematics Education Seminar at the University of Goteborg* (pp. 61–67). Spokane, WA: People to People Ambassador Programs.

Artigue, M. (2002). Learning mathematics in a CAS environment: The genesis of a reflection about instrumentation and the dialectics between technical and conceptual work. *International Journal of Computers for Mathematical Learning, 7*, 245–274.

Artigue, M., Cerulli, M., Haspekian, M., & Maracci, M. (2009). Connecting and integrating theoretical frames: The TELMA contribution. *International Journal of Computer for Mathematical Learning, 14*, 217–240.

Artigue, M., & Mariotti, M. A. (2014). Networking theoretical frames: The ReMath enterprise. *Educational Studies in Mathematics, 85*, 329–355.

Arzarello, F., Bairral, M., Danie, C., & Yasuyuki, I. (2013). Ways of manipulation touchscreen in one geometrical dynamic software. In E. Faggiano & A. Montone (Eds.), *Proceedings of the 11th International Conference on Technology in Mathematics Teaching* (pp. 59–64). Bari: University of Bari.

Arzarello, F., Bartolini-Bussi, M., Leung, A., Mariotti, M. A., & Stevenson, I. (2012). Experimental approach to theoretical thinking: Artifacts and proofs. In G. Hanna & M. de Villers (Eds.), *Proof and proving in mathematics education: The 19th ICMI Study* (New ICMI Study Series, pp. 97–137). Berlin: Springer.

Arzarello, F., Olivero, F., Paola, D., & Robutti, O. (2002). A cognitive analysis of dragging practices in Cabri environments. *ZDM: The International Journal on Mathematics Education, 34*(3), 66–72.

Balacheff, N., & Sutherland, R. (1994). Epistemological domain of validity of microworlds: The case of Logo and Cabri-geometre. In R. Lewis & P. Mendelsohn (Eds.), *Proceedings of the IFIP TC3/WG3.3 Working Conference on Lessons from Learning* (pp. 137–150). Amsterdam: Elsevier; North Holland: IFIP.

Bartolini Bussi, M. G., & Mariotti, M. A. (2008). Semiotic mediation in the mathematics classroom: Artifacts and signs after a Vygotskian perspective. In L. English (Ed.), *Handbook of international research in mathematics education* (2nd ed., pp. 746–783). New York: Routledge.

Bokhove, C. (2013). Using crises, feedback and fading for online task design. In C. Margolinas (Ed.), *Task design in mathematics education: Proceedings of ICMI Study 22* (pp. 17–24), Oxford, UK. Available from http://hal.archives-ouvertes.fr/hal-00834054

Bottino, R. M., & Kynigos, C. (2009). Mathematics education & digital technologies: Facing the challenge of networking European teams. *International Journal of Computers for Mathematical Learning, 14*, 203–215.

Brousseau, G. (1998). *Theory of didactical situations in mathematics*. Dordrecht, The Netherlands: Springer.

Bruner, J. S. (1966). *Toward a theory of instruction*. Cambridge, MA: Belkapp.

Buchbinder, O., & Zaslavsky, O. (2013). A holistic approach for designing tasks that capture and enhance mathematical understanding of a particular topic: The case of the interplay between examples and proof task. In C. Margolinas (Ed.), *Task design in mathematics education: Proceedings of ICMI Study 22* (pp. 25–34), Oxford, UK. Available from http://hal.archives-ouvertes.fr/hal-00834054

Cerulli, M., Pedemonte, B., & Robotti, E. (2007). An integrated perspective to approach technology in mathematics education. In M. Bosch (Ed.), *Proceedings of the Fourth Congress of the European Society for Research in Mathematics Education* (pp. 1389–1399). Sant Feliu de Guixols, Spain: IQS Fundemi Business Institute.

Chan, Y. C., & Leung, A. (2013). Rotational symmetry: Semiotic potential of a transparency toolkit. In C. Margolinas (Ed.), *Task design in mathematics education: Proceedings of ICMI Study 22* (pp. 35–44), Oxford, UK. Available from http://hal.archives-ouvertes.fr/hal-00834054

Chevallard, Y. (1994). Ostensifs et non-ostensifs dans l'activité mathématique. *Intervention au Séminaire de l'Associazione Mathesis (Turin, 3 février 1994). Actes du Séminaire pour l'année 1993-1994* (pp. 190–200).

Clark-Wilson, A., & Timotheus, J. (2013). Designing tasks within a multi-representational technological environment: An emerging rubric. In C. Margolinas (Ed.), *Task design in mathematics education: Proceedings of ICMI Study 22* (pp. 45–52), Oxford, UK. Available from http://hal.archives-ouvertes.fr/hal-00834054

Confrey, J., Hoyles, C., Jones, D., Kahn, K., Maloney, A., Nguyen, K., et al. (2010). Designing software for mathematical engagement through modeling. In C. Hoyles & J.-B. Lagrange (Eds.), *Mathematics education and technology: Rethinking the terrain – the 17th ICMI Study* (pp. 19–46). New York: Springer.

DiSessa, A., Hammer, D., & Sherin, B. (1991). Inventing graphing: Meta-representational expertise in children. *The Journal of Mathematical Behavior, 10*, 117–160.

Drijvers, P., Boon, P., Doorman, M., Bokhove, C., & Tacoma, S. (2013). Digital design: RME principles for designing online tasks. In C. Margolinas (Ed.), *Task design in mathematics education: Proceedings of ICMI Study 22* (pp. 53–60), Oxford, UK. Available from http://hal.archives-ouvertes.fr/hal-00834054

Engeström, Y. (1987). *Learning by expanding: An activity-theoretical approach to developmental research*. Available online at http://lchc.ucsd.edu/MCA/Paper/Engestrom/expanding/toc.htm. Accessed August 18, 2014.

Fauconnier, G., & Turner, M. (1998). Conceptual integration networks. *Cognitive Science, 22*(2), 133–187.

Fauconnier, G., & Turner, M. (2002). *The way we think: Conceptual blending and the mind's hidden complexities*. New York: Basic Books.

Freudenthal, H. (1973). *Mathematics as an educational task*. Dordrecht: Reidel Publishing.

Freudenthal, H. (1991). *Revisiting mathematics education*. Dordrecht: Kluwer Academic.

Geiger, V., & Redmond, T. (2013). Designing mathematical modelling tasks in a technology rich secondary school context. In C. Margolinas (Ed.), *Task design in mathematics education: Proceedings of ICMI Study 22* (pp. 119–128), Oxford, UK. Available from http://hal.archives-ouvertes.fr/hal-00834054

Gueudet, G., & Trouche, L. (2009). Towards new documentation systems for mathematics teachers. *Educational Studies in Mathematics, 71*(3), 199–218.

Healy, L. (2000). Identifying and explaining geometrical relationship: Interactions with robust and soft Cabri constructions. In T. Nakahara & M. Koyama (Eds.), *Proceedings of the 24th Conference of the International Group for the Psychology of Mathematics Education* (Vol. 1, pp. 103–117). Hiroshima, Japan: PME.

Healy, L., Fernandes, S. H. A. A., & Bolite-Frant, J. B. (2013). Designing tasks for a more inclusive school. In C. Margolinas (Ed.), *Task design in mathematics education: Proceedings of ICMI Study 22* (pp. 61–68), Oxford, UK. Available from http://hal.archives-ouvertes.fr/hal-00834054

Joubert, M. (2013). Using computers in classroom mathematical tasks: Revisiting theory to develop recommendations for the design of tasks. In C. Margolinas (Ed.), *Task design in mathematics education: Proceedings of ICMI Study 22* (pp. 69–78), Oxford, UK. Available from http://hal.archives-ouvertes.fr/hal-00834054

Kaput, J. (1986). Information technology and mathematics: Opening new representational windows. *The Journal of Mathematical Behavior, 5*(2), 187–207.

Kieran, C., Drijvers, P., Boileau, A., Hitt, F., Tanguay, D., Saldanha, L., et al. (2006). The co-emergence of machine techniques, paper-and-pencil techniques, and theoretical reflection: A study of CAS use in secondary school algebra. *International Journal of Computers for Mathematical Learning, 11*, 205–263.

Kynigos, C., & Lagrange, J.-B. (2014). Cross-analysis as a tool to forge connections amongst theoretical frames in using digital technologies in mathematical learning. *Educational Studies in Mathematics, 85*, 321–327.

Leung, A. (2011). An epistemic model of task design in dynamic geometry environment. *ZDM: The International Journal on Mathematics Education, 43*, 325–336.

Leung, A., & Lee, A. M. S. (2013). Students' geometrical perception on a task-based dynamic geometry platform. *Educational Studies in Mathematics, 82*, 361–377.

Mackrell, K., Maschietto, M., & Soury-Lavergne, S. (2013). The interaction between task design and technology design in creating tasks with Cabri Elem. In C. Margolinas (Ed.), *Task design in mathematics education: Proceedings of ICMI Study 22* (pp. 79–88), Oxford, UK. Available from http://hal.archives-ouvertes.fr/hal-00834054

Margolinas, C. (Ed.). (2013). *Task design in mathematics education: Proceedings of ICMI Study 22*, Oxford, UK. Available from http://hal.archives-ouvertes.fr/hal-00834054

Mariotti, M. A. (2012). ICT as opportunities for teaching-learning in a mathematics classroom: The semiotic potential of artefacts. In T.-Y. Tso (Ed.), *Proceedings of the 36th Conference for the International Group for the Psychology of Education* (Vol. 1, pp. 25–42). Taipei, Taiwan: PME.

Mason, J., Graham, A., & Johnston-Wilder, S. (Eds.). (2005). *Developing thinking in algebra*. London: Sage.

Morgan, C., Mariotti, M. A., & Maffei, L. (2009). Representation in computational environments: Epistemological and social distance. *International Journal of Computers for Mathematical Learning, 14*, 241–263.

Noss, R., Healy, L., & Hoyles, C. (1997). The construction of mathematical meanings: Connecting the visual with the symbolic. *Educational Studies in Mathematics, 33*(2), 203–233.

Or, A. C. M. (2013). Designing tasks to foster operative apprehension for visualization and reasoning in dynamic geometry environment. In C. Margolinas (Ed.), *Task design in mathematics education: Proceedings of ICMI Study 22* (pp. 89–98), Oxford, UK. Available from http://hal.archives-ouvertes.fr/hal-00834054

Rabardel, P. (1995). *Les homes et les technologies, une approche cognitive des instruments contemporains*. Paris: Armand Colin.

Radford, L. (2013). Sensuous cognition. In D. Martinovic, V. Freiman, & Z. Karadag (Eds.), *Visual mathematics and cyberlearning* (pp. 141–162). New York: Springer.

Robotti, E. (2013). Dynamic representations for algebraic objects available in AlNuSet: How develop meanings of the notions involved in the equation solution. In C. Margolinas (Ed.), *Task design in mathematics education: Proceedings of ICMI Study 22* (pp. 99–108), Oxford, UK. Available from http://hal.archives-ouvertes.fr/hal-00834054

Rogoff, B. (1998). Cognition as a collaborative process. In W. Damon (Series Ed.) and D. Kuhn & R. S. Siegler (Vol. Eds.), *Handbook of child psychology* (Vol. 2, pp. 679–744). New York: Wiley.

Sfard, A. (2008). *Thinking as communicating*. Cambridge: Cambridge University Press.

Sinclair, N. (2013). TouchCounts: An embodied, digital approach to learning number. In E. Faggiano & A. Motone (Eds.), *Proceedings of the 11th International Conference on Technology in Mathematics Teaching* (pp. 262–267). Bari, Italy: University of Bari.

Sinclair, N., & Pimm, D. (2014). Number's subtle touch: Expanding finger gnosis in the era of multi-touch technologies. In P. Liljedahl, C. Nical, S. Oesterle, & D. Allan (Eds.), *Proceedings of the 38th Conference of the International Group for the Psychology of Mathematics Education and the 36th conference of the North American Chapter of the Psychology of Mathematics Education* (Vol. 5, pp. 209–216). Vancouver, Canada: PME

Thomas, M. O. J., & Lin, C. (2013). Designing tasks for use with digital technology. In C. Margolinas (Ed.), *Task design in mathematics education: Proceedings of ICMI Study 22* (pp. 109–118), Oxford, UK. Available from http://hal.archives-ouvertes.fr/hal-00834054

Trouche, L. (2004). Managing the complexity of human/machine interactions in computerized learning environments: Guiding students' command process through instrumental orchestrations. *International Journal of Computers in Mathematical Learning, 9*, 281–307.

Trouche, L. (2005). Instrumental genesis, individual and social aspects. In D. Guin, K. Ruthven, & L. Trouche (Eds.), *The didactical challenge of symbolic calculators: Turning a computational device into a mathematical instrument* (pp. 197–230). New York: Springer.

Turner, M., & Gilles Fauconnier, G. (2002). *The way we think. Conceptual blending and the mind's hidden complexities*. New York: Basic Books.

Verillon, P., & Rabardel, P. (1995). Cognition and artifacts: A contribution to the study of thought in relation to instrumental activity. *European Journal of Psychology of Education, 10*, 77–103.

Vygotsky, L. S. (1978). *Mind and society*. Cambridge, MA: Harvard University Press.

Vygotsky, L. S. (1981). The instrumental method in psychology. In J. V. Wertsch (Ed.), *The concept of activity in Soviet psychology* (pp. 134–144). Armonk, NY: M.E. Sharpe.

Whiteley, W., & Mamolo, A. (2013). Optimizing through geometric reasoning supported by 3-D models: Visual representations of change. In C. Margolinas (Ed.), *Task design in mathematics education: Proceedings of ICMI Study 22* (pp. 129–140), Oxford, UK. Available from http://hal.archives-ouvertes.fr/hal-00834054

Wozniak, F. (2013). Pragmatic value and semiotic value of ostensives and task design. In C. Margolinas (Ed.), *Task design in mathematics education: Proceedings of ICMI Study 22* (pp. 141–150), Oxford, UK. Available from http://hal.archives-ouvertes.fr/hal-00834054

Yerushalmy, M. (2015). E-textbooks for mathematical guided inquiry: Design of tasks and task sequences. In A. Watson & M. Ohtani (Eds.), *Task design in mathematics education: An ICMI Study 22*. New York: Springer.

Part III
Plenary Presentations

Chapter 7
E-Textbooks for Mathematical Guided Inquiry: Design of Tasks and Task Sequences

Michal Yerushalmy

7.1 Textbook Culture: Traditions and Challenges

A textbook is a message from the professional community to students about *what* they should learn. It also represents the ideas of the author about *how* the content should be taught and learned. It plays a central role in school pedagogy and classroom norms, and its authoritative image has been the dominant aspect of the common classroom culture, often identified as *textbook culture*. Love and Pimm's (1996) "text on texts" is an important resource for understanding the features that textbooks are usually assumed to have: textbooks are closed in the sense that both text and images have been created in the past; they include problems for exercising but not aimed at questioning the content; they are linear and follow the "linear textual flow of reading" (p. 381), and usually they consist of cycles of expositions, examples, and exercises.

Traditional textbook culture assumes that teachers make sure their students learn the content of the book, usually in the specified order, because the book acts as a model for standards and for the way standards are assessed. The textbook is supposed to provide guidance and present opportunities for students to learn, making the objectives and ideas of the curriculum more readily apparent. For teachers, it also provides guidance in bringing their teaching in line with the expectations of the external authority, which may be the school, the syllabus, or some central assessment. In this function, the textbook serves as syllabus and timekeeper, and its author is considered to be the authorized entity charged with delivering content and pedagogy. Note the direct etymological link between *author* and *authority* (further described by Young, 2007), underscoring the authoritarian position of the textbook as written by a recognized expert author or group of authors. As an *author(iz)ed*

M. Yerushalmy (✉)
University of Haifa, Haifa, Israel
e-mail: myerushalmy@univ.haifa.il

This chapter has been made open access under a CC BY-NC-ND 4.0 license. For details on rights and licenses please read the Correction https://doi.org/10.1007/978-3-319-09629-2_13

object, the textbook is considered to be a solid personal resource for the learner. Learning with a textbook in this manner is often referred to as *learning by the book*, that is, a passive type of learning. *Teaching by the book* refers to teaching that treats the textbook as an authority that should be fully accepted. Both are known components of textbook culture, which varies only slightly across different contexts worldwide. Herbel-Eisenmann's analysis of the sources of textbook authority (2009) distinguishes between the textbook as an objective representation of knowledge, where the authority is an intrinsic property of the text, and the participatory relations between textbook and teaching. Teachers use various practices to confer authority onto the text and simultaneously onto themselves.

Developments across nations attempt to change the mathematics teaching standards and to shift the teacher's authority toward *guided inquiry* teaching. Although mathematics classrooms have increasingly adopted various formats of constructivism, reviews of post-reform math textbooks do not find that the practice of using textbooks has changed. Post-reform studies that examined the teachers' adoption of new curricula found that, although teachers attempt to reform their pedagogy to emphasize guided inquiry, the student-teacher-book relations have not changed and the textbook remains a central formal authoritative resource for the teacher and the learner. Learning mathematics with understanding has become a shared objective, but teachers seldom have opportunity and time to develop mathematical tasks, textbooks, or teaching sequences in which richness of tasks and learning for understanding are emphasized (Pepin, 2012). Analyzing newly developed textbooks, Nathan, Long, and Alitalia (2007) found that many new books have remained similar to the traditional ones.

Reyes, Reyes, Tarr, and Chavez (2006) studied for 3 years how newly developed textbooks supported by the National Science Foundation in the USA affected math teaching and learning in a US middle school. According to the authors, half the teachers declared that "My math book is my bible," and the other half were influenced more by the state-determined curriculum and assessment materials. The teachers covered 60–70 % of lessons of the specially designed-to-reform textbook, just as they did when using non-reform textbooks. McNaught (2009) presented similar findings in a study of the Core-Plus curriculum project (used as an example of an integrated content textbook), showing that around 60 % of the content of the textbook was taught, but not necessarily from the textbook, and about one third of the teaching was based on other supplementary materials.

Studying Swedish teachers as they guided students solving tasks from the textbook, Johansson (2007) showed that perception of the textbook as a resource of ultimate correctness has affected the teachers' decision not to question a textbook solution. In sum, (a) the textbook remains the single printed and bound object that acts as authoritative pedagogic guideline for what should be learned and for how it should be taught and assessed; (b) new textbooks that attempt to adopt a somewhat less authoritative tone create a didactic challenge and are not likely to be fully adopted by experienced teachers; and generally speaking, (c) although the textbook is assumed to provide devices for actively involving the students in examples and exercises, the studies focus on engagements with textbooks that are reserved mostly for the teachers.

The studies on interactions with textbooks present a somewhat more complex picture of the relations between mathematics, the text, teachers, and students than traditionally assumed. Teaching does not depend on a single textbook: approximately 30 % is accomplished using other resources, mainly previous textbooks that the teachers feel are more likely to achieve their goals. Moreover, different nations have different cultures, traditions, and expectations from teaching, all reflected in the textbook. The comparison between Norwegian and French textbooks by Pepin, Gueudet, and Trouche (2013) is illustrative. Despite the authority of textbooks, an increasing number of studies report on teachers looking for new models and views on teaching aimed at understanding the use of textbooks. Remillard (2012) examined how teachers are positioning themselves with the textbook and found that standard-based resources present a challenge to many teachers and require orientation, and that many teachers use them in ways not intended by the designers. Rezat (2012) analyzed six phases describing how teachers mediate textbook use. Patterns that Rezat identified in students' uses of textbooks are strongly related to the choice of order of expositions, order of pages, and of tasks used by the teacher. It suggests that the power of indirect teacher mediation of the textbook is underestimated.

Technology has been an important resource in several textbooks attempting reform. But such technological resources as the Open Education Resources, which accompany the newer books, are reported to be considered enrichment, while the textbook remains the core external authority. Textbook publishers are addressing a wide range of expected changes in the affordances of the digital object, including the material aspects of weight and cost, the quality and attractiveness of the material, the richness of the modes of presentation, and the opportunities for personalization. As textbooks rapidly change from print to digital formats, it is assumed that the ways in which they will be used will also change. Publishers allow teachers to personalize digital textbooks for their courses, emphasizing flexibility and inexpensive dynamic changes. Schoolteachers can personalize the textbook by selecting from existing chapters and content and even individualizing the book for the student.

Korea, considered to be one of the leading countries in math and science achievement, became one of the foremost innovators in the area of e-textbooks, especially in school math and sciences. Korea holds an integrative view in which textbooks remain the central learning resource, surrounded by other types of facilitating media. Paper textbooks are being digitized, integrating live resources such as hyperlinks, multimedia, dictionaries, references, and other data sources into a learning management system connecting students, assessment, etc. Other educational systems are adopting a similar view of the new textbook (Taizan, Bhang, Kurokami, & Kwon, 2012). The Israeli education system requires that each textbook appear in at least one of three formats: a digitized textbook, a digitized textbook that is enriched with external links and multimodal materials, or a textbook that is especially designed to work in a digital environment and includes online tools for authoring, learning, and management. An assumption of the integrative view is that the digital textbook can also function as printed text and assume the traditional format of the textbook. Thus, the change of the *object* prompted changes in the publishing process (editions, price, attractiveness) and in the ways of integrating digital add-ons with the instructional materials. The different formats of multimodal interactive textbooks in mathematics have yet to be thoroughly studied.

The previous literature review addresses the studies of printed textbooks, but the findings regarding mediation and patterns of use by students and teachers point to challenges that should be considered when using non-sequential and multimodal interactive texts. An important issue is that of authority; Gueudet and Trouche (2012) argued that the notion of author and authorship is often less transparent in online sources than in printed material, as does Usiskin (2013) when discussing the disappearance of transparency in e-Textbooks. Another is that of order or sequence. It is commonly believed that digital books enable changes of sequence and flexibility, but at the same time such flexibility may cause lack of clarity. In this situation, the main challenge is to rethink the sets of concepts and images used to guide us in thinking about the structure of traditional printed textbooks and to consider the consequences of interactivity, multimodality, and personalization on the design and structure of use—primarily the teacher.

7.2 The Design of an Interactive Unit

Yerushalmy, Katriel, and Shternberg (2002) designed an interactive digital textbook characterized by the extensive roles assigned to visual semiotic means to interactivity between the reader and the visual mathematical objects and processes, and by a conceptual order of digital pages that could be rearranged to serve a variety of instructional paths (for more details, see Yerushalmy, 2013). The design of the *VisualMath* textbook is situated in the larger view field of mathematical guided inquiry and school algebra, taking the function (on real number field) to be its mathematical and pedagogical root. We have long been seeking less formal control structures that attempt to respond to or to signal subjective schemes and views that teachers and students bring to their engagement with the text. Two fundamental principles directed the design:

- A unit, a pedagogical structure for a collection of algebra tasks, should be regarded as a gallery in an interactive museum. The design principles and views of the units were borrowed from a museum setting and were consistent with the distinction Kress and van Leeuwen (1996) made to describe linear and nonlinear texts. They described nonlinear texts as an "exhibition in a large room which visitors can traverse any way they like… It will not be random that a particular major sculpture is placed in the center of the room, or that a particular major painting has been hung on the wall opposite the entrance, to be noticed first by all visitors entering the room" (p. 223). Just as a curated exhibit, a unit should present opportunities for readers to focus on a concept through multi-sensual experience, making the objectives of the collection apparent. Each unit attempts to be a coherent collection of tasks, which, although they can be used in any flexible order according to the decision of the teacher or reader, is constructed to deliver the mathematical lesson by (a) offering a balanced collection of multimodal mathematical activities that include inquiry tasks, problems, and exercises (in general, all the tasks require non-routine thinking and reasoning by interacting and experiencing the mathematics) and (b) addressing various modes of learning and teaching, as described by Chazan and Schnepp (2002) and Lobato, Clarke, and Ellis (2005). Although the tasks are written for the student, it is assumed that each task can be used in different modes, including independent study by students, group collaboration, and teacher-conducted whole-class discussion demonstrating a process of inquiry that supports systematizing and institutionalizing experimental results: (c) a design that meets the institutional demands, including tasks that support transforming inductive results into consolidated mathematics (Barzel, Leuders, Predinger, & Hußmann, 2013 categorize textbook tasks along these terms). Each unit addresses three types of tasks (Kieran, 2004): *transformational activities* that are mostly rule-based manipulations, *generational activities* focused on representing and interpreting situations from outside or within the mathematics, and *integrative activities* that require the use of manipulations and generational actions but also go beyond these, to meta-processes such as generalizations, predictions, etc.

- Although the organization of each unit resembles a traditional set of textbook tasks—expositions, exercises, and problems—the principle that guided the design of tasks and characterization of activities is rooted in the interactivity of tools and diagrams. Each type of interactive element has its semiotic and pedagogical meaning. Tools are artifacts designed to carry a specific element of mathematics: function-based algebra. An interactive diagram (ID) is a relatively small and simple software application (applet) built around a *pre-constructed example* that serves as a basis for change. Both tools and diagrams were designed to support conjecturing and argumentation by providing various degrees of control to the user. Interactive elements are designed to support the systematic generation of examples in linked multiple representations, to accommodate various entry points, and to provide nonjudgmental mirror feedback that should be interpreted subjectively. Indeed, the challenge in constructing a task around an ID is to design opportunities for action (Yerushalmy, 2005).

This section outlines and exemplifies the design rationale and structure of a unit by analyzing the design of expositions, tools, problems, exercises, and essay tasks included in the *Transformations of Linear Functions* unit (http://visualmath.haifa.ac.il/en/linear_functions/transformation_of_graphs) (in Yerushalmy et al., 2002; Yerushalmy, Shternberg, & Katriel, 2014).

7.2.1 Interactive Exposition

Freisen's (2013) analysis of the structures that characterize the advent of modern textbooks describes the expository nature of textbooks as the main obstacle to considering newer modes of textbooks along the history as an environment supporting the personal construction of knowledge. Following Kuhn (1962), Freisen questioned the role of the exposition in traditional textbooks and the attempt to present the universal truths and to expect students to later recite them back to the teacher. In more recent textbooks, the terminology of the *scripted* exposition has changed to knowledge that is *prompted* by the textbook. Our assumption was that learners need to construct their own examples and to construct the concept image and definition by creating their own example space. The widespread use of examples in mathematics textbooks serves as a means of communication and mediation between learners and ideas. If examples are well selected, the variations between examples are the means by which students can distinguish between essential and redundant features (Bills et al., 2006; Goldenberg & Mason, 2008; Watson & Shipman, 2008). Our challenge was to design expositions that can be worked on and personalized by providing interactive illustrations to be controlled by the reader. Its design was centered on the *illustrating diagram* (see Fig. 7.1).

Exposition: Illustrating Diagrams	
Translations and Reflections of Linear Functions The figure shows how different students can attempt different examples with the exposition. 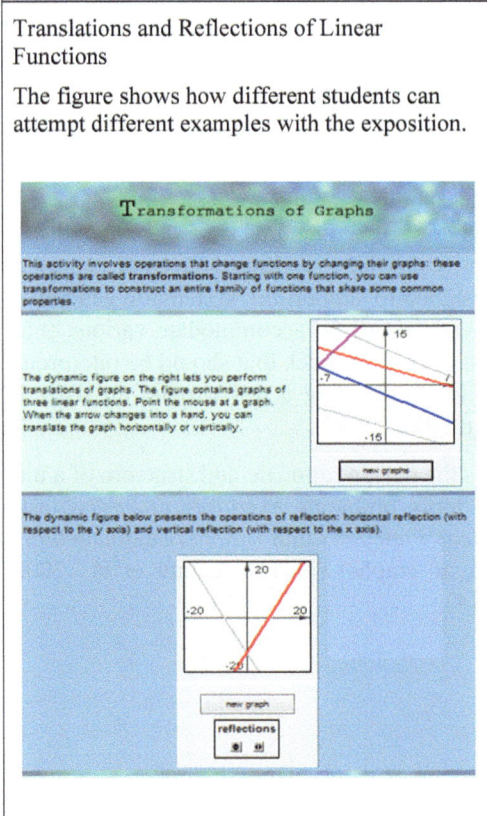	The role of the *illustrating interactive diagram* that opens the Transformations unit is to capture the central properties and quality of the translation of a linear function on a Cartesian system. It is designed to support the development of an initial working definition to be reconstructed while working on the unit. It is based on a known representation: a line in a Cartesian system and an assumed common, embodied understanding of dragging. The ID appearing in the exposition page are limited in types of representations (graphs only) and in tools used to manipulate the given examples (generation of examples and the translation and reflection tools). Thus, students cannot create counterexamples or turn an example into a non-example. The role of the interactive exposition that affords limited user control is to allow subjective development of awareness of the relevance of the unit features.

Fig. 7.1 Description of the role of illustrating interactive diagram (http://visualmath.haifa.ac.il/)

7 E-Textbooks for Mathematical Guided Inquiry

The exposition was designed to start the example generation, usually by offering a single graphic representation and relatively simple actions, such as viewing an animated example or dragging an interactive shape. The reader was expected to use these to create examples that represent new ideas with known means of control individually, in cooperative groups or in whole-class discussions.

7.2.2 Toolbox and Unit Tools

Tools are artifacts designed to carry specific elements of mathematics. Artifacts become purposeful tools in response to subjective needs and personal actions. The assumption in the design of the *VisualMath* e-textbook was that tools such as the microscope, the calculator, or any software become a way of thinking and knowing and have an epistemological role as they change the traditional assumptions of what we mean by knowing mathematics. A toolbox of ten tools for doing mathematics is part of the e-textbook. The unit tools (or activity tools) are special cases of the tools in the Toolbox, designed to explore specific concepts mainly by limiting the generated types of example spaces through supporting a smaller range of actions. Although they make exploration and inquiry possible, they are designed to call for action in a specific manner that supports the construction of the principal idea of the unit. Toolbox and unit tools are part of the learning environment, and they are always available (see Fig. 7.2).

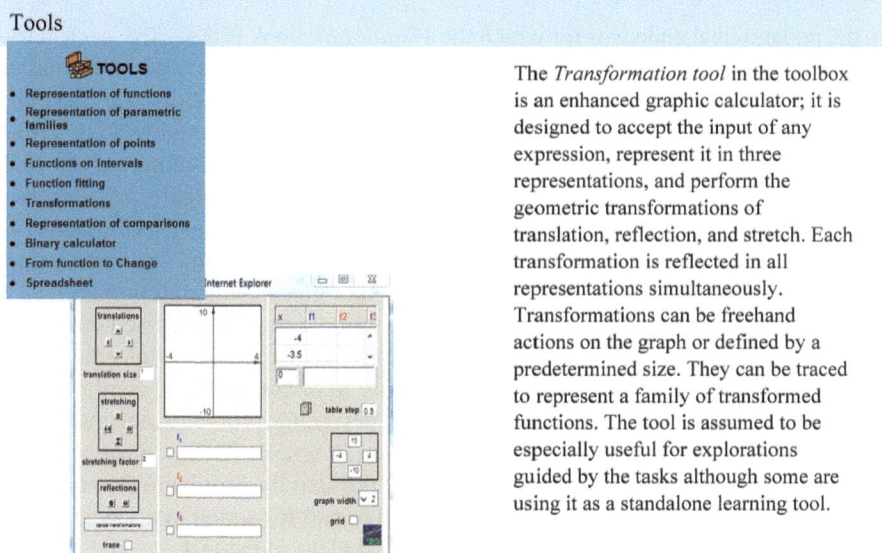

Fig. 7.2 Descriptions of toolbox and unit tools (http://visualmath.haifa.ac.il/)

Unit Tools

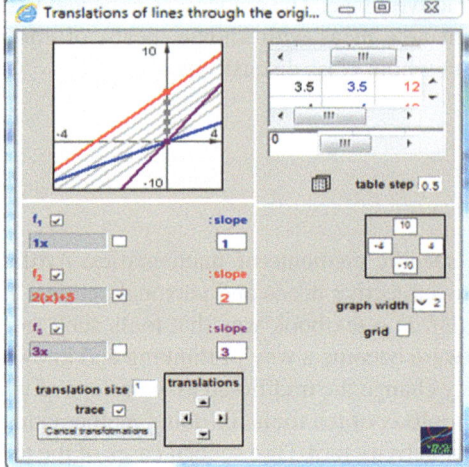

The *Translations of lines through the origin* is a transformation tool designed to focus on an object $f(x) = mx$ and an operation (translation). Symbolic input is not allowed. New examples can be generated only by changing the original given functions using translations, or by free input of the size of the slope. Thus, by limiting the example space, it demonstrates how terms in linear expressions reflect or correspond to parallel lines. The tool is on one hand general, as one can present any linear function and line; on the other hand it is limited to translations and starts with a specific generic example, creating boundaries for the story to be learned.

Fig. 7.2 (continued)

7.2.3 Tasks or Problems and Exercises

In the pedagogical endeavor for which the *VisualMath* book is designed, each activity was intended to grant opportunities for students to explore in order to formulate ideas. The traditional order of direct teaching of procedures followed by drilling, practice, and word problems (applications of the taught algorithms) was not a relevant consideration. The objectives of the tasks were to use explorations out of which conjectures grow, are discussed, are explained, and are informally and formally proved or rejected. The tasks require making sense of problems and, as the new core standards state, require the students to "persevere in solving them," spending longer on analyzing givens, constraints, relationships, and goals. Although problem solving can always be helped by use of appropriate tools, it should be carried out strategically, constructing viable arguments and critiquing the reasoning of others. The problems are designed to provide an opportunity to obtain peculiar, conflicting, or unexpected examples. The primary means of designing problems are interactive diagrams (IDs). Especially useful are the *guiding diagrams* (GIDs), designed to be the principal delivery channel of the message of the activity. Similar to the narrator's voice in Goldenberg (1999), GIDs are designed to call for action in a specific manner that supports the construction of the principal ideas of the task.

As in many algebra books, exercises call for writing an expression, equation, or inequality. In function-based algebra, the expressions are functions and thus represent a procedure. Rather than asking to simplify or to solve, most exercises ask for equivalent expressions and encourage more than a single correct answer, or they require the construction of a function or of an equation that meets the specified conditions. The design assumption regarding exercises was that they are performed after students have adopted a correct (if not complete) concept image and concept definition and have practiced with understanding the principal processes related to the concept. The primary means used to design exercises are *elaborating diagrams*, which include a wide range of representations and controls within the representations. Elaborating diagrams leave it up to the learner to solve the task by offering a variety of general-purpose tools. The exercise can be solved without the support of the interactive diagram, using known procedures mentally as a paper exercise. Alternatively, the answer can be reached by trials with multiple representation reflecting feedback. Students can use openly available input, watch the feedback, compare the linked results, and gradually improve their guesses. See examples in Figs. 7.3 and 7.4.

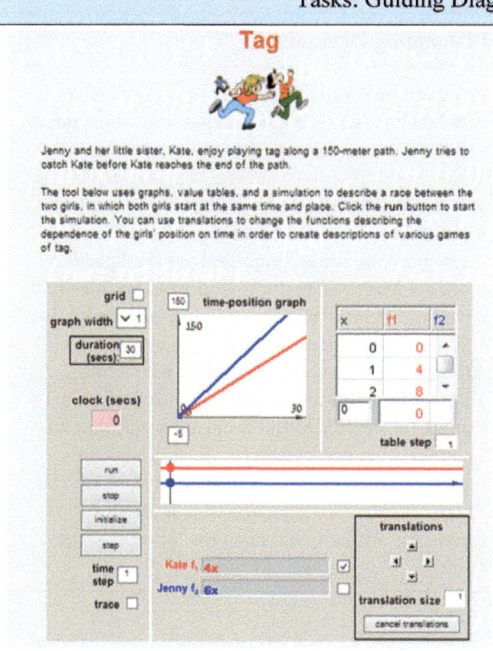

Fig. 7.3 Guiding diagrams and elaborating diagrams (http://visualmath.haifa.ac.il/)

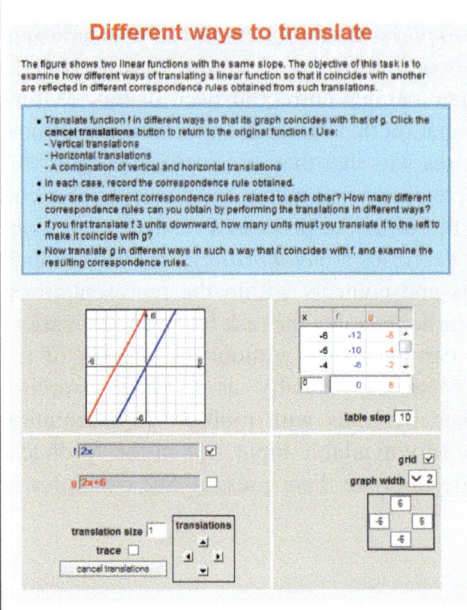

Different ways to translate problem focuses on the expression as a mathematical story. For any two parallel lines, there are many ways of translating one to coincide with the other:

For example, translating $2x + 6$ by 3 to the right is expressed symbolically as $2(x - 3) + 6$, which is equivalent to $2x$.

Translations are geometric manipulations; input or editing of expressions is not allowed. This restriction, which is a matter of principle in the design of guiding diagrams, helps focus the exploration on the connections between the geometric changes of the object and its symbolic algebraic description.

Exercises: Elaborating Diagrams

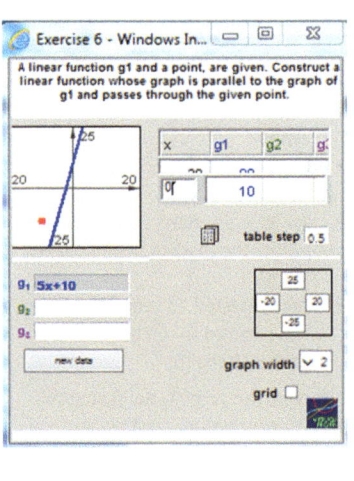

The exercise (or the infinite number of exercises generated by the restricted randomness of the point and the line – in this case it is always a point that is not on any axis) is a traditional task that requires an answer. It is designed as a graphic calculator tool. Input is free, trials can be attempted, and feedback is given by the linked graph and values. Specific tools, such as the transformation tools, which could have been helpful here, are missing on purpose. It is assumed that students can solve the exercise on paper without the tool, and use the interactive diagram for checking, or that they are exploring the exercise as beginners and need, at least at first, to find their own way of solving the task with an interactive that does not limit such exploration. The library of exercises is presented in an order of increasing difficulty based on the location of the point.

Fig. 7.3 (continued)

The task: Write an expression describing the given line graph containing the two points marked on the diagram.

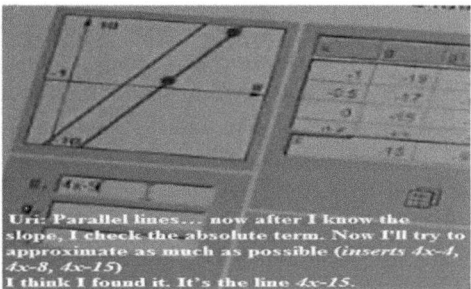

"Their attempts demonstrate diverse working strategies in using the free symbolic input and the three function representations in the diagram. One student (Uri) used the free symbolic input to write an expression and systematically change its parameter values. We interpret his work as building a generic example. The second student (Ilana) used the free input only after the interviewer pointed out this option. She started her solution by entering various non-explicit function expressions and other invalid syntax inputs that were coincidental combinations of variables and coordinates of intersection points. Ilana's process of trial and error reached a turning point when she incidentally plotted and interpreted the constant function graph. Once she was able to link between a constant term and a graph, she was able to generalize the meaning of the b parameter in $f(x) = ax + b$, but she misunderstood the meaning of the parameter a. Dani looked for numeric values on the graph and in the table, which he then used for his paper and pencil computations. When he started using the free symbol input, he realized that he had made a mistake in his work on paper. He went back to calculate the slope and the constant term on paper, and rewrote an expression with the ID to check it. Once the line of the input expression looked parallel with the original graph, Dani's focus changed from finding the Y-intercept to comparing the two lines and achieving a correct constant term by making his graph overlap the given line.

The possibility of entering various expressions to be viewed in three linked function representations in the elaborating ID offers a range of working strategies. At the initial stage of the work with expressions using the ID, all three students used the elaborating ID only to check data for the given example. After the initial stage, they proceeded along different paths: systematic search for meanings of parameters in the expression, experimentation with numerical changes in the expression by trial and error, and elaborative work on the expression with paper and pencil, with the ID being used only for checking."

Fig. 7.4 An example of students working on an exercise (adapted from Naftaliev & Yerushalmy, 2011)

7.2.4 Write an Essay Task

Returning to the museum image, the role of the main *integrative task* of writing an *essay* is analogous to choosing the piece of art to be positioned in the center of the gallery. It is placed in such a way as to be visible and considered by all, but it is not necessary to stop by this exhibit for a long time on a first visit. Most units offer a math concept essay. The essay should guide students' exploration or summarize and demonstrate ideas that students already explored with tools, talked about in groups, and tried in other tasks of the unit. The essay can also be regarded by the teacher and students as suggestions for directions to follow (see Fig. 7.5). The suggested directions are the gallery "centerpiece" or the main goals that the unit wants to bring to the fore using its tools, problems, and exercises. In this sense, the suggested outline for the exploration tasks is also a suggestion for the teacher or for any reader interested in what would constitute "knowing the concept" upon completing the unit.

Essay Integrative Task: Use available Tools, Unit Tools, and Interactive Tasks

How are the various transformations expressed in the different representations of a function?
- In the graph
- In the correspondence rule
- In the value table

How do the various transformations affect the properties of a function (such as increase or decrease, points of intersection with the axes, or slope)? Which of the properties change and which do not?

How can transformations help in describing phenomena? When a transformation is applied to a graph that describes a story, what is the meaning of the transformation in the terms of the story?

Fig. 7.5 Example of an essay integrative task (http://visualmath.haifa.ac.il/)

Writing an essay task can serve as a significant tool in supporting the organization of knowledge in a constructivist classroom because it asks to structure divergent results and consolidate the mathematics learned, to generalize results to other settings, and to be skillful. It creates opportunities for students to raise questions that cut across objects, operations, and the terms of the subject. In this sense, essay tasks are considered "expert problems." Expert work consists of solving problems when they arise, searching for the foundations of the task, learning necessary definitions, searching for examples, looking for similar problems, performing meta-level heuristics, and instrumenting artifacts to serve as mathematical and psychological tools. Expert problems in a textbook "are rich tasks, each presented in a form it might arise in mathematics, science or daily life. They require effective use of problem solving strategies, as well as concepts and skills. Performance on these tasks indicates how well a person will be able to do and use the mathematics beyond the mathematics classroom" (Burkhardt & Swan, 2013, p. 440). Standing as the central exhibit of the gallery, the essay can be visited and revisited in various scenarios of exploration: an evolving essay that the student is responsible to complete in the course of the learning period or an individual project that can be developed and expanded to connect with other units (e.g., the exact same essay task appears in the transformations unit of the quadratic function section of the textbook).

7.2.5 Restructuring a Unit Along the Space of Interactive Elements

Our initial attempt in constructing a balanced unit was to support readers, mainly teachers, by keeping the structure of the unit based on more or less traditional components. At the same time, the interactive tools and diagrams became the main consideration of the design and of the pedagogical implementation. The exercise room of the transformation gallery was designed to support practice with meaning, but the simple and open design of the exercises as elaborating IDs caused teachers to often use it as expositions in guided class discussions or as an exploration activity. The unit tools that initially were designed to support specific tasks were used as a learning environment, replacing expositions. And the essay tasks, which were initially designed to support an open-ended project for students or groups to pursue on their own, were assigned as reviews upon completion of the unit, and teachers often used it as a map to guide them in teaching the unit. These pedagogical decisions made it clear that specifying the functions of the gallery rooms along the lines of the structure of textbook units as traditionally described needs to be rethought. We suggest, therefore, changing our perspective and considering designing the interactive units from the vantage point of the semiotics of interactive elements.

We used this semiotic framework to examine the aspects of interactive tasks from the point of view of presentational, orientational, and organizational functions. The *organizational function* refers to the connection between all the components of the task: verbal text, representations, tools, examples, etc. To describe the process of design with IDs, it is useful to look at three types of organization: illustrating, elaborating, and guiding. These types were most apparent in the description of the components of the unit in this section. The example that initially appears in the ID determines the nature of its *presentational function*. Three types of examples are widely used in IDs: specific, random, and generic. Specific examples that present the exact data of the activity of which they are part provide a dynamic illustration of the text without altering the information. In random examples, a specific example is generated within given constraints, presenting different information at various times and for different users. To serve as a generic example, the diagram must be structured to be representative; it must present a situation that could be part of the given task, but its focus is not on the specific data of the activity. The tone in which the text addresses the learner is subject to design decisions having to do with the *orientational function*. The "sketchiness" or "rigorousness" of the diagrams is an important aspect of reader orientation. An example that appears in a diagram can have an accurate appearance and speak in a strict, distant tone, or it can include a more subtle description and adopt a non-authoritative tone. IDs can function both as sketches and as accurate diagrams.

The elements of this framework were valuable in explaining student learning with various interactive diagrams in different contexts of algebra tasks (Naftaliev & Yerushalmy, 2013). They were also valuable in guiding our design of new instructional resources, and it remains to be explored whether this framework is valuable and productive as a guide for instruction.

7.3 Sequencing in Non-sequential Textbooks

Brown (2009) compares teaching to a jazz player's use of the notes and argues that rather than designing curriculum materials as one-size-fits-all documents, designers should support different modes of use according to their pedagogical design capacity (p. 31). I support this call and suggest that transparent design of the mathematical idea and in the mapping of the objects and of actions of the mathematical subject matter can be a helpful tool for teachers who design or modify curricular documents and resources. For a collection of semi-ordered materials and multimodal digital "pages," which to a certain extent stand on their own, to be considered a textbook, the deep structure of the concepts and the interrelations between them must be simple and visible. Two principles guided the design of the *VisualMath* e-textbook. Our first decision was to organize the content along a single view of the algebra, focusing on the algebra of functions: *VisualMath* was designed to use functions as the foundation for mastering algebraic skills with understanding by all students. This is not the common structure of mathematics textbooks that usually represent a progression along various themes and views of the algebra (Rezat & Straser, 2013). The second decision was to organize the materials along a relatively small number of mathematical objects and operations that can mathematically and pedagogically support a variety of progressions and sequences. These two principal design considerations are described in detail in Yerushalmy (2013).

We therefore prepared two lists. One consisted of the mathematical objects involved. In its current form, the *VisualMath* e-textbook accommodates the linear and the quadratic "museum exhibits," each one appearing as a row on the map: a mathematical object. An additional row represents "any" or generic examples of functions. The other list consisted of operations on the objects and with them. The six operations included do not form an exhaustive list. Rather, they are what Schwartz had called the "interesting middle" (Schwartz, 1995): operations that represent important mathematical concepts and are appropriate and useful to learn as part of function-based school algebra. The operations are represent (a function), modify (reforming the view or structure without changing the function), transform (using operations to transform a function into families of functions), analyze the function and its change, operate with two functions (synthesizing new functions out of two different or identical functions), and compare two functions. The two lists are distinct and were therefore placed in an orthogonal organization in a 2D matrix map, where each cell represents the opportunities for learning resulting from the corresponding operation and object. Each operation with an object can take place in symbolic, graphic, or numeric representations (Fig. 7.6).

Symbolic/ Graphic/ Numeric						
Operation Objects	Represent	Modify	Transform	Analyze (function change)	Operate (with 2)	Compare
Generic						
Linear						
Quadratic						

Fig. 7.6 An organizational map for the *VisualMath* e-textbook (http://visualmath.haifa.ac.il/)

The next example illustrates design intentions aimed at supporting sequencing decisions. Assuming that students are already familiar with the three representations of functions, three conceptual "guided tours" are suggested for teaching the quadratic unit: the *analyzing tour*, the *solving tour*, and the *algebraic structure* (modifying and operating) *tour*. Although the three are complementary, each of the sequences by itself may respond to necessary foundational knowledge of quadratics.

Teachers can plan the course of the *analyzing tour* based on the concept of constant or nonconstant rate of change. This can be achieved by emphasizing the Analyze column in the map, which is elaborated in three units of the book (marked in Fig. 7.7): Rate of Change unit in the Linear part, Quadratic Growth unit, and Motion at Changing Speed unit, where motion is modeled by changing speed and constant acceleration.

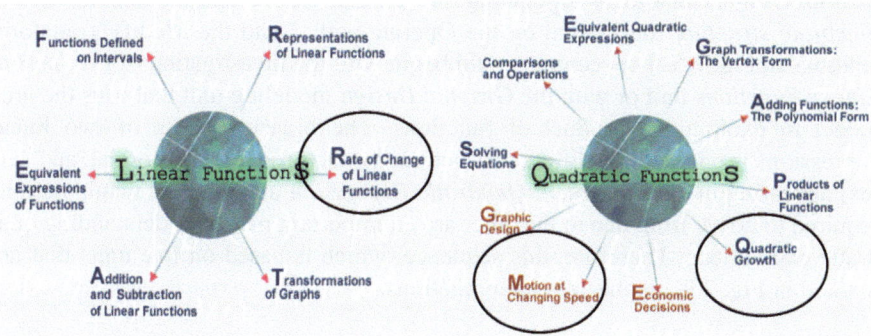

Fig. 7.7 The function analysis sequence based on three units: Rate of Change, Quadratic Growth, and Motion (http://visualmath.haifa.ac.il/)

An alternative focus places *equations and inequalities* in the center. Students have already learned to solve linear equations and probably view them as comparisons of linear functions. They have also experienced the idea of comparison of two functions representing equation or inequality and equivalent comparisons as the conceptual view of the process of solving equation or inequality in function-based algebra. Therefore, another challenging sequence aims at teaching along the two rightmost columns in Fig. 7.6: *Operate with 2* and *Compare* (comprising of 3 units: *Addition and Subtraction of Linear Functions*, *Comparisons and Operations*, and *Solving Equations* units). This sequence, marked in Fig. 7.8, is based on knowledge in the Modify column, but the focus is on the algebraic object of equation and on the concept of equivalence. Many tasks in various other units require solving and can be used to enrich the sequence.

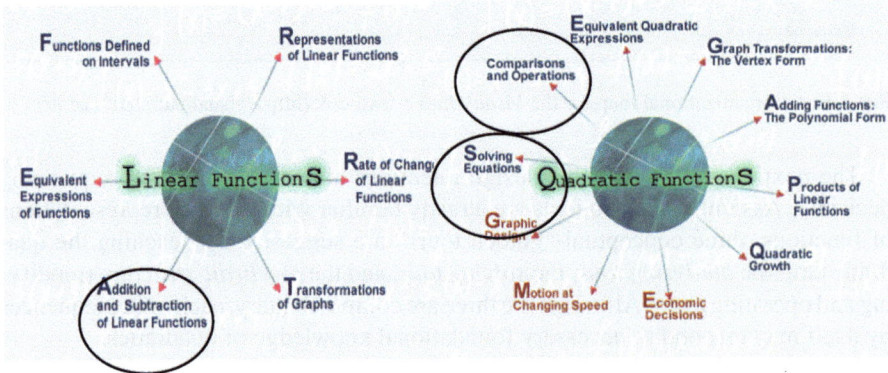

Fig. 7.8 The solving sequence based on three units: *Addition and Subtraction of Linear Functions*, *Comparisons and Operations*, and *Solving Equations* (http://visualmath.haifa.ac.il/)

A third choice of a focus for delving into the algebra of quadratic functions is the *algebraic structure* tour, based on the Operate with 2 and the Modify/Transform columns in Fig. 7.6. The scenario could begin with the investigation of *Products of Linear Functions* unit or with the *Graphic Design* modeling unit that uses the area model to explore the product of functions. The binary product of two linear expressions is one of three different structural forms (product, polynomial, and vertex) appearing in the *Equivalent Quadratic Expression* unit, and the manipulations required to arrive from one to the other are an important part of understanding quadratic expressions. Therefore, this sequence, which is based on five units that are marked in Fig. 7.9, emphasizes manipulations.

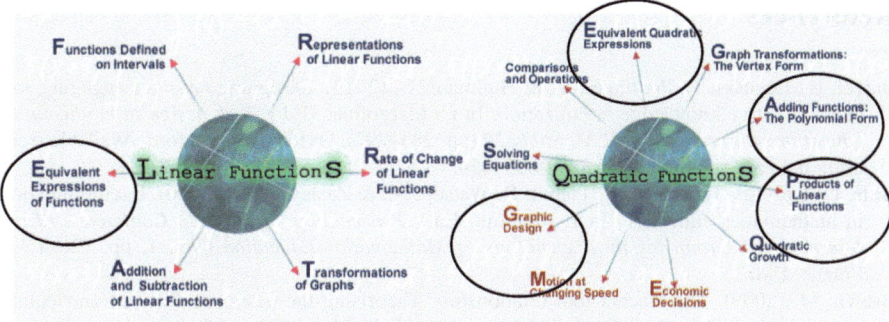

Fig. 7.9 The algebraic structure and manipulation sequence based on five units: *Equivalent Expressions*, *Products of Linear*, *Adding Functions to Polynomials*, *Equivalent Quadratics*, and *Graphic Designs* (http://visualmath.haifa.ac.il/)

7.4 Summary: Concept-Driven Navigation in the Space of Interactive Tasks

We started with a brief review of the traditional roles and images of order and authority of textbooks. We analyzed the difficulties that arise when teachers try to respond to student needs and ways of understanding based on an approach according to which the textbook remains an important resource that determines what should be taught and how. We then proceeded to describe common notions concerning the interactivity of digital textbooks and questioned what the Web and other resources appended to the digitized versions of textbooks can provide to support guided inquiry in the mathematics classroom. Finally, we discussed several design principles of interactive digital textbooks and reviewed examples of central design decisions reflected in the *VisualMath* algebra textbook. The three principles included in the discussion were (a) designing interactive diagrams that provide students with ways to explore within curricular boundaries, (b) suggesting a visual semiotic framework for typifying the conceptual components and terms inherent to the design of interactions within technology-based textbooks, and (c) organizing the textbook into units that respond to the principal objects and operations of the mathematics to be learned and sequenced to achieve personalization. I tried to address the difficulties expressed by many teachers who are deeply committed to changing the way in which mathematics is being taught today but are frustrated in their efforts by the authority of textbooks that dictate their teaching agenda. I argue that the organization of the e-textbook should make it possible for teachers to sequence the curricular material in a way that serves the needs of the classroom and of the students. I therefore suggest that, in the era of digital textbooks, it is necessary that such organizational/design principles should be part of teachers' knowledge.

References

Barzel, B., Leuders, T., Predinger, S., & Hußmann, S. (2013). Designing tasks for engaging students in active knowledge organization. In C. Margolinas (Ed.), *Task design in mathematics education: Proceedings of ICMI Study 22* (pp. 283–292). Oxford, UK: Oxford. Available from http://hal.archives-ouvertes.fr/hal-00834054.

Bills, L., Dreyfus, T., Mason, J., Tsamir, P., Watson, A., & Zaslavsky, O. (2006). Exemplification in mathematics education. In J. Novotna (Ed.), *Proceedings of the 30th Conference of the International Group for the Psychology of Mathematics Education* (Vol. 1, pp. 126–154). Prague: PME.

Brown, M. (2009). The teacher-tool relationship: Theorising the design and use of curriculum materials. In J. Remillard, B. Herbel-Eisenmann, & G. Lloyd (Eds.), *Mathematics teachers at work: Connecting curriculum materials and classroom instruction* (pp. 17–36). New York: Routledge.

Burkhardt, H., & Swan, M. (2013). Task design for systemic improvement: Principles and frameworks. In C. Margolinas (Ed.), *Task design in mathematics education: Proceedings of ICMI Study 22* (pp. 431–440). UK: Oxford. Available from http://hal.archives-ouvertes.fr/hal-00834054.

Chazan, D., & Schnepp, M. (2002). Methods, goals, beliefs, commitments, and manner in teaching: Dialogue against a calculus backdrop. In J. Brophy (Ed.), *Advances in research on teaching: Social constructivist teaching* (Vol. 9, pp. 171–195). Greenwich, CT: JAI Press.

Friesen, N. (2013). The past and likely future of an educational form: A textbook case. *Educational Researcher, 42*(9), 498–508.

Goldenberg, P. E. (1999). Principles, arts, and craft in curriculum design: The case of connected geometry. *International Journal of Computers for Mathematical Learning, 4*(2–3), 191–224.

Goldenberg, P., & Mason, J. (2008). Shedding light on and with example spaces. *Educational Studies in Mathematics, 69*(2), 183–194.

Gueudet, G., & Trouche, L. (2012). Teachers' work with resources: Documentational geneses and professional geneses. In G. Gueudet, B. Pepin, & L. Trouche (Eds.), *From text to 'lived' resources: Mathematics curriculum materials and teacher development* (pp. 23–42). New York: Springer.

Herbel-Eisenmann, B. A. (2009). Negotiating the "presence of the text": How might teachers' language choices influence the positioning of the textbook? In J. T. Remillard, B. A. Herbel-Eisenmann, & G. Lloyd (Eds.), *Mathematics teachers at work: Connecting curriculum materials and classroom instruction* (pp. 134–151). New York: Routledge.

Johansson, M. (2007). Mathematical meaning making and textbook tasks. *For the Learning of Mathematics, 27*(1), 45–51.

Kieran, C. (2004). The core of algebra: Reflections on its main activities. In K. Stacey, H. Chick, & M. Kendal (Eds.), *The future of the teaching and learning of algebra* (The 12th ICMI study, pp. 21–34). New York: Kluwer Academic.

Kress, G., & van Leeuwen, T. (1996). *Reading images: The grammar of visual design*. London: Routledge.

Kuhn, T. (1962). *The structure of scientific revolutions*. Chicago, IL: University of Chicago Press.

Lobato, J., Clarke, D., & Ellis, A. B. (2005). Initiating and eliciting in teaching: A reformulation of telling. *Journal for Research in Mathematics Education, 36*(2), 101–136.

Love, E., & Pimm, D. (1996). 'This is so': A text on texts. In A. Bishop, K. Clements, C. Keitel, J. Kilpatrick, & C. Laborde (Eds.), *International handbook of mathematics education* (Vol. 1–2, pp. 371–409). New York: Springer.

McNaught, M. D. (2009). Implementation of integrated mathematics textbooks in secondary school classrooms. In S. L. Swars, D. W. Stinson, & S. Lemons-Smith (Eds.), *Proceedings of the 31st Annual Meeting of the North American Chapter of the International Group for the Psychology of Mathematics Education*. Atlanta, GA: Georgia State University.

Naftaliev, E., & Yerushalmy, M. (2011). Solving algebra problems with interactive diagrams: Demonstration and construction of examples. *Journal of Mathematical Behavior, 30*(1), 48–61.

Naftaliev, E., & Yerushalmy, M. (2013). Guiding explorations: Design principles and functions of interactive diagrams. *Computers in the Schools, 30*(1–2), 61–75.

Nathan, M. J., Long, S. D., & Alitalia, M. W. (2007). *The symbol precedence view of mathematical development: An analysis of the rhetorical structure of textbooks.* Institute of Cognitive Science, University of Colorado, Bolder. http://www.colorado.edu/ics/sites/default/files/attached-files/00-07.pdf. Accessed June 4, 2013.

Pepin, B. (2012). Working with teachers on curriculum materials to develop mathematical knowledge in/for teaching: Task analysis as 'catalytic tool' for feedback and teacher learning. In G. Gueudet, B. Pepin, & L. Trouche (Eds.), *Mathematics curriculum material and teacher development: From text to 'lived' resources* (pp. 123–142). New York: Springer.

Pepin, B., Gueudet, G., & Trouche, L. (2013). Investigating textbooks as crucial interfaces between culture, policy and teacher curricular practice: Two contrasted case studies in France and Norway. *ZDM: The International Journal on Mathematics Education, 45*(5), 685–698.

Remillard, J. T. (2012). Modes of engagement: Understanding teachers' transactions with mathematics curriculum resources. In G. Gueudet, B. Pepin, & L. Trouche (Eds.), *Mathematics curriculum material and teacher development: From text to 'lived' resources* (pp. 105–122). New York: Springer.

Reyes, B. J., Reyes, R. E., Tarr, J. L., & Chavez, O. (2006). *Assessing the impact of standards-based middle school mathematics curricula on student achievement and the classroom learning environment.* Washington, DC: National Center for Education Research. http://mathcurriculumcenter.org/PDFS/MS2_report.pdf. Accessed June 4, 2013.

Rezat, S. (2012). Interactions of teachers' and students' use of mathematics textbooks. In G. Gueudet, B. Pepin, & L. Trouche (Eds.), *Mathematics curriculum material and teacher development: From text to 'lived' resources* (pp. 231–246). New York: Springer.

Rezat, S., & Straser, R. (2013). Mathematics textbooks and how they are used. In P. Andrews & T. Rowland (Eds.), *Masterclass in mathematics education: International perspectives on teaching and learning* (pp. 51–62). London: Bloomsbury.

Schwartz, J. L. (1995). The right size byte: Reflections on educational software designer. In D. Persinks, J. Schwartz, M. West, & S. Wiske (Eds.), *Software goes to school* (pp. 172–182). New York: Oxford University Press.

Taizan, Y., Bhang, S., Kurokami, H., & Kwon, S. (2012). A comparison of functions and the effect of digital textbook in Japan and Korea. *International Journal for Educational Media and Technology, 6*(1), 85–93.

Usiskin, Z. (2013). Studying textbooks in an information age – a United States perspective. *ZDM: The International Journal on Mathematics Education, 45*(5), 713–723.

Watson, A., & Shipman, S. (2008). Using learner generated examples to introduce new concepts. *Educational Studies in Mathematics, 69*(2), 97–109.

Yerushalmy, M. (2005). Functions of interactive visual representations in interactive mathematical textbooks. *International Journal of Computers for Mathematical Learning, 10*(3), 217–249.

Yerushalmy, M. (2013). Designing for inquiry in school mathematics. *Educational Designer, 2*(6). Retrieved from: http://www.educationaldesigner.org/ed/volume2/issue6/article22/.

Yerushalmy, M., Katriel, H., & Shternberg, B. (2002). *The functions' web book interactive mathematics text.* Israel: CET –The Centre of Educational Technology. Available from http://www.cet.ac.il/math/function/english, with *Explorer* only.

Yerushalmy, M., Shternberg, B., & Katriel, H. (2014). *The VisualMath functions and algebra e-textbook.* http://visualmath.haifa.ac.il/. Accessed October 12, 2014.

Young, S. (2007). *The book is dead: Long live the book.* Sydney: University of New South Wales Press.

Chapter 8
Didactic Engineering as a Research Methodology: From Fundamental Situations to Study and Research Paths

Berta Barquero and Marianna Bosch

8.1 Didactic Engineering as a Research Methodology

The notion of didactic engineering (DE) has been at the core of the project of a science of didactics founded by Guy Brousseau in the 1970s along with the theory of didactic situations (TDS). In a recent paper presenting the origin of DE, Brousseau (2013) explains its necessity and locates it in the interface between research and teaching:

> Didactic engineering was a necessary and 'concrete' domain between a poorly invested activity, teaching mathematics, and an absent science, Didactics. The latter was supposed to, on the one hand, newly define both of them and, on the other hand, find its contingency in their confrontation and complementarity. 'Do not content yourself only with evidence', 'systematically reproduce', 'analyze in order to save experiences', 'only accept exogenous concepts under their testing in didactic engineering'—those have been the guiding principles [of Didactics]. (Brousseau, 2013, p. 4, our translation)

In the entry *didactic engineering* of the new *Encyclopaedia in Mathematics Education*, Michèle Artigue tries to clarify this intermediate role between the reality of classrooms and the science of didactics:

> The idea of didactical engineering (DE) [...] contributed to firmly establish the place of design in mathematics education research. Foundational texts regarding DE such as (Chevallard, 1982) make clear that the ambition of didactic research of understanding

---

Supported by the Spanish R&D projects EDU2012-39312-C03-01, EDU2012-39312-C03-03, and EDU2012-32644

B. Barquero
Universitat de Barcelona, Barcelona, Spain

M. Bosch (✉)
Universitat Ramon Llull, Barcelona, Spain
e-mail: mariannabosch@gmail.com

This chapter has been made open access under a CC BY-NC-ND 4.0 license. For details on rights and licenses please read the Correction https://doi.org/10.1007/978-3-319-09629-2_13

and improving the functioning of didactic systems where the teaching and learning of mathematics takes place cannot be achieved without considering these systems in their concrete functioning, paying the necessary attention to the different constraints and forces acting on them. Controlled realizations in classrooms should thus be given a prominent role in research methodologies for identifying, producing and re-producing didactic phenomena, for testing didactic constructions. (Artigue, 2014, p. 159)

It is important to keep in mind that, in the theory of didactic situations (TDS), didactic engineering was part of a collective project, led by Brousseau, to build an empirical science of didactic phenomena where the issue of the empirical validation of results was to be carefully taken into account. This is how he remembers those beginnings, in the same text quoted above:

My contribution was to design, project and start creating a proper science, which has to be responsible for the original theoretical concepts needed by engineering and for submitting them to the exigencies of any mature science, enriched by its scientific peer to peer relationships with other educational approaches. (*ibid.*, p. 4, our translation)

In this context and as Artigue (2008, p. 4) explains, didactic design was called to fulfill two different needs: to take into account the complexity of classrooms, at a time when research mainly relied on laboratory experiments and questionnaires; and to articulate the relationships between research and teaching innovation. She also highlights five main characteristics of DE as a theory-based intervention: the central role given to the notion of *situation* in both the modeling of mathematical knowledge and the organization of its teaching; the crucial attention paid to the epistemology of knowledge and the need to rebuild any mathematical content as the answer to an issue raised within a social situation; the importance given to the characteristics of the empirical *milieu* of the situation and of the students' interaction with this *milieu*; the three different functionalities assigned to mathematical knowledge, *action—formulation—validation*; and the vision of the teacher's role as the organizer of the relationships between the adidactic[1] and the didactic dimensions of situations (*devolution, institutionalization*).

As we shall see, DE appears as a research methodology to be closely related to the TDS, although it exceeds this initial framework:

As a research methodology, DE emerged with this ambition, relying on the conceptual tools provided by the Theory of Didactical Situations (TDS), and conversely contributing to its consolidation and evolution (Brousseau, 1997). It quickly became a well-defined and privileged methodology in the French didactic community, accompanying the development of research from elementary school up to university level […] (Artigue, 1990, 1992). From the nineties, DE migrated outside its original habitat, being extended to the design of teacher preparation, and professional development sessions, used by didacticians from other disciplines […] and also by researchers in mathematics education in different countries. (Artigue, 2014, pp. 159–160)

What are then the main characteristics of DE that are preserved in the evolution of TDS and the approaches sharing its main epistemological principles, such as the

---

[1] In an adidactic situation, students interact with a milieu only considering the logic of the problem approached, without taking into account the teachers' didactic intentions.

Anthropological Theory of the Didactic (ATD) we are considering here? We are answering this question using the four-phase structure of DE as a research methodology proposed by Michèle Artigue (2008). It will help us distinguish the theoretical assumptions underlying all DE works and emphasize its internal role in didactics research as a *phenomenotechnique*, that is, as a tool to produce didactic phenomena.

At the starting point of a DE process, we are locating a concrete content or issue to be taught and learned and usually a didactic problem related to it. The first phase, called *preliminary analysis*, mainly includes an epistemological questioning of the mathematical content at stake and of the necessity to introduce it at school, and a study of the conditions and constraints offered by the institutions where the teaching and learning process is to take place. This is an essential first step where research hypotheses are formulated and the content to be taught and learned is questioned, usually considering different kinds of hypothetical didactic phenomena involved. It is also in this phase where previous research results can be reinvested.

The second phase concerns the *design and a priori analysis*. This phase corresponds to the statement of how the content at stake is considered or modeled within didactics research. A mathematical and a didactic level may be distinguished here, to first "define" or "characterize" the content (mathematical analysis), and then to propose how to make it emerge from problematic questions within a sequence of concrete situations (didactic analysis). In the theoretical frames here considered, these analyses are carried out in terms of mathematical and didactic situations (TDS) or mathematical and didactic praxeologies (ATD).

The third phase includes the implementation of the previously designed didactic process, its observation, and data collection. At this experimental level, an *"in vivo" analysis* is usually developed, when interpreting in real time (or straight after) what is taking place in the classroom. Finally, the a posteriori *analysis* culminates the DE process. It is organized in terms of the contrast, validation, and development of the research hypotheses and design proposals of the previous phases, usually often leading to the formulation of new problems, related to both fundamental research and teaching development (Fig. 8.1).

It needs to be highlighted that, even if the a priori analysis precedes the in vivo and a posteriori analyses, there is always a constant interaction between the outcomes of the different phases: results from the a posteriori analysis may not only suggest introducing changes in the design of the teaching process, but also developing the characterization of the content at stake (preliminary analysis). It may also contribute to the *science of didactics* with the results obtained and the open problems raised, leading to new theoretical or methodological developments. In this sense, DE is not a development practice where previously established research results are transformed into teaching proposals. It is a way to empirically contrast assumptions about the possibilities of the diffusion of mathematical knowledge and the phenomena hindering it. As Brousseau said:

> My ambition has been to turn didactic engineering not into a socio-professional cover, but a scientific activity based on a coherent and 'proper' body of scientific knowledge. (2013, p. 6, our translation)

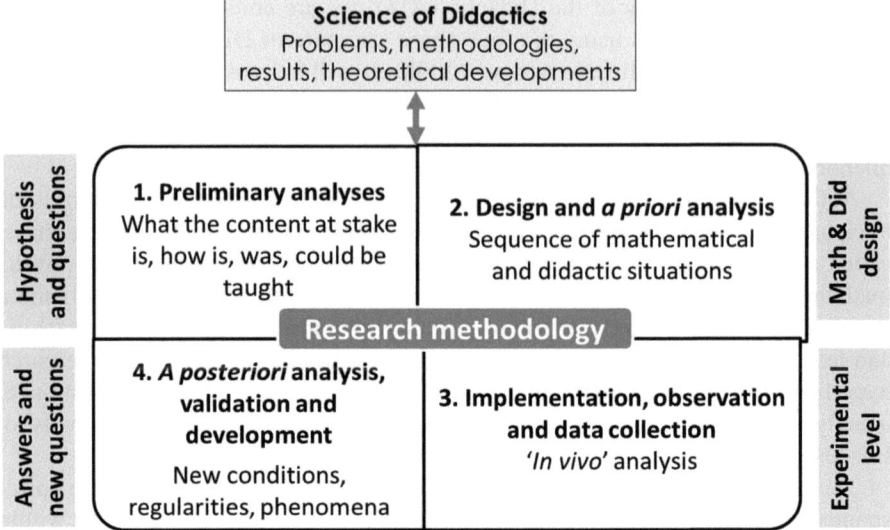

Fig. 8.1 Phases of the DE research methodology within TDS

DE is one among many other empirical methods elaborated and used by didactic research. We will not refer to it for instance when what is investigated is not directly a teaching and learning process but results from evidence coming from other sources gathered through naturalistic observation of institutions and their outcomes (including classes, historical documentation, etc.) or through direct intervention via interviews, questionnaires, etc.

8.2 Didactic Engineering Within the Theory of Didactic Situations

After this brief introduction to the notion of DE, we are presenting two examples of research based on DE processes, one from the TDS about the measure of quantities at primary school level, and another one from the ATD and the teaching of modeling processes at university level. In spite of the initial difference between both investigations, strong commonalities are being stressed, relying on what we propose to conceive as the mainstream of DE in Didactics.

8.2.1 An Example: Measuring Quantities at Primary School

We are using the case of the measurement of quantities at primary school to illustrate the four phases of the DE methodology within the TDS and, especially, their interactions. This example corresponds to a crucial issue in elementary mathematics education and has been the object of many investigations in the TDS that have not been widely disseminated in the international community. We will describe it in a brief and necessarily simplified way. More details can be found in Bessot and Eberhard (1983), Brousseau and Brousseau (1987, 1991–1992), Brousseau (2002), Douady and Perrin-Glorian (1989), Perrin-Glorian (2002, 2012), and Sierra (2006).

8.2.1.1 Preliminary Analysis

The starting point of the research is not the teacher's problem, *How to introduce the measurement of quantities in primary school?* Rather, it is the insertion of this problem into a broader questioning including epistemological as well as social issues, such as: *Why is it necessary to teach the measurement of quantities at primary school? What mathematical entities and practices are related to it? What social activities? How is it related to other mathematical notions, such as numbers, ratios, relationships, areas, volumes?*

To answer these questions, one should take into account the processes of didactic transposition (Chevallard, 1982) and the analysis of the activities that have been, are, and could be taught at school, an analysis that usually leads to the identification of *didactic phenomena*. For instance, it can be shown (Brousseau, 1997; Chambris, 2010; Perrin-Glorian, 2002, 2012) that, with the introduction of New Maths into the French curriculum in the 1970s, magnitudes and quantities disappeared from school mathematics, where they supported the construction of numbers. Only some basic practical measures and the metric system remained. Curricula have changed a lot since then, but the synthesis between quantities and sets to support the construction of numbers has still not been solved. Some indicators of this phenomenon are the fact that the choice of the unit of measure (gauge) is never raised, the blurry role played by units in modeling strategies and calculations, and the frequent situation that mathematical work is dominated by "abstract" numbers instead of "concrete" ones, that is, those directly representing physical quantities. Many years ago, Hans Freudenthal described this absence in the following terms:

> To count people and eggs there are natural units. To measure quantities, one needs gauges; the result of the measuring procedure is a number, which measures the quantity. There is a variety of gauges, because there is a variety of magnitudes; length, area, volume, height, mass, work, current intensity, air pressure, and monetary value are notions that become magnitudes by measuring procedures. Sometimes it is not clear why some magnitudes need different gauges. […] A few of these gauges are learned in arithmetic instruction, and as far as he needs it, the physicist develops a rational measure system. In between a large domain is no man's land. This is the fault of the mathematician. (1973, pp. 197–198)

During the same period, Hassler Whitney (1968) developed a mathematical theory of physical quantities to justify calculations, not between numbers but between quantities (such as $6\,m \div 2\,s = 3\,m/s$ or $5.25€/m \times 0.8\,m = 4.20€$), thus trying to build a bridge between engineering or science practices and mathematical ones. However, his proposals remained in the "scholarly mathematics" and have not permeated the prevailing school mathematical culture where calculations are very often done with abstract numbers and where units appear (if at all) only at the end at the process.

8.2.1.2 A Priori Analysis: Design of Mathematical and Didactic Situations

In order to face the complex problem quickly outlined previously, research in didactics needs to elaborate its own vision about measure and quantities or, more precisely, a *reference epistemological model* (Bosch & Gascón, 2006). In the TDS, reference epistemological models are formulated in terms of fundamental situations defined as games of action, communication, and validation, in interaction with an experimental *milieu*. The situation proposed by Brousseau (2002) defines the measure and quantities in terms of three intertwined *universes* and different situations between them. The first universe is the world of concrete measurable objects and their material comparison (putting objects side by side, on the two plates of a weighing scale, into a liquid, etc.). The second one is the universe of quantities (lengths, weights, areas, volumes, prices, etc.) as equivalence classes of objects considering analogical measures, where objects do not need to be manipulated but can be compared through some intermediate measures (gauges). The third is the universe of units, numbers, and change of units, obtained after defining a single privileged gauge for each magnitude (Fig. 8.2). We can thus obtain a general definition of measure in terms of triplets, including two universes and a situation to link them: something to measure (objects); a way to put objects into correspondence (adding specific conditions to get a measure application); and a positive numerical structure to express the measure (also with specific conditions depending on the number of units considered and other requirements). Usually, in school culture, only the first and third universes are considered, and only the third acquires a mathematical status.

To be operative, this definition needs to be specified in terms of sequences of games or adidactic situations passing through phases of action (solving a problem through empirical interaction with a *milieu*), communication (explaining the answer so that another person can follow and even reproduce the solution), and validation (justifying the solution without referring to the contingency of the *milieu*). Depending on the educational level and institution considered, the types of situations may obviously vary. Their design is part of both the delimitation of the reference epistemological model (mathematical situations) and their concrete realization under specific conditions (didactic situations).

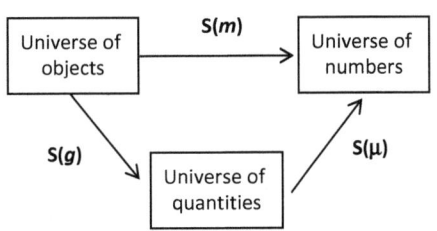

Fig. 8.2 Universes of measure (Brousseau, 2002)

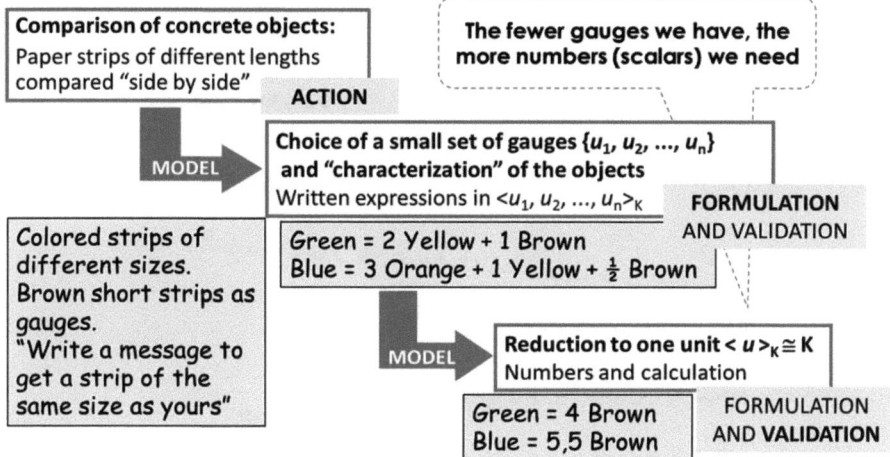

Fig. 8.3 Measuring situations: a priori analysis

Let us take a short example from Brousseau and Brousseau (1987) which is part of a larger DE design: a situation for grade 4 of primary school where it is proposed to introduce the measure of "length" through the following situation of communication. The *milieu* is composed of similar strips of different lengths and colors, with repetitions: short brown strips of the same length, medium red strips of different lengths, and long blue strips of different lengths. In this *milieu*, two strips can be compared by putting them side by side (action). The communication game, played by teams, consists in, given a blue strip, asking another team to bring as many smaller strips as necessary to build a new strip of the same size (Fig. 8.3). The aim of the activity is to raise the need for gauges to simplify the comparison of objects (common units for the messages) and, since there is no simple relationship between long and short strips, to move to the choice of a single unit and its fractions to simplify the messages without decreasing the precision of the measure.

8.2.1.3 Implementation, Observation, and Data Collection

In the first part of the sequence related to the strips communication game (where small brown strips are called "u"), it can be seen how the initial messages may fail and students learn how to make more precise messages to get a strip of the same size as theirs. The types of messages produced are "2 u plus 3 quarters of a u", "5 strips and fold the small u strip in 2", "3 times u, half, half of the half, half of the half of the half", "2 strips and another one with a small part missing", etc. (Fig. 8.4).

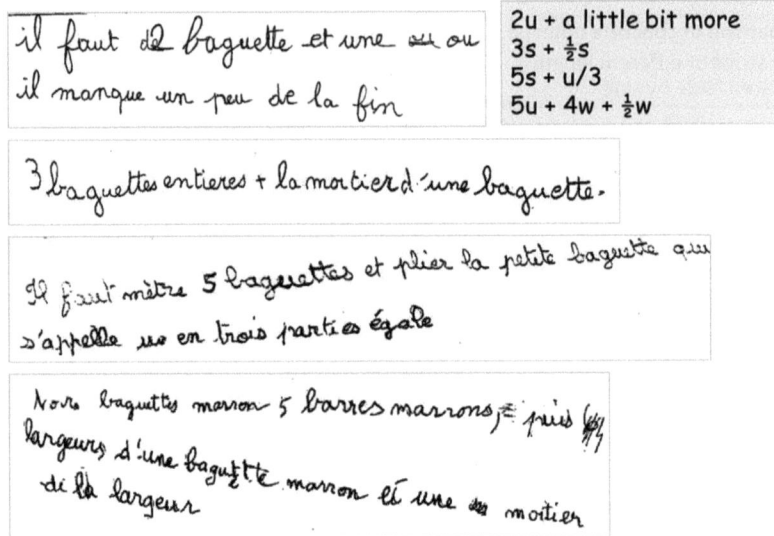

Fig. 8.4 Measuring situations: experimentation (students' productions)

The width of the strips was also used, thus including a new gauge that was not forecasted in the a priori analysis: "5 brown strips plus the width of a brown strip and half the width" (Brousseau & Brousseau, 1987, pp. 3–6).

During the very first experimentation of the sequence, an interesting problem appeared that partially discredited the a priori analysis. Because the students were familiar with the ruler and with measuring lengths in cm, some of them did not feel the need to choose a small strip as a gauge and started writing their messages using cm: "2 brown strips, one small strip and not a whole one, 3 or 4 cm have to be eliminated". It was then very difficult, and artificial, to move the students back to the single brown strip as unit, keeping cm aside.

Due to this unexpected event, after the first two sessions of working with lengths, it was decided to change lengths for weights, less familiar to the students, and avoid the use of metric units. Strips were replaced by small objects (pencil cases, small glasses, exercise books, etc.); different sizes of nails and small plates were introduced as gauges, and the comparison was made with a two-plate scale. This shows how the experimentation and in vivo analysis can make the design and a priori analysis evolve. We can consider that the reference epistemological model was also enriched through the experimentation, showing new conditions for the construction of the process of measuring quantities, as for instance the difficulties for the second universe (quantities) to exist without being directly absorbed by the third one (numbers), and also the relationships between the set of scalars needed in this second universe and the number of generators (gauges) used (Sierra, 2006).

8.2.1.4 A Posteriori Analysis: Results, New Phenomena, New Research Questions

The situations about comparing lengths and weights are part of a long sequence of 30 activities which form the DE work described in (Brousseau & Brousseau, 1987) to introduce the measuring of quantities in grade 4 of primary school. It contains the following activities:

- Measurement of lengths: communication game; studying the messages
- Measurement of weights: communication game; messages; work on the writings; comparing expressions; conversions; adding weights; comparing sums and total weight; transformation in basis 60
- Measurement of time: time and duration; calculation with numbers in basis 60
- Legal units of weight: presentation; conversions
- Finding the weight of an empty recipient: first part; second part (challenges)
- Measurement of lengths: adding lengths; decimal measures
- Writing decimal measures: length measures; decimal length and weight measures; comparison of decimal measures; order in decimal measures
- Operations with decimal measures: addition; multiplication by an integer; subtraction

In a later work, Brousseau and Brousseau (1991–1992) present some crucial issues derived from this research and describe some of the related phenomena. As we have seen before, there is, for instance, the fact that familiar *milieus* (such as lengths) are not always didactically productive, even if they may initially facilitate the devolution of a situation. A similar didactic phenomenon occurs with the teaching of rational or negative numbers, when the fractional or directional measures that are used to introduce them become a didactic obstacle when defining their multiplication or division.

There is also another example related to a very interesting experience with one of the reported activities, *the weight of the receptacle*. In spite of the errors of measure of the full and half-full receptacle, the students postulate and confirm an affine relationship between the volume of water and the total weight of the receptacle, which enables them to deduce the weight of the empty receptacle. It thus shows a complex relationship between the students' reasoning in a validation situation and the empirical *milieu* used, because taking into account the errors of measure appears as a mathematical necessity. And it submerges students at the core of scientific activity: "[Due to the measure errors], children became aware that when a theory or a method is made to forecast or obtain a result, the fact that its application happens once or twice is not enough for it to be accepted as true or valid. It has to 'work' in all cases, something which can only be established through reasoning" (*ibid.*, p. 80).

8.2.2 Didactic Engineering in the Science of Didactics

8.2.2.1 The COREM as a *Didactron*

Even if DE can be understood as a general research methodology closely related to the constitution of Didactics as a scientific domain, its existence cannot be separated from the COREM, *Centre d'Observations et de Recherches pour l'Enseignement des Mathématiques*. It was created in 1973 by Guy Brousseau as a research laboratory of the University of Bordeaux 1 and was integrated in the elementary school Jules Michelet in Talence (Bordeaux, France).[2] Till its closure in 1999, the COREM functioned as what Brousseau amusingly called a *didactic accelerator* or *didactron*.

In the COREM, new teaching proposals based on the TDS were regularly experienced by researchers, in close cooperation with the teachers of the school, who participated in the design, a priori analysis, teaching, observation, and a posteriori analysis of the lessons. Furthermore, all didactic engineering components, from the conception of situations to their setting up, managing, and observation, were the concern of all the staff, teachers, and researchers (Brousseau, 2013, p. 7). In fact, Michelet School was (and still is) a normal public elementary French school with 4 classes of preschool level and 10 classes of primary level (2 groups per grade), with pupils from the neighborhood and the same curricular and administrative requirements as any other French school. The teachers at the school were also normal ones, in the sense that no specific educational training was required, even if they were asked to participate in research activities. The peculiarity is that they worked in teams of 3 teachers per 2 classes, devoting one third of their time to the COREM, where they attended seminars and meetings with researchers, made observations, and had teaching preparation sessions with the other teachers of the team. According to Gresland and Salin, "The complexity of the [COREM] functioning is due to the fact that the creators of the project wanted to avoid the educational vocation of the school being altered by the investigations, and that these could later be carried out in the best possible methodological conditions" (1999, p. 30, our translation). It also supposed a detailed regulation of the interactions between researchers, teachers, and the classes observed.

Usually, in the development of a didactic engineering process, researchers presented a teaching proposal partially including the a priori analysis (goals expected, problems addressed, strategies forecast) to the team of teachers. Then they jointly elaborated the details of the sequence of lessons up to the preparation of a *didactic card* (*fiche didactique*) for each lesson. Researchers prepared the observation and decided on the kind of data to be gathered. During the lessons, observers had to try to be as invisible as possible, and teachers were supposed to forget that they were observed, taking their own decisions about the teaching of the lessons. Immediately after each lesson, a short meeting took place for the teacher, researchers, and other possible observers to share impressions, starting with the teacher's

---

[2] A detailed presentation can be found at http://faculty.washington.edu/warfield/guy-brousseau.com/index.html and http://guy-brousseau.com/le-corem/

8 Didactic Engineering as a Research Methodology

report of the experience. The observation of "ordinary lessons" which had not been organized through a didactic engineering process also took place at the COREM on a regular basis, although with a less structured procedure.

Throughout the existence of the COREM, all materials related to the teaching of mathematics were gathered, including class preparation activities (both the normal ones and those specially designed for research), students' productions, video recording of the lessons, reports from teachers and researchers after the lessons, annual planning of the courses, research seminars, etc. There was also a classroom especially designed for observations, with an extra surrounding area for observers, a windowed cubicle for the observers in an outer room, and technicians doing video and audio registration. Since 2010, the COREM archives have been made available by the Centre of Resources in Didactics of Mathematics Guy Brousseau (CRDM-GB) of the Spanish university Jaume 1 of Castelló (Valencia) (http://www.imac.uji.es/CRDM). Video recordings can also be accessed at visa.inrp.fr/visa. The list of didactic engineering realizations observed is very long, including teaching proposals about the main mathematical content from preschool to grade 5:

Reasoning and logic (preschool, grade 1): Designation, equality, lists, belonging to a list; Classing, sets, propositions, no-and-or, equivalence, equality; Comparisons, physical quantities ordering; number, length, mass, price, capacity; Order, <, >, next, previous; $P(E)$; Implicit theorems, demonstration (race to 20); Theorems, proofs (bigger number)

Quantities and measure (preschool, grades 1, 4, 5): Natural quantities (cardinal, lengths, masses, prices); Capacities; Sums, products, extractions, partitions; Volumes, capacities; Rational and decimal quantities (commensuration, unit partition); Events measure (statistics)

Discrete quantities and arithmetic: Operations on natural numbers (addition, multiplication, subtraction, division); Functions

Rational quantities, arithmetic, and algebra: Rational and decimal numbers, definition, writing, operations; Order (density); Linear applications, enlargements; Numerical dilations, ordering, composition; Structure of rational numbers

Space and geometry: Topology, figures; Fundamental situation of geometry; Congruencies; Dilations

Statistics and probability: Random walk; Confidence interval; Compose probability (two successive events); Approximation to the Law of Large Numbers. (Brousseau, 2013, p. 12)

8.2.2.2 An Experimental Epistemology

In the research program set up by the TDS, the experimental work carried out by DE processes is crucial, as it represents a way to empirically test epistemological and didactic proposals formulated in terms of sequences of adidactic and didactic situations. In a sense, the TDS appeared as a reaction to the New Mathematics reform of the 1960s and 1970s that Brousseau considered as "a utopia totally ignoring all the difficulties and laws of the dissemination of knowledge and practices in a society [...], which believed and died in the illusion of transparency of didactic facts" (2004, p. 23). This explains the importance given to the empirical contrast of teaching proposals before their dissemination, as well as the necessity to base them on a consistent and explicit framework of theoretical assumptions. That was the very precise role of DE:

> *Didactic engineering* became, de facto, a part of Didactics of mathematics where precise, observable and reproducible teaching devices, specific to different forms of knowledge of determined mathematical entities, were conceived and also empirically, experimentally and theoretically studied [...] (Brousseau, 2013, p. 6, our translation)

In this context, the notion of *situation* does not only appear as a way to describe teaching activities, but was also first used as a model to conceive mathematical activities specific to each mathematical content to be taught. It contains the requirement to characterize each piece of mathematical knowledge by a set of conditions making it progressively appear as the answer to a given problematic question. The adidactic situations are thus a way to show the functionality of mathematical knowledge in the institutional environment of the students who have to learn it.

This ambitious project requires a double rupture: researchers need to allow themselves to question mathematics as it is usually conceived and presented by mathematics scholars and by school institutions, elaborating their own alternative reconstructions of mathematical knowledge and activities (the *reference epistemological models*). They also need to have the same attitude towards other disciplines (psychology, pedagogy, sociology, etc.) concerning the effects of their proposals on mathematical practices and knowledge. This is why it is important that the results obtained are empirically based, protecting researchers from adopting unfounded ideologies or implicit institutional viewpoints on both educational facts and mathematical knowledge.

8.3 Didactic Engineering Within the Anthropological Theory of the Didactic

8.3.1 From Situations to Study and Research Paths

The TDS conception of DE is located in what Chevallard (2012) calls the *paradigm of questioning the world*: mathematical contents, just like the content of any other subject matter, should not be taught as if their value and importance were taken for granted. On the contrary, they need to be constructed and appear for the students as true answers to real questions. The search for a fundamental situation to represent, model, and rebuild any given piece of knowledge is in fact a way for didactics research to assume its own responsibility in the search for the possible *raisons d'être* of mathematical contents within the students' reach, and for the rationale of their teaching at school. For instance, the epistemological question, *what is the measure of quantities and how can it be constructed through a sequence of situations?*, includes the primary question: *what is the measure of quantities for, and why is it important to learn it?*

The Anthropological Theory of the Didactic, as it has been developed by Yves Chevallard (1992, 2006, 2007, 2012), shares the TDS essential epistemological questioning, the search for a rationale for any piece of knowledge to be taught, and the central place given to problematic issues in learning and teaching processes. The modeling in terms of fundamental situations is replaced by two main theoretical tools: the notion of *praxeology* used to describe any kind of human activity (Chevallard, 1999) and the *Herbartian schema*, named after the J. F. Herbart (1776–1841), into account the way praxeologies are built, taught, learned, or disseminated as the answer to a given problematic question (Chevallard, 2011).

If the starting point of the teaching and learning process is a given praxeology P a group of students X should learn under the supervision of a group of teachers Y, then the didactic process involving X, Y, and P can be described in terms of *study and research activities* structured in six didactic moments or dimensions closely linked to the structure of praxeologies (Barbé, Bosch, Espinoza, & Gascón, 2005; Bosch & Gascón, 2010; Chevallard, 1999). However, this is not the only possible pattern to represent teaching and learning. A didactic process does not necessarily begin with the delimitation of a given piece of knowledge to be taught, but can also be motivated by the need to consider a problematic question Q_0 a group of students X wants (or has) to answer with the help of a group of teachers Y. What then appears is a sequence of linked study and research activities called *study and research paths* (*SRP*), which can be formalized using the general "Herbartian schema" as follows:

$$\left[S(X;Y;Q_0) \to \{A^\diamond_i, O_j, Q_k, A_k\} \right] \to A^\heartsuit$$

The starting point of an SRP should be a "lively" question of genuine interest for the community of study, what we call a *generating question* referred to as Q_0. The question has to be taken seriously, not as a mere opportunity to cover some fixed a priori mathematical content. Elaborating answers to Q_0 must become the main purpose of the study and an end in itself.

The study of Q_0 evolves and opens many other *derived questions* Q_k that appear as the starting point of new SRP or new branches of the initial one. One needs to constantly ask whether these derived questions are relevant in the sense of being capable of leading *temporary answers* A_k that can be helpful in elaborating a *final answer* $A^\heartsuit$ for Q_0. As a result, the study of Q_0 and its derived questions Q_k leads to successive temporary answers A_k tracing out the *possible routes* to be followed in the effective experimentation of the SRP. The work of producing $A^\heartsuit$ can thus be described as a *tree of questions Q_k and temporary answers A_k* related to each other through a modeling process.

The implementation of the SRP usually requires resorting to external preestablished answers $A_i^\diamond$ to the derived questions Q_k, as well as some other objects O_j used to test the available answers, elaborate new ones, and formulate new questions. The preestablished answers $A_i^\diamond$ are accessible through different means of communication and diffusion called the *media* (in the sense of "mass media"). However, knowledge provided by the *media* corresponds to constructions that have usually been elaborated to answer other questions than those specifically approached. Thus, it has to be "deconstructed" and "reconstructed" according to the new needs. This is the main role of the *experimental milieu*, M, containing empirical objects O_j as well as other old, well-established answers $A_i^\diamond$. Milieu M evolves throughout the study process and becomes one of the main guarantees of a successful outcome. It is usual that, during the SRP, emerges the need to make a given $A^\diamond$ available to X because it is required or seems necessary to produce $A^\heartsuit$, The specific branch of the SRP starting in this case is called a *study and research activity* (SRA) focused on $A^\diamond$. In this sense, SRP together with SRA provides a general modeling tool to describe any kind of teaching and learning process, from those based on the direct transmission of knowledge to those centered on inquiry activities.

This broadened conception of didactic processes can be used to describe almost any form of teaching and learning strategy, and it prevents researchers from assuming any kind of specific form of school organization as normal or natural. Furthermore, it encourages taking into account a broad set of conditions and constraints affecting the teaching and learning processes that far exceed the limits of the classroom. At the same time, the border between mathematics and didactics (in the sense of teaching and learning) is blurred: doing mathematics includes study, research, and supervision; learning mathematics includes collectively carrying out an activity of study and research; and teaching mathematics corresponds to leading or supervising a research and study activity.

It is in this context that DE experimentations are carried out in the setting of ATD, in very different conditions than those established by the COREM, although they maintain the main methodological gestures exposed at the beginning of the chapter.

8.3.2 An Example: Teaching Modeling at University Level

The second case of DE we are presenting approaches the problem of teaching mathematical modeling at university level. This case illustrates some of the tools used in the framework of the ATD going through the four phases of a DE methodology process (see Fig. 8.5). More details about this particular case can be found in (Barquero, 2009; Barquero, Bosch, & Gascón, 2008). Some other research about the design and integration of SRP at different school levels, and even in teachers' professional development, has been established in the same framework and following similar methodologies (García, Gascón, Ruiz-Higueras, & Bosch, 2006; Hansen & Winslow, 2010; Rodríguez, Bosch, & Gascón, 2008; Ruiz-Munzón, Matheron, Bosch, & Gascón, 2012; Winsløw, Matheron, & Mercier, 2013).

8 Didactic Engineering as a Research Methodology

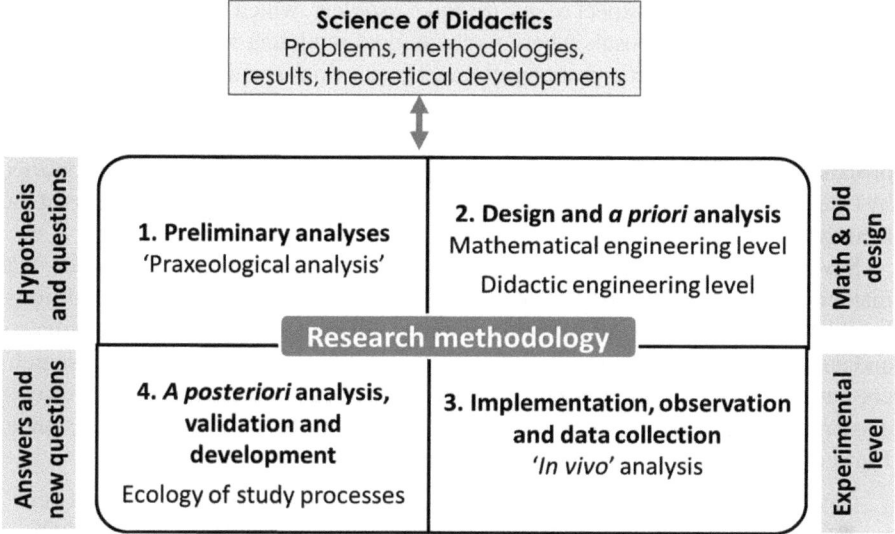

Fig. 8.5 Phases of the DE research methodology within the ATD

8.3.2.1 Preliminary Analysis

The starting point of the research here considered is the integration of mathematical modeling in first-year university courses of Mathematics for Natural Sciences. When analyzing what kind of mathematics is taught at this level, one could think that natural sciences university degrees would offer favorable institutional conditions to teach mathematics as a modeling tool, as mathematical models are becoming more and more essential to the understanding, use, and development of scientific disciplines. However, this seems far away from reality: despite the fact that mathematical models appear in the syllabi of almost all the courses, teaching mathematical models often comes at the end of the process, if there is time left for it. The dominant ideology is that modeling represents a mere application of some preestablished knowledge, leaving little room for the process of proposing, constructing, validating, and questioning mathematical models. We define as *applicationism* this spontaneous epistemology, which appears to be dominant in many university institutions (Winsløw et al., 2013).

According to the ATD and to the epistemological principles considered, if we start from the principle that intra-mathematical modeling is part of mathematical modeling, then many mathematical activities can be reformulated as modeling activities. It is considered that, in a modeling process, both the initial system considered and the models used have a praxeological structure. Mathematical modeling activity then appears as a process of (re)construction and articulation of mathematical praxeologies which become progressively broader and more complex, the main aim of which is to provide answers to problematic questions.

Thus, mathematical modeling cannot only be considered as an aspect or modality of mathematical activity but has to be placed at the core of it. This integration constitutes an essential aspect of our *research problem*, which opens the issue of the design of teaching proposals where mathematical modeling adopts an explicit and crucial role, emerging from initial problematic questions and able to link mathematical content that now appears as tools or models to provide answers to questions. Our working hypothesis is to suggest an SRP as one of the appropriate teaching proposals to move toward the (new) paradigm of *questioning the world* proclaimed by Chevallard, and which explicitly situates mathematical modeling problems at the heart of teaching and learning processes.

Several investigations from different theoretical perspectives have shown that mathematical modeling activities can exist at school under appropriate conditions, at all levels and in almost all curricular content. However, besides the good progress and encouraging results in research for the integration of modeling, many researchers have pointed out the existence of strong limitations hindering the large-scale dissemination of mathematical modeling practices in the classroom. For instance, Burkhardt makes the following harsh statement:

> [W]e know how to teach modelling, have shown how to develop the support necessary to enable typical teachers to handle it, and it is happening in many classrooms around the world. The bad news? 'Many' is compared with one; the proportion of classrooms where modelling happens is close to zero. (2008, p. 2091)

The research problem that has to be addressed is thus the study of the conditions that make the teaching of mathematical modeling possible at school, as well as the constraints that hinder its development as a normalized activity. Of course this problem depends on how mathematical modeling is conceived by both the research community (epistemological model) and the institutions where it is to be disseminated (the usual school and scholar epistemology, which takes here the form of applications). In ATD, it is referred to as the problem of the *ecology* of mathematical modeling in current educational environments. It can be specified with central questions about: what kinds of limitations and constraints exist in our current educational systems that prevent mathematical modeling from being widely incorporated in daily classroom activities? What kind of conditions could help a large-scale integration of mathematical modeling at school?

According to our previous analysis, the problem of the ecology of mathematical modeling becomes the problem of the ecology of SRPs and of their capacity to ensure the development of modeling activities. In the following section, we outline our partial answer to this enormous problem, focusing on the mathematical and didactic design of a particular SRP at university level with respect to the question of how to predict population dynamics.

8.3.2.2 A Priori Analysis: Mathematical and Didactic Design

Our generating question Q_0 that leads to the a priori mathematical and didactic design of the SRP, given the size of a population over previous periods of time, focuses on the following questions: *How can we predict the long-term behavior of its size?*

What sort of assumptions about the population, its growth, and its surroundings should be made? How can one create forecasts and test them? In all its implementations, Q_0 was introduced using different populations: first a pheasant population, then a fish population, and finally, a yeast population that was cultivated either in independent containers or mixed.

To provide answers to Q_0 and to the sequence of the derived questions that followed it, the construction of different mathematical models was required. Depending on whether time was considered as a *discrete* or *continuous* magnitude and if population generations were considered *independent* (x_t only depends on x_{t-1}) or *mixed* (x_t depends on $d>1$ past generations $x_{t-1}, x_{t-2}, \ldots, x_{t-d}$), a four-branch structure of the SRP can be delimited, giving rise to its a priori mathematical design (Fig. 8.6).

Looking into the derived questions opens a sequence of modeling activities that cover most of the content of a first-year course of mathematics for natural science students at university level: sequences and its convergence, one-variable calculus, linear algebra, and ordinary differential equations and their systems. This first mathematical design step is followed by the *didactic* a priori *design* of the SRP. It has inherited the structure defined in the mathematical a priori design and now includes questions about the *mesogenesis* (evolution of the experimental *milieus*), *chronogenesis* (evolution of the new questions and the knowledge introduced through the media), and *topogenesis* (sharing of responsibilities between teacher and students).

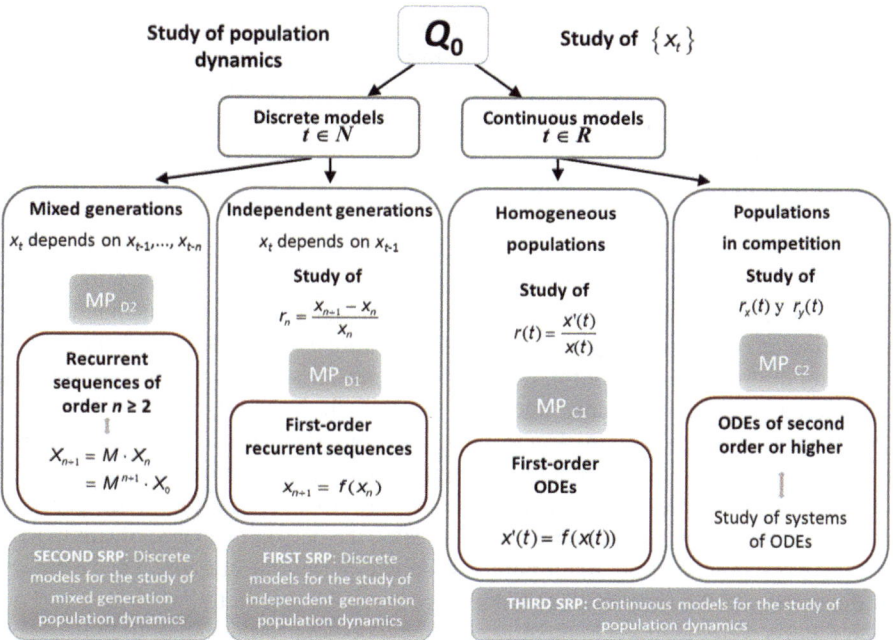

Fig. 8.6 General structure of the SRP branches derived from the study of Q_0 (Barquero, 2009)

Many important decisions are taken at this point to support the necessary change of students' and teachers' common strategies rooted in a dominant university teaching culture. For instance, students were constantly asked to assume new responsibilities so as to formulate new questions and approach them, to provide their own temporary answers to the successive derived questions, to plan the collective work, etc. In turn, the teacher has a new role to play as the *supervisor of the inquiry*, avoiding the temptation of imposing possible answers, inviting the groups of students to defend the successive answers they provide, to help decide on the questions to pursue, etc. Moreover, students should be able to introduce any external work or piece of knowledge they find appropriate in the *milieu*. The whole class will have the task to create the appropriate *milieu* for an internal validation of all those preestablished answers. All those new conditions in the implementation of the SRP required new teaching strategies and new devices: enabling the students to plan the work, elaborate new answers, compare data and models, write reports with temporary answers, validate final answers, defend them, etc.

8.3.2.3 Implementation and In Vivo Analysis

We tested the use of the SRP for five academic years (from 2005/06 to 2009/10) with first-year students of technical engineering degrees at the Autonomous University of Barcelona (Spain), who were attending a 1-year *Mathematical Foundations of Engineering* course. A special educational activity, called the "mathematical modeling workshop," was introduced in the general organization of the course; it was optional for students. The workshop ran in parallel with the lecture and problem practice sessions scheduled in the usual course. In the successive implementation of the SRP, a 2-h weekly session of the workshop took place as follows: students worked in teams of 2 or 3 members and had to develop their own study and propose their own "temporary" answers to the intermediate questions of Q_0.

Throughout the modeling workshop and its successive implementation year after year, the necessity arose to introduce several teaching and learning devices that were nonexistent in our usual university teaching settings. They had to evolve throughout the course and become accepted by the students. On the one hand, at the beginning of each workshop session, the teams were asked to deliver a report summarizing the work carried out in the previous sessions with respect to assumptions considered, main problematic questions dealt with, mathematical models used, "temporary" answers obtained, and new questions opened. In each session, one team was in charge of explaining and defending their report. A discussion followed to state the main progress and to agree on how to continue the study process. Moreover, there was the "secretary of the week", the person in charge of summarizing the work done and the main points of debate during the session. The secretary of the week and the team of the week played a crucial role in the workshop and all their reports were included in the diary of the workshop. At the end of the SRP, each individual student had to write a final report of the entire study: evolution of the main questions studied, work in and with different mathematical models, relationship between them, and so on. On the other hand, students were asked to search for any external information about the mathematical models they were building, and the answers they were providing in the media. Their findings were also explained in the workshop sessions and they discussed how these external mathematical objects could be useful (or not) and how they could be validated and used in relation to the questions they were dealing with.

Thanks to the several variations in the successive implementations of the SRP and to its *in vivo* analysis, several aspects could be improved every year. The a priori mathematical and didactic design of the SRP was gradually enriched, that is, after each implementation we had a more detailed description of the derived questions and temporary answers that were likely to appear in each of the SRP branches. Moreover, we obtained more details and got more control of the use and functionality of the different learning and teaching devices that were included throughout the workshop, especially those aspects related to the new conditions of mesogenesis, chronogenesis, and topogenesis:

- What responsibilities did students find more difficult to assume?
- What teaching strategies can help achieve the transfer of passing on more responsibilities to students?
- Do the weekly reports help students to formulate their own assumptions and to pose new questions?
- Does the debate generated at the beginning of each session help students to organize their own work?
- Does the workshop diary encourage the students and the teacher to have a broader perspective of the whole modeling process?

8.3.2.4 A Posteriori Analysis and Ecology

When considering the SRP as a whole, we verified from its first implementation that the sequence of derived questions arising from the generating question Q_0 led the students and the teacher to consider most of the main content of the entire mathematics course (sequences and their convergence, one-variable functions, derivative, ordinary differential equations, matrices, etc.). However, during the workshop, this content appeared in a structure that was completely different from the usual organization proposed in the main course. Instead of the classical "logic of mathematical concepts", the workshop was more guided by the "looking for answers to problematic questions" and "types of models" that progressively appeared. During its five courses of implementation, because the instructors were the same year after year, the first author of the paper and a lecturer who is an expert in applied mathematics, we found it easier to make SRP compatible with the standard formats of teaching (lectures and problem practice sessions). In the end, all these traditional devices were subordinated to the study of questions opened during the workshop. For instance, when questions appeared that needed some theoretical developments, such as "what was the relation between the relative rate of growth and the derivative or how to calculate the n-power of a matrix", we suspended the workshop sessions and spent several lectures and practice sessions on a *study and research activity* centered on the diagonalization of matrices before carrying on with the workshop.

However, it was not easy to preserve and transfer all these good conditions to the new teacher who came to replace us. Although he had all the material and descriptions from the previous SRP and all our assistance, the year after we left the course, the new implementation of the SRP only took 2 weeks. When we asked the teacher why it had taken such a short time, he told us that he only needed three sessions to show the students how to solve the questions and explain all the mathematical models they had to apply… In the end, the traditional ways, focused only on direct transmission and application, seem to have prevailed (Barquero, Bosch, & Gascón, 2013).

Other important constraints that could be identified were mainly related to the difficulties for keeping in mind the generating question of the SRP, given the fact that students were not used to pursuing a question for such a long period of time. SRP requires a strong modification of the usual didactic contract that currently exists at universities, where the teacher provides long lists of different small problems which the students have to solve. On the contrary, some other responsibilities that are usually assigned exclusively to the teacher were easily assumed by the students: searching information about models, discussing different ways of looking for an answer, comparing experimental data and reality, writing and defending reports with partial or final answers, etc. Others, however, were more difficult to share: choosing the relevant mathematical tools, criticizing the scope of the models constructed, posing new questions to continue with the study, planning the work to do, etc.

Last but not least, another strong constraint appeared in all the SRP implementations: the necessity of an *ad hoc mathematical discourse* available to describe the process that had just taken place. The work carried out in the workshop led to a need for new words, concepts, and discourses to talk about what was going on and to formalize it theoretically. The teacher and the students could no longer base their work on previously selected material, such as the one provided by textbooks or by previous lectures. In each case, they had to elaborate their own narrative of the process followed, a collective and original *mathematical text* indispensable to describe the dynamics of the work done and to provide material for the writing of the final answer $A^\vee$. This lack of mathematical discourses to express, describe, and formalize the dynamics of mathematical activity brings to light the necessity to develop new mathematical and didactic infrastructures to support self-sufficient modeling activities.

Following Hans Freudenthal's observation in the case of the mathematical work with quantities, we came across other "no man's lands" which appear to be crucial for mathematical modeling to live in our school institutions. The problem is not only "the fault of the mathematician"; it seems to affect the entire educational culture and the conceivable ways of making it evolve.

8.4 Open Questions

As was said in the introduction, this chapter focuses on the notion of DE as it was introduced in the TDS to empirically organize the study of didactic phenomena and new teaching proposals, and its later developments in the ATD with the problem of the ecology of teaching and learning processes. We have left aside other conceptions of DE which are more or less related to them (Margolinas et al., 2011), their contrast with other task-design works, and more general reflections about the role of design and theories in mathematics education (Burkhardt & Schoenfeld, 2003; Design-Based Research Collaborative, 2003; Godino et al., 2013).

In order to encourage the debate and nourish future comparative studies on this issue, we conclude by briefly addressing three main issues that, in our opinion, cannot be left aside in the research work of contrasting and trying to articulate different approaches. First of all, we have seen that, in the research program established by the TDS and developed by the ATD, the transition to the paradigm of questioning the world becomes crucial: mathematical content, as well as any other subject matter, needs to appear as true answers to real questions rather than mere *monuments to visit* (Chevallard, 2012). The necessity to move away from *monumentalism* is not new, but it has not always been considered in the same manner, especially when researchers' epistemological assumptions require a certain distance from assumptions which prevail in teaching and research institutions.

The TDS and the ATD locate the problem of the ecology of design realizations at the heart of DE research, didactics appearing as the scientific study of the conditions for mathematical knowledge (praxeologies) to disseminate in human institutions. Furthermore, the ATD proposes a considerable enlargement of the unit of analysis for research corresponding to the different levels of the scale of didactic codetermination (Chevallard, 2002). In the case of mathematical modeling, some approaches (Burkhardt, 2008; Kaiser & Maaβ, 2007; Lesh & Doerr, 2003; among others) have also highlighted the problem of the large-scale dissemination of new teaching proposals. However, this issue is still far from becoming central in the main stream of research in mathematics education. We need more insight about how other approaches have experienced and proposed to deal with this ecological problem.

It seems clear that the ecological problem needs to engage different partnerships of the educational community: researchers, designers, policy makers, teacher associations, mathematicians, editors, etc. In some approaches, the role of research is clearly distinguished from the teachers' role, even if they are in broad agreement on their tight cooperation. The problem of the roles assigned in mathematics education to the different partners of the education process appears as an unavoidable issue related to the problem of the ecology, especially at a moment when all efforts should be put together, while responsibilities are of course typically different among the partners and institutions involved.

References

Artigue, M. (1990). Ingénierie didactique. *Recherches en Didactique des Mathématiques, 9*(3), 281–308. (English translation: Artigue, M. (1992). Didactical engineering. In R. Douady & A. Mercier (Eds.), *Recherches en didactique des mathématiques, Selected papers*. La Pensée Sauvage, Grenoble, pp. 41–70.)

Artigue, M. (1992). Didactic engineering as a framework for the conception of teaching products. In R. Biehler, R. W. Scholz, R. Sträßer, & B. Winkelmann (Eds.), *Mathematics didactics as a scientific discipline* (pp. 7–39). Dordrecht: Kluwer Academic.

Artigue, M. (2008). Didactical design in mathematics education. In C. Winslow (Ed.), *Nordic research in mathematics education: Proceedings from NORMA08* (pp. 7–16). Copenhagen. Available from https://isis.ku.dk/kurser/blob.aspx?feltid=212293. Accessed on January 14, 2013.

Artigue, M. (2014). Didactic engineering in mathematics education. In S. Lerman (Ed.), *Encyclopedia of mathematics education* (pp. 159–162). New York: Springer. Available from http://www.springerreference.com/docs/navigation.do?m=Encyclopedia+of+Mathematics+Education+(Humanities%2C+Social+Sciences+and+Law)-book188. Accessed on January 14, 2013.

Barbé, J., Bosch, M., Espinoza, L., & Gascón, J. (2005). Didactic restrictions on the teacher's practice. The case of limits of functions in Spanish high schools. *Educational Studies in Mathematics, 59*, 235–268.

Barquero, B. (2009). *Ecología de la modelización matemática en la enseñanza universitaria de las matemáticas*. Ph.D. dissertation. Barcelona: Universitat Autònoma de Barcelona. Available from http://www.tdx.cat/handle/10803/3110. Accessed on January 03, 2015.

Barquero, B., Bosch, M., & Gascón, J. (2008). Using research and study courses for teaching mathematical modelling at university level. In D. Pitta-Pantazi & G. Pilippou (Eds.), *Proceedings of the Fifth Congress of the European Society for Research in Mathematics Education* (pp. 2050–2059). Larnaca, Cyprus: University of Cyprus.

Barquero, B., Bosch, M., & Gascón, J. (2013). The ecological dimension in the teaching of mathematical modelling at university. *Recherches en didactique des mathématiques, 33*(3), 307–338.

Bessot, A., & Eberhard, M. (1983). Une approche didactique des problèmes de la mesure. *Recherches en Didactique des mathématiques, 4*(3), 293–324.

Bosch, M., & Gascón, J. (2006). Twenty-five years of the didactic transposition. *ICMI Bulletin, 58*, 51–64.

Bosch, M., & Gascón, J. (2010). Fundamentación antropológica de las organizaciones didácticas. In A. Bronner, M. Larguier, M. Artaud, M. Bosch, Y. Chevallard, G. Cirade, et al. (Eds.), *Diffuser les mathématiques (et les autres savoirs) comme outils de connaissance et d'action* (pp. 49–85). Montpellier: IUFM de l'Académie de Montpellier.

Brousseau, G. (1997). *Theory of didactical situations in mathematics*. Dordrecht: Kluwer Academic.

Brousseau, G. (2002). Les grandeurs dans la scolarité obligatoire. In J. L. Dorier, M. Artaud, M. Artigue, R. Berthelot, & R. Floris (Eds.), *Actes de la XIe Ecole d'été de didactique des mathématiques* (pp. 331–348). Grenoble: La Pensée Sauvage.

Brousseau, G. (2004). L'émergence d'une science de la didactique des mathématiques. *Repères IREM, 55*, 19–34.

Brousseau, G. (2013). *Introduction à l'Ingénierie Didactique* (non-published lecture). Available from http://guy-brousseau.com/2760/introduction-a-l%E2%80%99ingenierie-didactique-2013/

Brousseau, N., & Brousseau, G. (1987). La mesure en CM1. Compte rendu d'activités. *Publications de l'IREM de Bordeaux*.

Brousseau, G., & Brousseau, N. (1991–1992). Le poids d'un récipient. Etude de problèmes du mesurage en CM. *Grand N, 50*, 65–87.

Burkhardt, H. (2008). Making mathematical literacy a reality in classrooms. In D. Pitta-Pantazi & G. Pilippou (Eds.), *Proceedings of the Fifth Congress of the European Society for Research in Mathematics Education* (pp. 2090–2100). Larnaca, Cyprus: University of Cyprus.

Burkhardt, H., & Schoenfeld, A. H. (2003). Improving educational research: Toward a more useful, more influential, and better-funded enterprise. *Educational Researcher, 32*(9), 3–14.

Chambris, C. (2010). Relations entre grandeurs, nombres et opérations dans les mathématiques de l'école primaire au 20e siècle: théories et écologie. *Recherches en Didactique des Mathématiques, 30*(3), 317–366.

Chevallard, Y. (1982). *Sur l'ingénierie didactique*. IREM d'Aix Marseille, Marseille. Available from http://yves.chevallard.free.fr/spip/spip/IMG/pdf/Sur_l_ingA_c_nierie_didactique_-_YC_-_1982.pdf. Accessed on January 14, 2013.

Chevallard, Y. (1992). Fundamental concepts in didactics: Perspectives provided by an anthropological approach. In R. Douady & A. Mercier (Eds.), *Research in didactique of mathematics. Selected papers* (pp. 131–167). Grenoble: La Pensée Sauvage.

Chevallard, Y. (1999). L'analyse des pratiques enseignantes en théorie anthropologique du didactique. *Recherches en didactique des mathématiques, 19*(2), 221–266.

Chevallard, Y. (2002). *Organiser l'étude: 3. Ecologie & régulation. XIe école d'été de didactique des mathématiques* (pp. 41–56). Grenoble: La Pensée Sauvage.

Chevallard, Y. (2006). Steps towards a new epistemology in mathematics education. In M. Bosch (Ed.), *Proceedings of the IVth Congress of the European Society for Research in Mathematics Education* (pp. 22–30). Barcelona: Fundemi IQS.

Chevallard, Y. (2007). Readjusting didactics to a changing epistemology. *European Educational Research Journal, 6*(2), 131–134.

Chevallard, Y. (2011). La notion d'ingénierie didactique, un concept à refonder. Questionnement et éléments de réponse à partir de la TAD. In C. Margolinas, M. Abboud-Blanchard, L. Bueno-Ravel, N. Douek, A. Flückiger, P. Gibel, F. Vandebrouck, & F. Wozniak (Eds.), *En amont et en aval des ingénieries didactiques* (pp. 81–108). Grenoble: La Pensée Sauvage.

Chevallard, Y. (2012). *Teaching mathematics in tomorrow's society: A case for an oncoming counterparadigm*. Plenary talk at 12th International Congress on Mathematical Education, Seoul, Korea.

Design-Based Research Collaborative. (2003). Design-based research: An emerging paradigm for educational enquiry. *Educational Researcher, 32*(1), 5–8.

Douady, R., & Perrin-Glorian, M. J. (1989). Un processus d'apprentissage du concept d'aire de surface plane. *Educational Studies in Mathematics, 20*(4), 387–424.

Freudenthal, H. (1973). *Mathematics as an educational task*. Dordrecht: Reidel.

García, F. J., Gascón, J., Ruiz-Higueras, L., & Bosch, M. (2006). Mathematical modelling as a tool for the connection of school mathematics. *ZDM: The International Journal on Mathematics Education, 38*(3), 226–246.

Godino, J., Batanero, C., Contreras, A., Estepa, A., Lacasta, E., & Wilhelmi, M. (2013). Didactic engineering as design-based research in mathematics education. *Proceedings of CERME 8*. Available at from http://www.ugr.es/~jgodino/eos/Godino_CERME_2013.pdf. Accessed on January 14, 2013.

Greslard, D., & Salin, M. H. (1999). La collaboration entre chercheurs et enseignants dans un dispositif original d'observation de classes: Le centre d'observation et de recherche sur l'enseignement des mathématiques (COREM). In F. Jacquet (Ed.), *Proceedings of CIEAEM 50* (pp. 24–37).

Hansen, B., & Winslow, C. (2010). Research and study courses diagrams as an analytic tool: The case of bidisciplinary projects combining mathematics and history. In M. Bosch (Ed.), *Un panorama de la TAD. An overview of ATD* (pp. 257–263). Bellaterra: Centre de Recerca Matemàtica.

Kaiser, G., & Maaβ, K. (2007). Modeling in lower secondary mathematics classroom – problems and opportunities. In W. Blum, P. L. Galbraith, H. Henn, & M. Niss (Eds.), *Modelling and applications in mathematics education: The 14th ICMI Study* (pp. 99–107). New York: Springer.

Lesh, R., & Doerr, H. M. (2003). *Beyond constructivism: Models and modeling perspectives on mathematics teaching, learning, and problem solving*. Mahwah, NJ: Lawrence Erlbaum Associates.

Margolinas, C., Abboud-Blanchard, M., Bueno-Ravel, L., Douek, N., Fluckiger, A., & Gibel, P. (Eds.). (2011). *En amont et en aval des ingénieries didactiques*. Grenoble: La Pensée Sauvage.

Perrin-Glorian, M. J. (2002). Problèmes didactiques liés à l'enseignement des grandeurs. In J. L. Dorier, M. Artaud, M. Artigue, R. Berthelot, & R. Floris (Eds.), *Actes de la XIe Ecole d'été de didactique des mathématiques* (pp. 299–315). Grenoble: La Pensée Sauvage.

Perrin-Glorian, M. J. (2012). Quelques réflexions sur l'enseignement des nombres et grandeurs au long de la scolarité obligatoire. *Educmath*. Available from http://educmath.ens-lyon.fr/Educmath/dossier-manifestations/conference-nationale/contributions/conference-nationale--perrin-glorian

Rodríguez, E., Bosch, M., & Gascón, J. (2008). A networking method to compare theories: Metacognition in problem solving reformulated within the anthropological theory of the didactic. *ZDM: The International Journal on Mathematics Education, 40*, 287–301.

Ruiz-Munzón, N., Matheron, Y., Bosch, M., & Gascón, J. (2012). Autour de l'algèbre: les entiers relatifs et la modélisation algébrico-fonctionnelle. In L. Coulange, J. P. Drouhard, J. L. Dorier, & A. Robert (Coordinators), Enseignement de l'algèbre élémentaire. Bilan et perspectives (pp. 81–101). *Recherches en Didactique des Mathématiques,* Special issue.

Sierra, T. A. (2006). *Lo matemático en el diseño y análisis de organizaciones didácticas. Los sistemas de numeración y la medida de magnitudes* (Doctoral dissertation) Universidad Complutense de Madrid. Available from http://eprints.ucm.es/tesis/edu/ucm-t29075.pdf

Whitney, H. (1968). The mathematics of physical quantities, Part II: Quantity structures and dimensional analysis. *The American Mathematical Monthly, 75*(3), 227–256.

Winsløw, C., Matheron, Y., & Mercier, A. (2013). Study and research courses as an epistemological model for didactics. *Educational Studies in Mathematics, 83*(2), 267–284.

Chapter 9
The Critical Role of Task Design in Lesson Study

Toshiakira Fujii

9.1 Introduction

Lesson Study, the Japanese approach to improving classroom teaching, came to the attention of educators outside of Japan primarily through the publication of *The Teaching Gap* (Stigler & Hiebert, 1999). Though most of the book focuses on findings from the 1995 TIMSS Video Study, Chap. 7 of the book, based on work by Makoto Yoshida (Fernandez & Yoshida, 2004; Yoshida, 1999), describes Lesson Study in detail. Since then, many mathematics teachers and teacher educators have been involved in Lesson Study, and many books and research papers have been written on various aspects of Lesson Study and the typical structure of Japanese problem-solving mathematics lessons.

It is becoming clear that there are aspects of Lesson Study that are implicitly understood by Japanese teachers that have not transferred easily to other countries. For Lesson Study to be successful, these aspects should be made explicit. This chapter tries to clarify an embedded key aspect of Lesson Study: task design. In particular, the chapter discusses how a task for a lesson is designed and evaluated in the context of Lesson Study.

The author would like to thank Thomas McDougal for reading and editing numerous revisions and for his invaluable comments on this chapter

T. Fujii (✉)
Tokyo Gakugei University, Tokyo, Japan
e-mail: tfujii@u-gakugei.ac.jp

This chapter has been made open access under a CC BY-NC-ND 4.0 license. For details on rights and licenses please read the Correction https://doi.org/10.1007/978-3-319-09629-2_13

9.2 Japanese Lesson Study

The history of Lesson Study in Japan spans more than a century. For Japanese educators, Lesson Study is like air, felt everywhere because it is implemented in everyday school activities, and so natural that it can be difficult to identify the critical and important features of it.

Catherine Lewis (2011) characterized the Lesson Study cycle as follows (see also Fig. 9.1):

1. *Study curriculum and formulate goals*: Consider long-term goals for student learning and development. Study curriculum and standards; identify topic of interest
2. *Plan*: Select or revise research lesson. Write instruction plan that includes long-term goals, anticipated student thinking, data collection, model of learning trajectory, and rationale for the chosen approach
3. *Conduct the research lesson*: One team member teaches the lesson; others observe and collect data
4. *Reflect*: In a formal lesson colloquium, share data from the lesson to illuminate student learning, discrepancies in content, lesson, and unit design, and broader issues in teaching-learning. Document the cycle to consolidate and carry forward learning as well as new questions into the next cycle of lesson study (Lewis & Hurd, 2011, p. 2)

Lesson Study in Japan takes place at three different levels: the individual school level, the district or regional level, and the national level. The Lesson Study cycle is basically the same at each level and usually spans 1 year. At the school level, the typical Lesson Study cycle begins at the end of one academic school year—i.e., in February or March—when the faculty decides upon a research theme for the next school year which starts in April. Several research lessons are scheduled from, say, May to November. Each research lesson and its post-lesson discussion occupy only 1 day, but the teachers reflect on what they learned at the research lessons and usually write a booklet or long summary report by the end of the school year.

Although the Lesson Study cycle is the same at all levels, the purposes are different, and these different purposes impact the task design. National-level Lesson

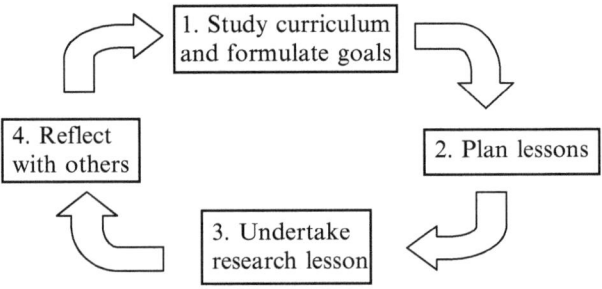

Fig. 9.1 The Lesson Study cycle (adapted from Lewis & Hurd, 2011, p. 2)

Study is usually research oriented: an academic or veteran teacher may take primary responsibility for the Lesson Study and teach the research lesson. It emphasizes the use of materials or tasks never seen before, and the goal is usually to demonstrate that the materials or task has a good mathematical and educational value. The goal of school-based Lesson Study, in contrast, is usually to accomplish the school theme or mission. School-based teams usually use familiar tasks from a textbook, perhaps with slight revision.

In any case, Japanese teachers involved in Lesson Study spend at least a few months, but sometimes more than half a year, designing a task and planning a lesson. The long-term period of planning a lesson crystallizes into a detailed lesson proposal or lesson plan. The lesson proposal includes the task for the lesson and the reason why it is used, described in detail. Therefore, we can analyze the framework of a lesson proposal to gain insights into the nature of task design in Lesson Study in Japan. We do that in the next section. Then, in the following section, we discuss the common policy of focusing a research lesson on a single task.

9.2.1 The Detailed Lesson Proposal

One of the characteristics of Japanese research lessons is that they are based on a *gakushushido-an*. *Gakushushido-an* is usually translated as "lesson plan." Because "plan" misleadingly implies a fixed script, however, the term "proposal" is better.

Japanese teachers spend a lot of energy and time crafting a lesson proposal. The contents of the typical lesson proposal give clues about the task design process. Although the details vary from school to school, and even from teacher to teacher, Lewis (2002) notes that a typical proposal for a research lesson in Japan consists of the following:

1. Name of the unit
2. Unit objectives
3. Research theme
4. Current characteristics of students
5. Learning plan for the unit, which includes the sequence of lessons in the unit and the tasks for each lesson
6. Plan for the research lesson, which includes:
   - Aims of the lesson
   - Teacher activities
   - Anticipated student thinking and activities
   - Points to notice and evaluate
   - Materials
   - Strategies
   - Major points to be evaluated
   - Copies of lesson materials (e.g., blackboard plan, student handouts, visual aids)
7. Background information and data collection forms for observers (e.g., a seating chart).

The explicit inclusion in the proposal of the tasks for each lesson in the unit indicates how important the tasks are believed to be, and that the authors of the proposal think carefully about their sequence within the unit. Their role or function in the unit or even their position in the whole curriculum are studied by teachers and clearly stated in the lesson proposal. In other words, the lesson proposal shows that task design involves the explicit linking among tasks within the unit and across units in Lesson Study. Connections among tasks are revealed also when a research lesson is implemented; Shimizu (2010) showed evidence that tasks are connected to each other within the teaching unit through the teacher's explicit efforts to link students' ideas and experiences.

To link tasks within the unit and across related units in previous and later grades, teachers need to understand the scope and sequence of the curriculum. This requires a reasonably precise curriculum. Teachers also have to consider the learning trajectory of students, considering the mathematical and educational value of each task not only for the current lesson but also for the future. The learning trajectory is a critical consideration in constructing the detailed lesson proposal and, therefore, is critical in task design.

9.2.2 Structured Problem Solving

Almost every research lesson in mathematics follows a certain form, referred to by Stigler and Hiebert (1999) as *structured problem solving*. A structured problem-solving lesson focuses on a single task and contains four phases:

1. Presenting the problem for the day (5–10 min)
2. Problem solving by the students (10–20 min)
3. Comparing and discussing (*neriage*) (10–20 min)
4. Summing up by the teacher (*matome*) (5 min)

This type of lesson imposes certain demands on the task design. In Japan, "presenting the problem" means helping students understand the context of the task and what it will mean to solve the task, but it specifically excludes any exposition by the teacher about how to solve the task. Instead, students are expected to work independently on the task for 10–20 min, during which time at least some students should solve it. The third phase, *neriage*, assumes that students will arrive at different solution methods, which are then compared and discussed for the purpose of helping all students learn new mathematics and ways of thinking. Thus, the task should be understandable by the students with minimal teacher intervention; it should be solvable by at least some students (but not too quickly), and it should lend itself to multiple strategies.

In the fourth phase, *matome*, the teacher may say something about which strategy may be the most sophisticated and why, but it should go beyond that to include comments by the teacher concerning the mathematical and educational values of the task and lesson (Fujii, 2008).

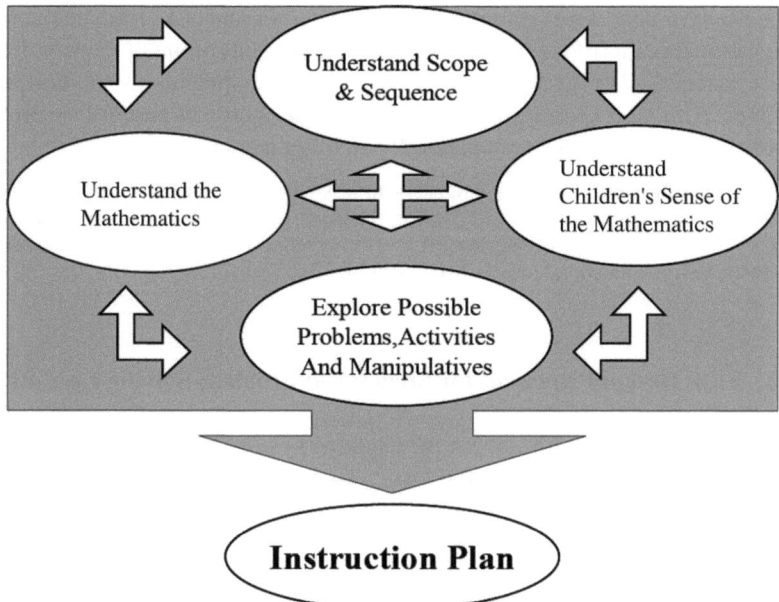

Fig. 9.2 *Kyozaikenkyu* process (adapted from Watanabe et al., 2008, p. 140)

9.2.3 Designing the Task as Part of Kyozaikenkyu

The activities or factors involved in creating a research lesson proposal can be categorized based on whether they relate primarily to (1) the curriculum, (2) the students, (3) mathematics, or (4) tasks. Ultimately, however, the lesson requires a task, and so all activities eventually focus on investigating appropriate tasks consistent with the aim of the lesson. Watanabe, Takahashi, and Yoshida (2008) identified four core steps involved in constructing an instruction plan for a lesson: (1) understand the scope and sequence; (2) understand children's mathematics; (3) understand mathematics; and (4) explore possible problems, activities, and manipulatives (Fig. 9.2). Japanese teachers routinely do this as part of preparing a detailed lesson proposal; the process is called *kyozaikenkyu*.

9.2.3.1 The Meaning of *Kyozaikenkyu*

Kyozaikenkyu literally means the study of, or research on, teaching materials. For Japanese educators, designing the task is the essential activity of *kyozaikenkyu*. The word *kyozaikenkyu* and the activity to which it refers may be unfamiliar to non-Japanese, but it is a common educational term used in academic journals in Japan. In fact, the *Journal of the Japanese Society of Mathematics Education* has a section devoted to *kyozaikenkyu*.

Kyozaikenkyu involves examining teaching materials and tasks from mathematical and educational points of view as well as from the students' point of view. Moreover, Japanese teachers also investigate ways to encourage students to solve a task by themselves. Although *kyozaikenkyu* is recognized as a critical part of Lesson Study by Japanese educators, teachers outside Japan often neglect it:

> Japanese educators place a strong emphasis on task selection, [but] this effort is largely ignored by non-Japanese adapters of Lesson Study, possibly because the effort involved may be almost invisible, in the way that 90 % of an iceberg is invisible, with all of our attention going to its visible tip. (Doig, Groves, & Fujii, 2011, p. 182)

9.2.3.2 Task Design Principles for Structured Problem-Solving Lessons

Japanese educators distinguish between "teaching how to solve the task" and "teaching mathematics through solving the task". This is why most structured problem-solving lessons focus on a single task. If chosen well, a single task allows for the important new mathematical ideas to emerge in the discussion, and additional tasks are unnecessary.

But the ultimate aim of a structured problem-solving lesson is not just to promote students' mathematical understanding or skill; the aim is also to deepen and widen their wisdom and thinking as human beings. This might sound strange or unrealistic, but consider the following problem:

> Squares are made using matchsticks as shown in the picture. When the number of squares is eight, how many matchsticks are there?

This problem lends itself to many solution strategies, including these:

(a) Consider a *C*-shape (3 match sticks) as a unit 8 times, and add the final side: 3 * 8 + 1
(b) Consider a *C*-shape as a unit 7 times, and finish with a full square: 3 * 7 + 4
(c) Draw the entire figure and count matchsticks one by one.

A comparison and discussion of these strategies might help students see the merits of strategy (a) relative to (b), because that solution (based on counting 8 *C*-shapes) is more directly related to a condition given in the problem (8 squares). If the condition is changed to 100 squares, adapting the solution is very simple. Meanwhile, strategy (c), though mathematically primitive, is nonetheless quite powerful: you are certain to arrive at the answer. So this problem makes it possible for students to gain at least two general bits of wisdom: (1) one should think about the conditions of a problem and look for a solution in terms of those conditions; and (2) even if you cannot come up with a "clever" solution, you may still be able to solve a problem through hard work. Thus, we have the following principles for an ideal task:

- It is appropriate and mathematically valuable in terms of the aims of the lesson
- It interests the students
- It is at the appropriate level of difficulty
- It can be solved in several ways
- It can apply to other mathematical problems or real-life problems
- It has a potential to elicit valuable basic wisdom

9.3 Designing Tasks Using *Kyozaikenkyu* in Lesson Study

Doig et al. (2011) illustrate four types of tasks typically used in Lesson Study: tasks that either

1. Directly address a concept
2. Develop mathematical processes
3. Are chosen based on a rigorous examination of scope and sequence
4. Address a common misconception

In this section, we focus on an example of designing a task based on a rigorous examination of scope and sequence, using *kyozaikenkyu* in Lesson Study in Japan.

9.3.1 The Topic: Subtraction with Regrouping

Japanese first-grade textbooks contain a unit concerning subtraction of one digit numbers from two digit numbers (less than 20) using regrouping. There are a total of 36 such possible subtractions: 18–9, 17–9, 17–8, 16–9, 16–8, 16–7, …, 11–3, and 11–2. This is regarded as an important area of content, and which of these 36 subtractions should be the first for children to learn is hotly contested.

9.3.1.1 Teachers Know There Are Reasons for the Numbers Used

Chapter 7 of *The Teaching Gap* follows a teacher team as they engage in Lesson Study focusing on this specific unit. Upon examining different textbooks, the teachers realize that almost all textbooks start with 13–9 or 12–9, and after reading the teacher's manuals, they understand why.

This activity, that is, investigating and studying textbooks and teachers' manuals, is a typical early step in the design task for teachers engaged in Lesson Study. Teachers may decide to use a task that is in one of the textbooks, or they may not. But they know that the specific choice of numbers influences students' solutions and

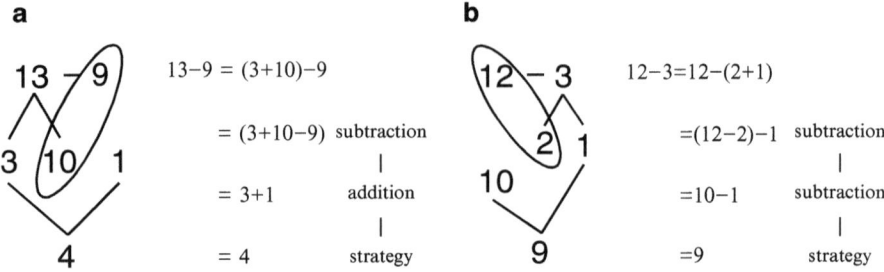

Fig. 9.3 The *subtraction–addition strategy* (**a**) and the *subtraction–subtraction strategy* (**b**)

that there are reasons for the numbers in the textbooks. Therefore, the decision to deviate from the textbooks, or not, is made carefully.

In the textbooks, the reason why 13–9 is the first subtraction problem with regrouping is that the subtrahend, 9, is close to 10. It is easy for the student to separate 13 into 10 and 3, subtract 9 from 10, and then add the difference to 3: 13–9 = (10+3)–9 = (10–9)+3. This strategy is referred to as the *subtraction–addition strategy* (see Fig. 9.3a). Consistent with this sequence of tasks for subtraction, the addition part of the textbook uses 9+4 as the first task.

In contrast, with the subtraction 12–3, because 2 and 3 are close to each other, it is easy to break 3 down to 2 and 1 and subtract them sequentially: 12–3 = 12–(2+1) = (12–2)–1. This strategy is called the *subtraction–subtraction strategy* (see Fig. 9.3b).

With a subtraction like 14–8, the two strategies, *subtraction–addition* and *subtraction–subtraction*, tend to be used by students with approximately equal likelihood. Therefore, teachers who wish to promote argumentation in their classes sometimes use a problem like 14–8 as the first task for children, while textbook companies adopt a more conservative stance based on their desire to make it easy for teachers to anticipate student responses and to be sure that there will be enough children who use the subtraction–addition strategy.

The teachers in *The Teaching Gap* decided not to use 12–9 from the textbooks because "it's not very interesting". One teacher suggested 15–8 or 15–7, and then a teacher suggested 11–6: "Because kids can conceptualize in their heads about up to the number 6 at this age, I thought we should go with numbers like 11–6". Another teacher proposed 12–7, because "one of her students, who was a low achiever, happened to have seven family members. Everyone agreed that this was a good idea" (p. 118). So, the teachers decided to use 12–7, which seemed likely to provoke the subtraction–addition and subtraction–subtraction strategies equally, allowing for a discussion that would compare the relative merits of these two methods.

Such careful scrutiny of the sequencing of tasks is unusual by Western norms. "Western observers are often astonished… by the order of presentation being the subject of so much study and debate. However, Japanese Lesson Study is frequently used to investigate sequences of tasks that are different from those traditionally used" (Doig et al., 2011, p. 194).

9.3.2 Why Teachers Begin Kyozaikenkyu *with Textbooks*

Kyozaikenkyu with textbooks is a typical activity of Japanese teachers in Lesson Study. Japan has a National Course of Study, and textbooks must be authorized by the Ministry of Education, Culture, Sports, Science and Technology. So, all textbooks treat the same topics in each grade. But the six publishing companies that publish mathematics textbooks each have their own philosophy. Therefore, it is natural for teachers to compare each textbook's treatment of the same content. Investigating the textbooks often includes study of the teacher's manuals, which include not only suggestions about how to teach the content but also the reason for teaching the content and the reason behind the textbook's approach.

9.3.3 *Exploring Possible Manipulatives*

In the case of the Lesson Study on subtraction, the teachers implemented two research lessons focused on the subtraction problem 12–7. In the first lesson, children seemed to struggle with decomposing 12 into 10 and 2. Therefore at the second lesson, teachers modified the manipulative from a single piece of tape representing 12 to two pieces: a longer tape representing 10 and a shorter piece representing 2, and scissors were available for cutting the tape representing 10. The teachers spent a lot of time coming up with the new manipulatives. This is a good example of *kyozaikenkyu* in terms of exploring possible and appropriate manipulatives. And, it is a good example of how task design in Lesson Study includes consideration of the materials and manipulatives that should be provided to students.

In the case presented here, one can see that, both in terms of choosing specific numbers to use and in terms of choosing suitable manipulatives to provide, good task design must involve considerations of likely student thinking and strategies, which is why anticipating student responses to a task is a standard part of Lesson Study.

9.3.4 *Evaluating the Task in Action*

In Lesson Study, the quality or functionality of the task is evaluated through the research lesson and the post-lesson discussion. At the research lesson, observers collect evidence from students' activities of whether the task worked well or not in terms of aims of the lesson. During the post-lesson discussion, teachers talk about the effect of the task on students' learning in accordance with the aims of the lesson by citing concrete evidence from the research lesson.

Because the role of the task and anticipated solutions are described in the lesson proposal, observers will typically watch to see if the anticipated solutions emerge or not.

The proposal for the subtraction lesson identified four approaches that students might use to subtract 7 from 12:

1. Counting–subtraction, i.e., starting with a group of 12 objects, or a group of 2 and 10 objects, and eliminating 7 objects while counting one by one
2. Supplement–addition, i.e., counting on from 7 to 12 while keeping track of the number of counts ("8, 9, 10, 11, 12")
3. Subtraction–addition (Fig. 9.3a)
4. Subtraction–subtraction (Fig. 9.3b)

At the research lesson, four children presented their methods at the blackboard to the whole class. The four solutions included two that were anticipated: counting–subtraction and subtraction–subtraction. The supplement–addition method and the subtraction–addition method were not presented, but two unanticipated methods were presented. One was to subtract 2 from 12 to 7, and then subtract 5 from 10. This method could be expressed (not for first-graders) as $12-7=(12-2)-(7-2)=10-5=5$. Only one child used this strategy in the class, although the whole class eventually seemed to understand it. The other was to partition the number 12 as 5, 5, and 2. Then, as the student explained, "because seven is five and two, I moved the five and two of the number twelve." Only one child used this strategy although "many of the students said that her solution was good" (Fernandez & Yoshida, 2004, p. 165).

Taking a high-level view of the discussion that followed the research lesson, the school faculty raised 23 points of discussion. Ten of them, or about half, concerned the task:

- Two concerned the specific numbers used in the task, such as "12–8 or 13–7 would be better"
- Four concerned the manipulatives, such as "if all 12 tiles had been lined up in one straight line, students might have cut the 10-strip into 7 and 3 to use subtraction–addition strategy" (p. 176)
- The other four were (1) the way the problem was presented and how to present word problems in general, (2) the reason why only one child—who was not asked to present his solution—used the subtraction–addition strategy, (3) how the handout and manipulatives had been improved, and 4) why the supplement–addition method was unlikely to emerge in the lesson

These 10 points of discussion provide examples of how a task is evaluated in Lesson Study.

Taking a closer look at the post-lesson discussion, the teacher who implemented the research lesson with the problem 12–7 confessed that she was very disappointed because she could not get a variety of student solutions on the board; in particular, she had hoped to see the subtraction–addition method presented (Fernandez & Yoshida, 2004, p. 171). These comments at the post-lesson discussion show that a task cannot be evaluated solely on its mathematical merits, but should be judged based on its actual effects on student thinking and learning. This is characteristic of task design and task evaluation in the context of Lesson Study in Japan.

The faculty did not discuss the unexpected solution method that used the subtraction rule (i.e., $12-7=(12-2)-(7-2)$). This was reasonable, because the focus of the lesson was on using regrouping. But the rule is interesting: another useful variant is $(12+3)-(7+3)=15-10=5$. The task $12-7$ created an opportunity to learn the rule, and the teachers could have discussed the possibility of including the subtraction rule in the elementary school curriculum.

9.4 Discussion

The case study from *The Teaching Gap* points to two important features of task design in Lesson Study. First, task design involves anticipating students' solutions. Second, the task is evaluated in the post-lesson discussion based on concrete evidence collected during the lesson.

9.4.1 Task Design in Lesson Study Always Involves Anticipating Students' Solutions

For the subtraction lesson, the teachers seriously considered which numbers to use, because they know that the choice of numbers will affect which strategies the students will use when solving the task. Furthermore, the teachers recognized that each strategy has both mathematical and educational values.

Such close attention to the specific numbers does not mean that teachers are sticking to a concrete level of thinking and encouraging students to think about things concretely. On the contrary, teachers consider the general aspect of the number—its *quasi-variable* aspects. A *quasi-variable* is a number deliberately used in a general way so that it serves as a representative of many numbers, just as a variable would (Fujii & Stephens, 2001, 2008). Numbers are often chosen, then, based on their quasi-variable power, or how well they demonstrate a general truth.

For instance, the tasks 13–9 and 12–9 are likely to lead to the subtraction–addition strategy; thus they are not mere calculation problems, but lead to a particular general procedure for subtracting with regrouping in the base-ten system. Appreciating the base-ten system and place value notation system and its benefit for calculation is more important than getting an answer and gaining skill at calculating 13–9. To get such appreciation, however, students need to see alternative strategies, such as the subtraction–subtraction strategy or counting down, neither of which depends on the base-ten notation system. Therefore, a structured problem-solving lesson includes a *neriage* phase for students to compare or experience friends' methods and discuss similarities and differences among strategies in a whole class setting. Thus, when designing the task, there needs to be consideration of whether the task will elicit the alternative approaches needed for an effective *neriage*.

9.4.2 Task Design in Lesson Study Goes with Task Evaluation

The second feature of task design in the context of Lesson Study is that task evaluation is an inherent part of the process, wrapped into the evaluation of the lesson. The task is not judged based on some abstract determination about whether it is good or not for teaching a certain skill or concept, but based on concrete evidence from the lesson about how the students responded to it.

That evaluation often goes beyond the specific content of the lesson. For example, at the subtraction lesson, teachers discussed whether the task was appropriate or not and whether the manipulatives functioned well or not. But they also discussed more general issues. In fact, in 13 out of the 23 points of discussion, teachers discussed how to develop guidelines for fostering students' presentation skills. This was a concern not just for first-graders but all students. Sample comments included: "The skill like speaking in public and explaining what they think in a logical manner are the important things"; and "These skills are needed to do well beyond the subject of math" (Fernandez & Yoshida, 2004, p. 181). These show that the teachers evaluate a lesson and task in terms of a broader educational aim.

The final commentator also addressed broader educational values. He said that in order for students to pit solution strategies against each other, they must be given the opportunity to evaluate them based on their own attempts to solve the problem. He pointed out that there were no comments from students such as "Teacher, I found that this method is more convenient," or "This method is much faster," because each student had experienced only one way to solve the problem (p. 186). He gave specific examples from the lesson in proposing improvements to the lesson. But he seemed to be suggesting that this activity could make students think about classroom values, such as the importance of listening carefully to friends' opinions, of expressing ideas clearly to friends, of moving beyond "wrong or correct answers", of not underestimating friends' ideas, etc. Here is a good example from Lesson Study in Japan of how structured problem-solving can be a context in which to nurture students as human beings.

The final commentator explicitly addressed broad educational values in the very beginning of his comments. "He urged teachers to think carefully about what were the most important 'skills for living' that students should be learning from their mathematics instruction" (p. 182). Using as an example the formula for finding the area of a trapezoid, he said that "teachers should help students realize that moving from complicated to more simple forms is a convenient and a clever thing to do" (p. 183). This is an example of how Japanese Lesson Study concerns educational values.

This notion is related to the structured problem-solving type of lesson. A common misconception about such lessons is that solving the task is the main point. Such misconception leads to a focus on goals such as "students can do X" or "students understand X." But a structured problem-solving lesson is not just about finding the solution to a problem. It is well and good that students can do X, but X should

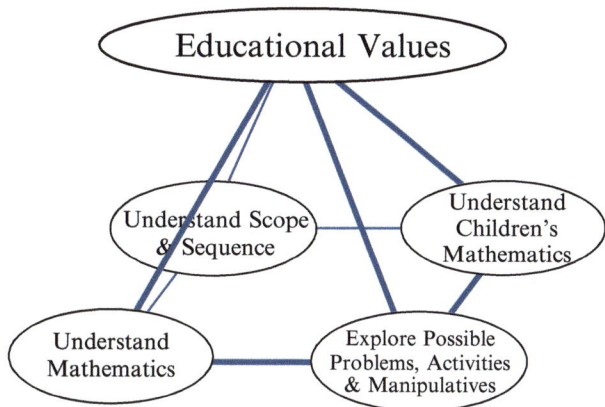

Fig. 9.4 A pyramid model of *kyozaikenkyu*, showing how the process aims to serve broad educational values

contain some value, and what that value is needs to be considered. To identify the educational values, the final commentator urged the teachers to do *kyozaikenkyu*.

Thus, we see that the flat model of *kyozaikenkyu* from Watanabe et al. (2008) needs to be extended to a three-dimensional model as shown in Fig. 9.4. In this revision of the model, the four goals of *kyozaikenkyu* collectively serve a larger goal, which is to develop tasks and lessons that bring broad educational values to life in the classroom.

It is hard to actually implement an ideal lesson with a rich task and a discussion that addresses broad educational values. Accomplishing this is, therefore, a lifelong goal of teachers. Lesson Study, an ongoing activity of Japanese teachers, both helps them develop such lessons and provides a testing ground for teachers.

9.4.3 Conclusion

Task design is the essential activity of *kyozaikenkyu*, which for Japanese educators is a critical part of Lesson Study. There are two sides to task design: anticipating students' solutions when writing the lesson proposal and evaluating the task during the post-lesson discussion in light of the actual students' responses in the research lesson.

We hope that by making the task design activity more visible, we can help teachers understand the *kyozaikenkyu* process more profoundly. Designing tasks as part of *kyozaikenkyu* will strengthen teachers' content knowledge, improve instruction, and deepen their understanding of Lesson Study itself.

References

Doig, B., Groves, S., & Fujii, T. (2011). The critical role of task development in lesson study. In L. C. Hart, A. S. Alston, & A. Murata (Eds.), *Lesson study research and practice in mathematics education* (pp. 181–199). Dordrecht: Springer.

Fernandez, C., & Yoshida, M. (2004). *Lesson study: A Japanese approach to improving mathematics teaching and learning*. New York: Routledge, Taylor & Francis Group.

Fujii, T. (2008). *Knowledge for teaching mathematics*. Plenary talk at the 11th International Congress on Mathematical Education, Monterrey, Mexico, July 6–13.

Fujii, T., & Stephens, M. (2001). Fostering an understanding of algebraic generalisation through numerical expressions: The role of quasi-variables. In H. Chick, K. Stacey, J. Vincent, & J. Vincent (Eds.), *Proceedings of the 12th ICMI Study Conference: The Future of the Teaching and Learning of Algebra* (pp. 258–264). Melbourne: University of Melbourne.

Fujii, T., & Stephens, M. (2008). Using number sentences to introduce the idea of variable. In C. Greenes & R. Rubenstein (Eds.), *Algebra and algebraic thinking in school mathematics* (pp. 127–140). Reston, VA: National Council of Teachers of Mathematics.

Lewis, C. (2002). *Lesson study: A handbook of teacher-led instructional change*. Philadelphia, PA: Research for Better Schools, Inc.

Lewis, C., & Hurd, J. (2011). *Lesson study step by step: How teacher learning communities improve instruction*. Portsmouth, NH: Heinemann.

Shimizu, Y. (2010). A task-specific analysis of explicit linking in the lesson sequences in three Japanese mathematics classrooms. In Y. Shimizu, B. Kaur, R. Huang, & D. Clarke (Eds.), *Mathematical tasks in classrooms around the world* (pp. 87–101). Rotterdam: Sense Publishers.

Stigler, J., & Hiebert, J. (1999). *The teaching gap: Best ideas from the world's teachers for improving education in the classroom*. New York: The Free Press.

Watanabe, T., Takahashi, A., & Yoshida, M. (2008). Kyozaikenkyu: A critical step for conducting effective lesson study and beyond. In F. Arbaugh & P. M. Taylor (Eds.), *Inquiry into mathematics teacher education* (AMTE monograph series, Vol. 5, pp. 131–142). San Diego, CA: Association of Mathematics Teacher Educators.

Yoshida, M. (1999). *Lesson study: A case study of a Japanese approach to improving instruction through school-based teacher development*. Unpublished Ph.D. dissertation, Department of Education, University of Chicago.

Chapter 10
There Is, Probably, No Need for This Presentation

Jan de Lange

10.1 Introduction

In the spring of 2013, *TIME* magazine published an article about the *Millennial* youth. The author, a well-known columnist Joel Stein, observed that the millennials are a generation mostly of teens and 20-somethings known for constantly holding up cameras, taking pictures of themselves, and posting them online. They are "narcissistic, overconfident, entitled, and lazy". Then comes a deep sigh: "Now, imagine being used to that technology your whole life… and having to sit through Algebra" (Stein, 2013).

Indeed, if they have to sit through algebra, is there any way that designers can connect the millennials with the learning of mathematics? Was all that we, the designers, did in the recent past without any merit? The reader will understand the title I chose for this presentation, especially if we accept Chris Schunn's observation: "The educational design community has no communal mechanisms for codifying craft knowledge" (Schunn, 2008). So, not only does the designer have a problem in dealing with the present youth, but we, as designers, do not even have tools and knowledge to describe our craft knowledge. You think that is bad news? Wait a second: we have no communal mechanisms for codifying craft knowledge, but that has an advantage if we take Collopy seriously (which I do): "Codifying design thinking threatens its central value of flexibility" (Collopy, 2009).

J. de Lange (✉)
Freudenthal Institute, University of Utrecht, Utrecht, The Netherlands
e-mail: J.deLange@uu.nl

This chapter has been made open access under a CC BY-NC-ND 4.0 license. For details on rights and licenses please read the Correction https://doi.org/10.1007/978-3-319-09629-2_13

10.1.1 *Educational Design: Is There a (Need for a) System?*

So, indeed the title of this chapter needs to be taken seriously; is there a *need* for a system in educational design? By my accepting the invitation to write this chapter, you might conclude (correctly) that I think there is indeed some need. Assuming there is research evidence to suggest we are experienced and proven designers, how do we help young and promising designers to become better, based on information that might contribute in the future to some kind of a system or framework?

Looking through the literature, we can find some definitions of educational design. One of the more helpful definitions is:

Design

- To create and execute according to plan
- To conceive and plan out in the mind
- Is very ego involving (own mark)
- To make drawing or sketch: process of design (Merriam Webster, 2008)

To *create* and *execute* according to plan seems like a trivial remark, but, reflecting on my past experiences, it is easy to interpret this as a warning: creative processes often tend to go on indefinitely; there is always room for improvement. It might even be true that *especially* in educational design, this is a problem, because if you see design as a recursive process that has to be validated by classroom experiences or by research, sometimes one needs some courage to decide that the plan has finally been executed.

The second point feels more in line with what one would expect. Thought experiments, dreams, and wandering thoughts, all play out in the mind before you ever start the actual design. The start of the concrete design seems the sublimation of the thought and mind process, and indeed, it feels as if you really are executing a plan, for a while at least.

One would like to deny that design is or has to be "ego involving" as suggested in the definition. But indeed there is no need to deny this. A designer is always trying to create something beautiful, and it would be nice if people recognize it as something that has your mark all over the place. There are some examples around of innovative or challenging educational design that can easily be attributed to a particular creator.

During the process, there are several phases for which drawing and sketching or, more generally, *visualization* is essential. Imagination can enable design for longitudinal development over time, for identification of different levels of competencies, for visualizing concepts differently than just by words, and many more aspects.

So, yes it is possible to reflect on your own practices using definitions, but many aspects are missing from this short list, and maybe such a general list is not at all helpful to future designers. It is therefore no wonder that Schunn poses the question: "Is there a system to the madness of design?" (Schunn, 2008). We must also admit that it is hard to reflect on your own system (if there is one), because often people may imagine they have direct access to their mental processes, but they do not

(Anderson, Lebiere, & Lovett, 1998). So, if possible at all, the description of your own mental processes is very difficult, and combining this with the reality as observed by Schoenfeld: "Educational designers have few incentives to codify their reflections and present them publicly" (Schoenfeld, 2009). We see that, indeed, writing about the design process might be an almost impossible task, given these observations. But the challenge is even bigger than merely being descriptive of method. As Schön observes:

> [Design takes place in] the swampy lowlands, problems are messy. The irony ... is that the problems of the high ground [theory] tend to be relatively unimportant to individuals or society at large, ... while in the swamp lie the problems of greatest human concern. The practitioner is confronted with a choice. Shall he remain on the high ground where he can solve relatively unimportant problems ... or shall he descend to the swamp of important problems where he cannot be rigorous in any way he knows how to describe? (1995, p. 27)

Given all these observations, one needs to be careful when writing anything on the design process, and hence the reader needs to take the following ideas not too seriously. I cannot access the mental processes I follow, and there may be no system to my madness. I have to make difficult choices between the high and low ground, and the whole context of talking about task design in the abstract is confusing, but I'll try anyway.

10.2 Personal Reflections on the Design Process

10.2.1 Slow Design

10.2.1.1 Slow Design: The Principles

It is tempting to start with a design process that seems to have disappeared: a process based on developmental research that takes high quality and evidence of effectiveness as guiding principles. It takes a lot of effort and time and needs serious planning and execution, goals, an understanding when these goals have been reached, assessments, and more. Therefore, it is generally considered too expensive in time, money, and human effort.

Wouldn't it be interesting to compare the costs of a really innovative and evidence-based "programme" against the costs of a "programme" constructed under high pressure with marginal evidence that it appears to serve the purpose and goals but that will sell very well because of smart and professional marketing? The costs of a market-based process might in the end be much higher in terms of the final product: ineffective in enabling real improvement, not meeting any educational design "standard", and creating a need for "even better" materials. A *leading principle* in *slow* design is that we use integrated partnerships: a researcher/designer partnership and a researcher/designer with teacher/student partnership. To speak a different language, we take design from the high ground into the mess of the swamp, and we try to be of relevance for all levels.

Next, we have to describe a plan. What I say about planning is a description of a process that is the result of a wandering brain and many years of experience. Of course, it is just a sketch and should not be taken too seriously, but can act as a guide through the mess of educational design. The designer needs to be involved in every aspect of process.

10.2.1.2 Slow Design: The Process

- Select your subject of choice (especially as a junior designer), its duration, and level. As a designer, it helps a lot if you have an affection with the subject or concept you are supposed to design material for. It will not only help you gain confidence, but will offer you opportunities to leave the often-travelled road; you will be able to make your own mark, and chances are high that you might really "invent" something unusual. Try to keep it a shorter "unit"; it is easier to make a coherent and convergent product with a tight focus.
- Design a (mental) sketch of flow and educational/didactical vision. As I will illustrate elsewhere, the flow of the unit must be sketched. It will give an order, or structure, to the design. Of course, in order to be able to do this one needs a didactical view and educational vision. It makes quite a difference if your philosophy is "show and tell" and lots of practice, or if you start with an exploration, in the vein of this memorandum from the American Mathematical Monthly (AMM): "To know mathematics means to be able to do mathematics... and what may be the most important activity, to recognize a mathematical concept in, or to extract it from, a given concrete situation" (Ahlfors et al., 1962, p. 8). The informed reader will not be surprised that a designer sees many more possibilities in the latter approach; the starting point of the design process in this case is the whole world and the degrees of freedom are high, although the actual design process will be very challenging.
- Use intuition. This may be a trivial remark or maybe not. Intuition is a very underestimated aspect of design, in general, and educational design, in particular. Intuition can suggest where to start if you have a clear understanding of the concept. Intuition can suggest how to get from that starting point to the concept in a challenging and motivating way. Intuition will help identify connections to other fields and concepts. We will return to intuition somewhat later.
- Choice of concept: if we take the recommendation of the AMM seriously, it is clear that if you want to extract a mathematical concept from a given concrete situation, the situation has to offer the opportunity for *conceptual mathematization*—giving meaning to the mathematical concept (de Lange, 1987).
- Choice of context: if the context is used to apply learned concepts, we need a high degree of authenticity—no artificial, camouflage, or fake contexts (de Lange, 1987).
- Look for inspiration in a "random" search library using associative thinking. *Googling* is a well-established practice for designers of all sorts. My experiences are such that I first look for "Images" which can stimulate associative thinking,

an approach that has led to many new and surprising discoveries. In the not too distant past, this activity took place in a physical library. Browsing there in a random part of the scientific library, maybe just looking at the titles of the books, can be extremely rewarding. A designer often needs only one page. Sometimes, you get much more from books; I recently re-found in a dark corner the book *Scale Effects in Animal Locomotion* (Pedley, 1977), a gem for mathematical designers of all ages.
- Refine your initial design. Look for continuity, for balance, and also leave your "own mark". Check also if you have used visualization, where appropriate. Is the text straightforward? Are there any "gender" problems?

This is the end of the first phase of my design process. Now it is time to leave your comfort zone and meet the real world. The main problem now is to not be too defensive; take all comments from colleagues and others seriously, and avoid taking a final position until you have heard everything, but the design decisions are yours.

10.2.1.3 Meet the Real World

- Seek a "real" discussion with experts of all kinds, even "real" mathematicians, and have your design ripped apart—a very desirable but uncommon stage in the process. To become a successful designer, the real world starts with the critical remarks of your colleagues. In many institutions, these discussions are often very "friendly", because your intention is to remain coffee-mates with those colleagues. The challenge, therefore, is to find a real and honest discussion, based on mutual respect. These exchanges can be harsh at times. Freudenthal was well known for his direct remarks: at one meeting in the 1970s, he concluded, after 2 hours of "friendly" interaction: "From a didactical point of view this is all rubbish!" A good designer loves discussions like this, although one needs some strength of character and self-esteem, if possible.
- Write a new version that is about ready for classroom experiments. Given the sharp exchanges before you reach this stage, this revision is not always easy to carry out. If you have too much respect for senior members, you might follow the wrong track. If you think you are right anyway, you are not learning anything.

As we all know, there are many real worlds all with their own rules and culture. The interaction with experts of course is not the "real" real world. The ultimate experts are in, and in front of, the classroom. We meet the teachers in the next phase:

- Discuss the design with experienced teachers; change your design if you think it would be a real improvement. Teachers should be ready to try out the design; prepare them on critical points in the design.
- Revise, and have a teacher teach using your design. NEVER teach it yourself. One of the serious errors designers may make is to teach the design themselves. Of course, there is at least a fair chance that the lesson will go smoothly; the teacher (designer) is really teaching the intended curriculum. Every now and

then you can do this if you are mainly looking for what is possible and do not have students of your own, but this does not count as a real try-out. As a designer, you are designing for other teachers, not for yourself.
- Observe authentic classroom activities, without the use of video; you will observe much more, in more detail, and with the possibility of interventions on the spot. You get feedback on the micro level. You walk around and check students' reactions and understanding. Look at their notes, and when seeing mistakes, ask questions in order to get a feel for the reason for the mistake. Try to understand why they do what they do.
- Make notes for revision; take discrepancies between "intended", "implemented", and "achieved" learning seriously. You do not design by just making a lesson, a module, or a curriculum. When you do that, you have an "image" of how it has to play out in the classroom, and, more often than not, it will play out quite differently in the real classroom.
- Concentrate in the first place on essential conceptual development, not on details. This is sometimes hard to do; your attitude is probably to "help" the students whenever they make a mistake. The focus should be on the conceptual development: do you see the intended "learning trajectory" materialize in the practice? Or does the teacher miss the longer developmental lines?

It will be no surprise that the next steps are:

- Start the cycle again in a different classroom, if necessary after redesign. Try it out in a different school with visits and observations by designer and other experts.
- And do not forget the importance of designing assessments, both formative and summative, and indicate the opportunities for excellent feedback (Black & Wiliam, 1998).

10.2.1.4 An Example of Slow Design

Logarithms are often considered a difficult subject for students to learn. In the philosophy of Realistic Mathematics Education, contexts should be used in order to develop mathematical concepts. This process is called *conceptual mathematization*. The context in this example is the "growth" of water plants that spread over the surface; a graph is essential to answer the questions posed (Fig. 10.1).

A student's answer on the first question is: "after about 4.3 weeks", reading the graph. The next question is about the time needed to grow 40 m^2 of plants. This seems a bit challenging as the value 40 is *not* in the graph. In answering this question, it is essential to understand that one can find the answer *without* using the graph: the fact that in 1 week the total area is *doubled*, in relation to the previous answer, yields: "it takes one more week to grow to 40 m^2. Thus, $4.3 + 1 = 5.3$ weeks". Following similar reasoning gives that it takes 3.3 weeks to grow 10 m^2. It was striking to observe in the trials that many teachers were hindered by their "context isolated" readily available knowledge of logarithms. To many students, this "doubling in 1 week" was obvious. The design process took this difference of perception into account, and we introduce the idea of logarithms using a definition arising from the context (Fig. 10.2).

10 There Is, Probably, No Need for This Presentation

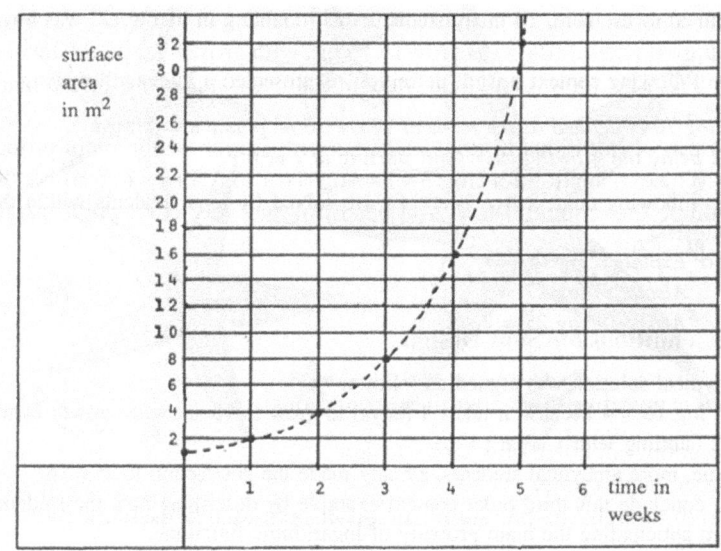

1. Estimate, using the graph, after how many days there are about 20 m² of plants.
2. Estimate, using the graph, after how many days there are about 40 m² of plants.

Fig. 10.1 Growth of water plants (de Lange, 1978a, p.71)

Next questions: Time to grow 40 m², 80 m²? 10 m² ?

²log 10 is defined as the moment where 10 m² plants are formed, 2 being the growth factor (and starting with 1 m²).

Explain:

²log 16 = 4   ³log 27 = 3   ⁵log 25 = 2

Explain: ²log 3 + 1 = ²log 6
                  ²log 7 + 1 = ²log 14
                  ²log 6 + ²log 2 = ²log 12

Fig. 10.2 Contextual definition (using the notation common in the Netherlands) (de Lange, 1978a, p.71)

A typical answer to the first question in Fig. 10.2 would be: "$^2\log 16$ (or $\log_2 16$) = 4 because it takes 4 (days) to grow to 16 m^2, with growth factor 2 (starting with 1 m^2 at $t=0$)". The more analytical students already make the formalization to "because $2^4 = 16$".

The last trio of questions in Fig. 10.2 is in preparation of the main property of logarithms. The explanation for $^2\log 3 + 1 = {}^2\log 6$ (or $\log_2 3 + 1 = \log_2 6$) is: "It takes 1 day to double the amount 3 to 6"—a beautiful and very conceptual insight that many students achieve.

10.2.1.5 Conditions for Slow Design

Slow design is possible only under these conditions:

- Some freedom of choice of what to design.
- Lots of freedom in time.
- Freedom of thought (no pressure from publishers, standards, etc.).
- Freedom to explore.
- There is restriction arising from your design philosophy; in my personal case, this is Realistic Mathematics Education (RME).
- Balance restrictions with freedom; in my case, it is a free interpretation of RME.

10.2.2 Fast Design

In the highly commercialized educational design community, a different set of rules applies. In the first place, we see a higher degree of separation: the designer designs, the researcher (if any) does the classroom observation, the technology part is outsourced to specialists, the teacher gets some preparation or a guidebook and then teaches, and the student does what the teacher asks. The main problems of this kind of process are obvious: a lack of communication between all the different components of the design, the designed coherence and convergence are under pressure, the designer is unaware of what happens with his/her design further on in the design cycle, and the risk is high that he/she does not feel any responsibility for the end product. In this mode of design, the design is not a process but a product. At the very least, opportunities for the designer to learn from feedback are missing. A typical situation is that a designer may have to deal with some national standards. Most curriculum standards have "mathematical practices" of one kind or another written in them, but the description usually leaves a lot to be desired from a designer's point of view, such as assumptions that a phrase like "problem-solving" has one agreed meaning.

Fast design needs products fast and is very market-driven. Therefore, the appearance of the design can outweigh the quality of content. Many books use full colour illustrations that have hardly any connection with the mathematics. Fast design makes it hard for an educational designer to have a high degree of satisfaction about the more "creative" and "out of the box" aspects of designing—the craft and art of design.

10.3 Design as Art

Design can be seen as art. Peter Hilton made an interesting observation about the relation between mathematics and applied mathematics. The essence is that mathematics is a science and applied mathematics is an art. His argument can be used, in my opinion, for the educational design in mathematics education. Hilton states: "Experience, intuition, inspiration makes it art" (Hilton, 1976). I use his words: "Educational design theory incorporates a systemic of knowledge and involves cumulative reasoning and understanding, it is to that extent a science. And since design and construction involves choices which must be made on the basis of experience, intuition, and even inspiration, it partakes the quality of art" (de Lange, 2012, inspired by Peter Hilton, 1976, p. 85).

It seems almost trivial to say that experience plays a major role in design and inspiration as well. We will not pay any more detailed attention to these two aspects. But the third component, and together with inspiration, the most important variable making design in education an art, is *intuition*. One can almost "feel" that many professional educational designers stay away, at a safe distance, from a discussion about the role of intuition. It is not easy to make it look more scientific, which is seen as a conditio sine qua non to write a refereed article; without a refereed article, the view of the designer is not seen to be relevant in the world of research. But one can find some interesting remarks about the role of intuition in educational design. And as the kernel of this presentation is a *reflection on my own practices*, my reflection was indeed helped by thinking about the role of intuition a bit more than casually.

10.3.1 The Role of Intuition

Rapid cognition is the sort of snap decision-making performed without thinking about how one is thinking. It takes place faster and often more accurately than the logical part of the brain can manage. Intuition is "rapid cognition" (as beautifully described in the book *Blink*, by Malcolm Gladwell, 2005). The secret is knowing which information to discard and which to keep. The brain is able to perform that work unconsciously. It can hardly come as a surprise that experience has something to do with intuition. According to Duggan (2007), brain science tells us there are three kinds of intuition: *ordinary*, *expert*, and *strategic*. In my own words, *ordinary* intuition is just a feeling, a gut instinct. *Expert* intuition is snap judgements, when you instantly recognize something familiar, the way a tennis pro knows where the ball will go from the arc and speed of the opponent's racket. The third kind, *strategic intuition*, is not a vague feeling, like ordinary intuition. Strategic intuition is a clear thought. And it's not fast, like expert intuition. It's slow. That flash of insight you had last night might solve a problem that's been on your mind for a month. And it doesn't happen in familiar situations, like a tennis match. Strategic intuition works in new situations. That's when you need it most.

Another way to make a clear distinction between the three kinds of intuition is that in ordinary intuition, it is just feeling, not thinking. In expert intuition, you use past experiences to make instant decisions, and past experiences make you an expert in this kind of intuition. Strategic intuition can be characterized by thinking, not feeling, informed by experience, leading to a deep insight that connects seemingly unrelated knowledge to create new insight and "gestalts" in unfamiliar, new, situations.

Strategic intuition is, in my opinion, a crucial factor in the art of educational design in mathematics and science. The problem seems to be that for obtaining strategic intuition, the crucial factor is time, especially reflective time. Reflective time is clearly available in slow design, both in the short-term and in the longer run. If we try to find lessons for young future designers from experienced designers who are the same age as the Rolling Stones, the problem is the same as for these rock and rollers: they can be seen as a thing of the past that refuses to fade away, or you can see their vitality and drive as an example for the future. The questions for designers therefore are: "how can you retain your drive, use fluid intelligence, keep on track with new media, use reflection in different time frames, try to beat your previous 'original' ideas (inspiration), and realize the roles of intuition?" Educational design is indeed art.

10.4 Design Examples

10.4.1 Central Concept Design

As the title suggests, a central mathematical concept or subject is the guiding principle in my design approach. In this example, a whole unit is designed as a sequence of tasks around the right triangle. It begins quite informally: "When you are standing on the rim of the Grand Canyon can you see the Colorado River or not, and explain why?" The concept of vision lines is not only explored, but also more or less formalized. Vision lines are also at the heart of the concepts of blind angle or blind area and the shadows.

Ladders have critical angles or steepness. So, the shift from the Canyon towards the new context of ladders is very natural, especially if combined with more on shadows. The right triangle, the object that connects all the different contexts, comes in handy again as the glide ratio, which is in a more formal sense the tangent. Figure 10.3 shows different contexts for the same concept.

10.4.2 Central Context Design

One design format that can be quite appealing is to choose, for use over a period of time, the same context with a variety of mathematical concepts that fit more or less authentically. It sometimes is a matter of "intuition" how far you can go in this

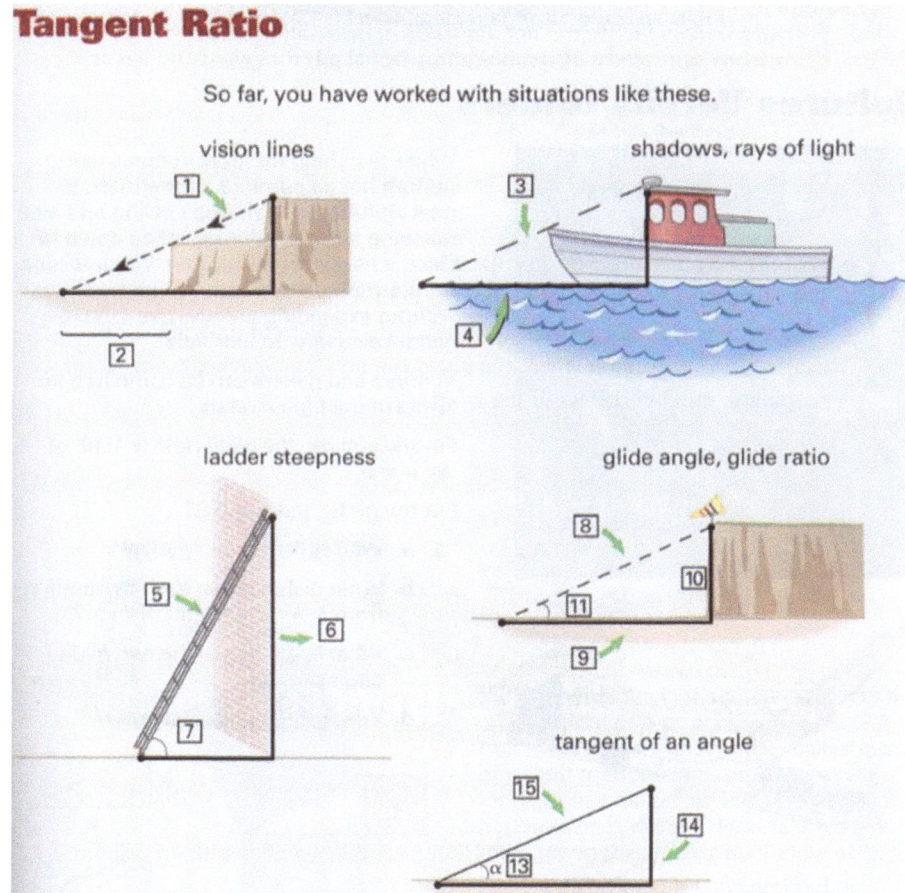

Fig. 10.3 Different contexts; same concept (Feijs et al., 2006, p. 45. Copyright Encyclopedia Britannica 2006)

respect. It is also possible to go too far. In the late 1970s, a request reached me from a teacher asking if I could "do something" about the problem of trigonometric ratios somewhere in the upper middle school region. She certainly knew how to challenge a designer, as this designer was a pilot as well. So in a relatively short time, a unit for a couple of weeks was designed about flying through trigonometry (see Fig. 10.4). Later in this chapter, you can see a very first draft of a unit, originally written with pencil and piloted in this format (Fig. 10.7). Within the flying context, the glide ratio was the central concept. The trial was declared a success, and a somewhat better format was used in the continuation of the process.

But the teacher was not satisfied; we could do better and treat another mathematical concept within this context. So a new design had vectors as an additional core concept—a subject that is easy to understand for young children, but often

30. a. Make a cross in the middle of a blank page. This represents an airport. Draw the region where the plane can be launched from and still land at the airport, given:

   (scale: 10 km = 1 cm)
   glide ratio 1:30
   altitude 2000 meters
   no wind.

   b. What is the glide angle of this plane?
   c. The plane flies at a speed of 60 km per hour. One day there is a bit of wind coming from the west at 20 km/h. Indicate in your drawing (started in a) the region where the plane can now be launched from, keeping in mind the effect of the wind.

31.

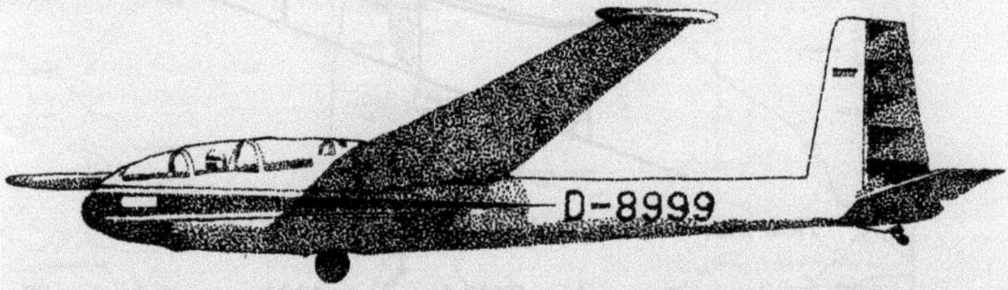

Blanik L-13

In a book on airplanes it says, "The Blanik L-13 has a glide ratio of 5%."
a. Explain what this means.
b. Compute the glide angle.

Fig. 10.4 Flying as a context for trigonometry (de Lange, 1978b, p. 20)

neglected in curricula. Eventually, we ended up with three lesson series about vectors (Fig. 10.5), triangles in the navigation context, and the definitions of trigonometric ratios.

Another example is the context of archaeology. In a Middle School Project (a collaboration between the Freudenthal Institute and The University of Wisconsin, funded by the US National Science Foundation), a unit was designed around this very intriguing context. The title of the unit is *Digging Numbers*. It started with the Mayan number system, but it dug much deeper. Classification and seriation of objects found in excavations are the main concepts that are treated with a high degree of authenticity. Two examples give the reader an impression of the task (Fig. 10.6).

Math Vectors

A **vector** has a *direction* and a *magnitude* (length).

Example: $\vec{a} = 140°/30$.
140° indicates the direction, 30 the magnitude.

1. Using grid paper, draw the vector $\vec{a} = 53°/50$. (1 grid unit equals ten km.)

2. Assuming that $\vec{a}$ represents a flight starting at airport A, how many km east of the airport does the airplane travel? How many km north of the airport?

3. Answer the same questions for the vector $\vec{b} = 14°/41$.

4. If the airplane travels a path equal to $\vec{a} + \vec{b}$, how far to the east and north does it travel?

A flight can also be described in the following way:

5. a. Draw the vector $\vec{c} = 50$ east/ 30 north.
   b. Add to $\vec{c}$ the following vector: $\vec{d} = 20$ east/ 40 north.
   c. Write the solution, using the same notation: $\vec{c} + \vec{d}$.

You can also write vectors such as "$\vec{c} = 50$ east/ 30 north" as "$\vec{c} = \begin{pmatrix} 50E \\ 30N \end{pmatrix}$."

6. Draw $\vec{e} = \begin{pmatrix} 20E \\ 40N \end{pmatrix}$ and $\vec{f} = \begin{pmatrix} 40W \\ 10N \end{pmatrix}$. Draw $\vec{e} + \vec{f}$ and write the sum.

Fig. 10.5 Flying as a context for vectors (de Lange, 1978b, p. 55)

10.4.3 Design over Time

It is quite helpful if the designer can reinvent himself or herself continuously. Even if a design proves to be quite successful, the challenge still exists to continue the development. Flying, for example, in the widest sense is indeed an excellent subject because you can easily go beyond the (sail)planes: flying squirrels and birds interest a whole group of different audiences and broaden the view of the students on the real world (Fig. 10.7). Does a sailplane soar better than an albatross?

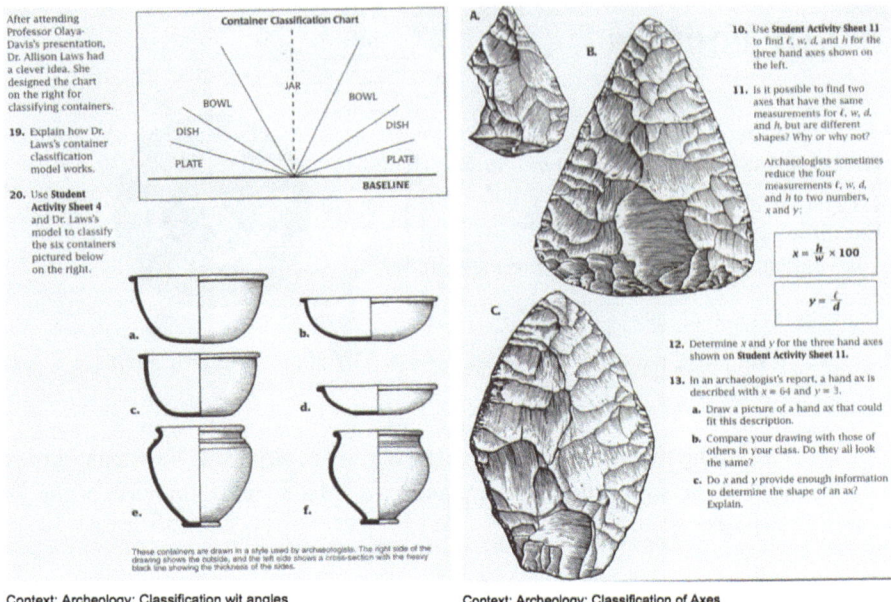

Fig. 10.6 Archaeology as context (de Lange et al., 2003, pp. 50, 84. Copyright Encyclopedia Britannica 2003)

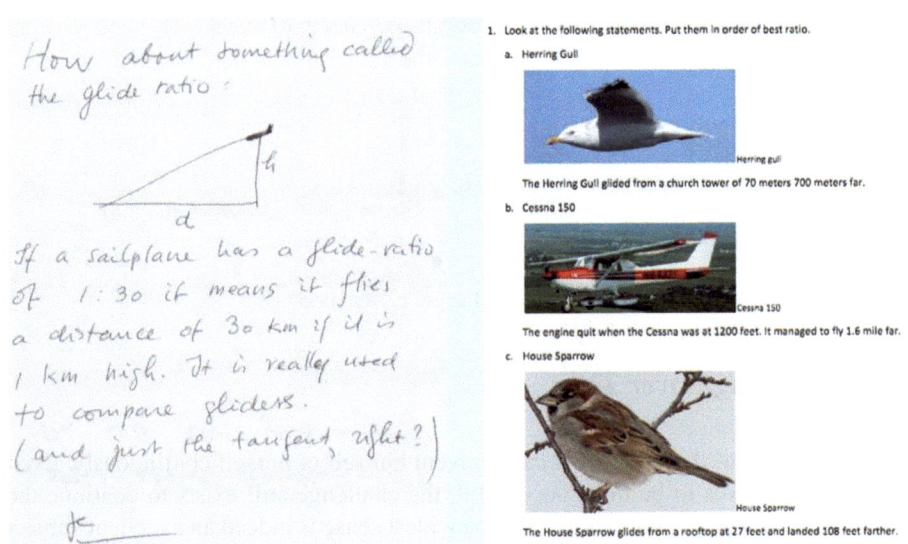

Fig. 10.7 Excerpts from the design process (de Lange originals)

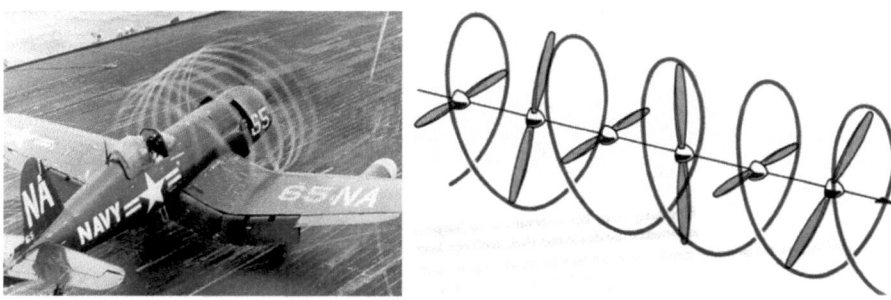

If you look from the side you will see a sinus:

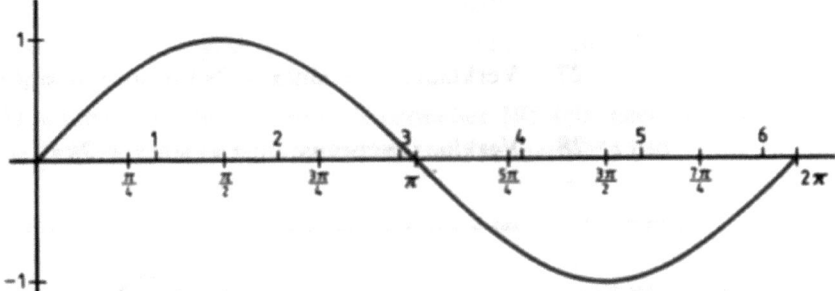

Fig. 10.8 Developments of the flying context (de Lange, 1978b, 1980, 1984; de Lange & Kindt, 1984)

But the concepts involved in flying can also be broadened, sometimes in surprising ways. The sine can be introduced in the flying context as a ratio in a triangle as indicated before (via glide ratio). But the graph of the sine function can also be introduced using the movement of a propeller tip as the sequence of illustrations in Fig. 10.8 shows.

10.4.4 Experimental Free Design

Sometimes a designer can be the happiest person in the world. The conditions for this mental stage are quite simple. The task was to design "something" for children in the age range from 6 to 12 that had to be challenging (of course), real world (of course), with some mathematics (of course) and science in it (de Lange, 2013). The context I used was simple: the weather. Children have seen weather maps, and the design goal was: ask them to prepare for the weather given a realistic weather map.

Weather maps are actually maps with contour lines. So as an exploration topic, students constructed an imaginary hike in a mountainous area: following hikes on a map can be made very realistic for students if you have video, graphics, and

Fig. 10.9 Contour lines (de Lange, 2013)

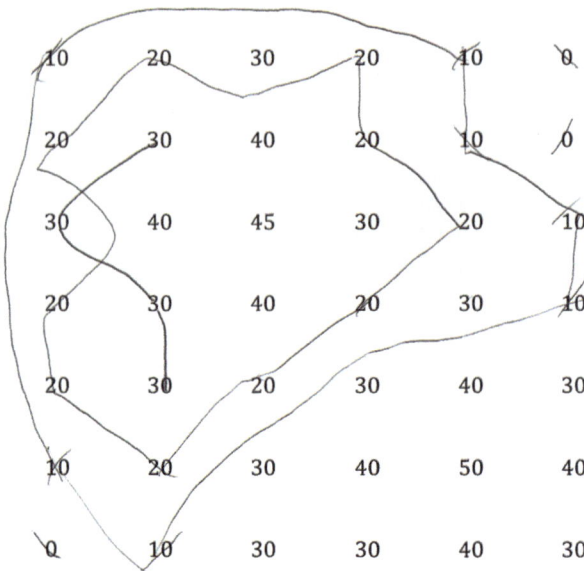

Fig. 10.10 Weather map (de Lange, 2013)

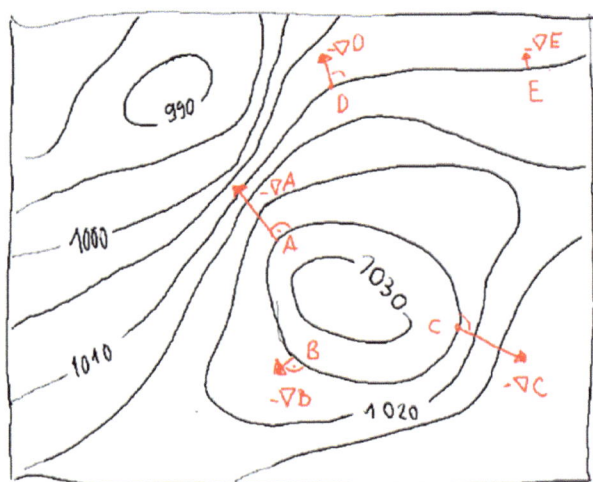

maps with and without contour lines. Next, we did the well-known problem that is illustrated in Fig. 10.9.

The students are given a lot of numbers on a grid and invited to draw the fitting contour lines. You can really get an interesting discussion if one of the students draws intersecting contour lines. A next step is to look at real weather maps. Just like a ball will follow the steepest path down a hill, the air will move in a direction perpendicular to the contour lines. The steeper the "hill", the faster the air will move. The red little arrows on the map tell the story convincingly (Fig. 10.10).

Fig. 10.11 Coriolis effect (de Lange, 2013)

The word *gradient* might be used here. Finally, the wind does not move in a direction perpendicular to the isobars (contour lines) but almost parallel to them. This is caused by the Coriolis effect (due to the rotation of the Earth). To understand this is quite a challenge for young children, until we used maple syrup on a slowly rotating globe (see Fig. 10.11). All of a sudden, the highlight of a whole series of lessons was reached with a plastic globe and maple syrup (I won't mention all the efforts in the kitchen to find which fluid substance gave the most convincing result; designing can be quite exhausting).

Free design can be extremely rewarding because it is a form of discovery learning; you can go to complex connections among topics that are only distantly connected to national standards, and you can show that well-designed problems can be solved by 6-year-olds or 12-year-olds. Of course, the answers and reasoning might differ quite a bit, but the results can be spectacular and the designer learns a lot.

10.5 Restrictions, Conditions, and Challenges

Educational design has many restrictions, conditions, and challenges. Lynn Steen made a point in 2007 that summarizes the main challenge: "There are many ways to organize curricula. The challenge, now rarely met, is to avoid those that distort mathematics and turn off students" (Steen, 2007, p. 93). Over the years, this problem has not yet been solved, and the present customer, I remind you, is the millennial described at the start of this chapter. It often seems that publishers are aiming more at selling their products to school districts and, hopefully, teachers. So the designer has to design for the student, while taking teachers and school boards into account.

I also remind you of Schön's view that the greatest problems lie in the swamp. Some may find great satisfaction in discriminating between educational design research and research-based educational design (McKenney & Reeves, 2012), but this is something from the "high ground". It does not help the educational designer who makes a living in the swamp while trying at the same time to use knowledge from the high ground *and* reflect on the experiences in the swamp, in order to give such theory more relevance.

In the 1980s, a brand new programme was developed in the Netherlands, *Mathematics A*. The process at the time took the traditional road: a national commission wrote a report exposing and explaining the philosophy of the programme. Next, the designers were asked to implement the vision into student materials. The first thing the designers did was construct a schema of possible modules. But, "Surprisingly, the commission had not mentioned the problems of achievement testing in their report" (de Lange, 1987, p. 164). So this task became a challenge for the designers. This task was indeed very challenging, and the struggle and results were published in de Lange (1987). In short, assessments are important and should be part of the design from the beginning, whether they are high stakes or monitoring, summative or formative, and portfolios or computer delivered.

More mundane restrictions and challenges are: externally imposed timelines that are always too short; the funding is never enough; the emphasis is on the final product rather than the dynamic design process or the students' learning. This problem has become even more serious because of the multidimensional design space we are living in; parallel processes seem to be inevitable: designer 1 designs the scope and sequence based on standards that have multiple interpretations; designer 2, at the same time, is designing animations; designer 3 does the assessments; designer 4 does the high stakes testing; designer 5 writes a script for video support; designer 6 … this system then requires another designer who coordinates and informs the whole system. Parallel design can be coherent if carried out by a well-managed team who shares interpretations and ideals, but often leaves the individual designer lost in the multidimensional space. Published curricula and standards are seldom as coherent and rigorous as their authors think.

Designers need to think of the children and students and of the beauty of mathematics as a discipline. Children and students are individuals, with a huge variety of capabilities, talents, and needs; in short, they are not standardized. The remarkable appearance and international growth of national standards suggest that individual learners have almost disappeared from the educational palette.

10.6 Task Design and Curricula Design

Task design is clearly an aspect of curricula design; one way of thinking about enacting a curriculum is as a series of clearly connected tasks. But even such a seemingly simple sentence has great consequences for design.

A curriculum can be seen as the planned interaction of pupils with instructional content, materials, resources, and processes, and that outlines the skills, performances, attitudes, and values pupils are expected to learn from schooling. The main problem is that there are many ways to start designing a curriculum; more often than not, the designer has to start with a framework that is not an optimal starting point from a design point of view. It may be a list of topics or concepts, or a list of assessment targets, or an exposition of a preferred way of teaching. In an ideal world, designers would already be involved in the initial phase of designing the curriculum as a whole, so they can influence the writing of the "standards" or "goals". This is important because curricula are not always written with either students in mind (e.g. as preferred hypothetical learning trajectories) or with an explicit view of what constitutes mathematics (e.g. integrating and connecting algebra and geometry, dealing with the imaginary 2D world and the realistic 3D world).

10.6.1 Left Out in the Cold

Most descriptions lack clarity on a variety of variables, for example, how important a "standard" or "goal" or "competency" is; the relative weight of, or how much time should be spent on, a subject; and how are topics connected, both horizontally or vertically through time or across experience. General guidance, such as "solve problems involving ratio", may be defendable if the designer was able to see the construction process in the curriculum. But more usually, the designer is left out in the cold. Without detailed interpretation, publishers can use the slogans "we meet the standards" with only superficial justification.

Designers have to make important choices. Should a unit of work start in the real world and develop mathematical concepts from there? Or should we use a more traditional model: first the math concepts and practice and then some applications? Or should we use both models, depending on the concept? And who makes this decision, the publishers, designers, curriculum writers, or teachers?

10.7 Summary

I shall summarize my chapter in terms of hints for young designers in mathematics education:

1. Ask yourself if you have a teaching background: did you teach to the book or did you "design" the lesson? If you fit the latter description, your chances of becoming a good designer are better.
2. Choose your first subject carefully: choose an obscure concept, e.g. *functions of two variables* for 16-year-olds. Arrange an experience that sheds new light on this subject, e.g. a trip to the Grand Canyon. Field experiences are almost indispensible. After visiting the Grand Canyon several times, the idea to use the horizontally structured layers of rock for introducing contour lines was almost inevitable (See Fig. 10.12).
3. If 2 was successful, choose as your next subject your least favourite topic at school, e.g. logarithms (See Fig. 10.13). Each chamber of the shell of a nautilus is an approximate copy of the next one, scaled by a constant factor. This gives rise to an approximation to a logarithmic spiral. Start to love understanding the concept; if you achieve this, you might design something useful.

Fig. 10.12 Contours in the Grand Canyon (de Lange, 1977, p.8)

Fig. 10.13 The logarithmic spiral: a joy for all ages (de Lange, 2013)

4. Insist on slow design whenever you can. Use the argument that you are doing developmental design research, including the latest results of cognitive science and developmental brain research (or some such story).
5. Use your design intuition as much as possible—design is also art. But do not mention it too much, until you're very old.
6. Keep teaching in a classroom yourself for two reasons:
    (a) To give you a feel for what is possible
    (b) To keep your feet in reality (the swamp)
7. Do not think primarily about writing for refereed journals; think of the students in the classrooms.
8. Accept curriculum constraints if you have to, but try to find and use the degrees of freedom.
9. Take initiatives for free design; it can be very rewarding, especially for the students.
10. Try to reflect on your practices; it may be very helpful for everyone in the process. I failed in that respect. This is actually my first reflection ever.
11. And reflect on the title of this paper; is it true that *there is, probably, no need for this presentation?*

References

Ahlfors, L. V., Bacon, H. M., Bell, C., Bellman, R. E., Bers, L., Birkoff, G., et al. (1962). On the mathematics curriculum of the high school. *American Mathematical Monthly, 55*(3), 191–195.
Anderson, J. R., Lebiere, C., & Lovett, M. (1998). Performance. In J. R. Anderson & C. Lebiere (Eds.), *Atomic components of thought* (pp. 57–100). Mahwah, NJ: Erlbaum.
Black, P., & Wiliam, D. (1998). Inside the black box: Raising standards through classroom assessment. *Phi Delta Kappan, 80*(2), 139–148. Available from http://weaeducation.typepad.co.uk/files/blackbox-1.pdf

Collopy, F. (2009). Lessons learned – Why the failure of systems thinking should inform the future of design thinking. *Fast Company blog.* http://www.fastcompany.com/1291598/lessons-learned-why-failure-systems-thinking-should-inform-future-design-thinking

de Lange, J. (1977). *Funkties van twee variabelen.* Utrecht: I.O.W.O.

de Lange, J. (1978a). *Exponenten en logaritmen.* Utrecht: I.O.W.O.

de Lange, J. (1978b). *Vlieg er eens in.* Utrecht: I.O.W.O.

de Lange, J. (1980). *Vlieg er eens in.* Utrecht: I.O.W.O.

de Lange, J. (1984). Fliegen, Papierflieger, Segelfliegen, Flugnavigation, Optimal fliegen, Point of no return. *Mathematik Lehren, 6.*

de Lange, J. (1987). *Mathematics, insight, and meaning: Teaching, learning and testing of mathematics for the life and social sciences.* Utrecht: O.W. & O.C.

de Lange, J. (2012). *Dichotomy in design: And other problems from the swamp.* Plenary address at ISDDE, Freudenthal Institute, Utrecht.

de Lange, J. (2013). http://jdlange.nl/category/meester-jan/

de Lange, J., & Kindt, M. (1984). *Sinus, culemborg.* Culemborg: Educaboek.

de Lange, J., Roodhardt, A., Pligge, M., Simon, A., Middleton, J., & Cole, B. (2003). *Digging numbers.* Chicago, IL: Encyclopaedia Britannica.

Duggan, W. (2007). *Strategic intuition: The creative spark in human achievement.* New York: Columbia Business School Publishing.

Feijs, E., de Lange, J., van Reeuwijk, M., Spence, M., Brendefur, J., & Pligge, M. (2006). "Looking at an angle". In *Mathematics in context.* Chicago, IL: Encyclopaedia Britannica.

Gladwell, M. (2005). *Blink: The power of thinking without thinking.* New York: Back Bay Books.

Hilton, P. (1976). Education in mathematics and science today: The spread of false dichotomies. In H. Athen & H. Kunle (Eds.), *Proceedings of the Third International Congress on Mathematical Education* (pp. 75–92). Karlsruhe, FRG: University of Karlsruhe.

McKenney, S., & Reeves, T. (2012). *Conducting educational design research: What it is, how we do it, and why.* London: Routledge.

Merriam Webster. (2008). http://www.learnersdictionary.com/definition/design

Pedley, T. J. (1977). *Scale effects in animal locomotion.* London: Academic.

Schoenfeld, A. H. (2009). Bridging the cultures of educational research and design. *Educational Designer, 1*(2). Available from http://www.educationaldesigner.org/ed/volume1/issue2/article5/pdf/ed_1_2_schoenfeld_09.pdf

Schön, D. A. (1995). The new scholarship requires a new epistemology. *Change, 27*(6), 27–34.

Schunn, C. (2008). Engineering educational design. *Educational Designer, 1*(1). Available from http://www.educationaldesigner.org/ed/volume1/issue1/article2/index.htm

Steen, L. (2007, January). Facing facts: Achieving balance in high school mathematics. *The Mathematics Teacher.* 100th Anniversary Issue, pp. 86–95.

Stein, J. (2013). Millennials: The me me me generation. *Time Magazine.* Available from http://time.com/247/millennials-the-me-me-me-generation/

Part IV
Commentaries

Chapter 11
Taking Design to Task: A Critical Appreciation

Kenneth Ruthven

11.1 The Evolution of Task

While task design has long been a central concern of mathematics education, it is only recently that an organized community has emerged in which task design has been linked with design research. Together, the editorial introduction to this book (Chap. 1) and Chap. 2 provide a useful historical sketch of the ground-laying for such activity, the emergence of several energetic groups, and the development of this international community, leading to the preparation of this book.

Using *Educational Studies in Mathematics* as a convenient historical section provides a simple means of tracing the penetration of talk about *tasks* into the mainstream of mathematics education research. One finds that *task*—used in the sense of a stipulation for some unit of mathematical activity—has been present since the inception of the field in the late 1960s (c.f. the use of *discovery task* in Scandura, Barksdale, Durnin, & McGee, 1969). More specialized terms followed, such as *task sequence* in the mid-1970s (c.f. Scandura, 1975) and *task-based interview* in the mid-1980s (c.f. Presmeg, 1986), but the term *task design* itself did not surface until the late 1990s (c.f. the title of one of the references in Noss, Healy, & Hoyles, 1997).[1] By comparison—as the traces revealed by this trail suggest—*task design* was already established in the psychological literature in the 1960s in relation to the design of diagnostic and other assessment tasks (c.f. its use in connection with multiple choice test items in Tversky, 1964).

---

[1] The next use of "task design" in *Educational Studies in Mathematics* occurred in 2001, and over 20 articles employed the term during the subsequent decade. By comparison, the earliest use of "task design" in the *Journal for Research in Mathematics Education* occurred in 1983, with only one further use before 2000.

K. Ruthven (✉)
University of Cambridge, 184 Hills Road, Cambridge CB2 8PQ, UK
e-mail: kr18@cam.ac.uk

This chapter has been made open access under a CC BY-NC-ND 4.0 license. For details on rights and licenses please read the Correction https://doi.org/10.1007/978-3-319-09629-2_13

The emergence of *task design* within the field of mathematics education has seen a reshaping of the originally psychological framing to accommodate a long-standing tradition of mathematical popularization. This tradition seeks to find relatively self-contained and accessible "activities" which are exemplary of some topic within mathematics and/or form of mathematical activity. For example, introducing an early paper on "geometrical activities for the upper elementary school", Engel wrote:

> In this paper I shall present a selection of activities which I have used in grades 5 to 7 for the past 16 years... At this level I prefer topics which are not treated later, but which are still interesting, important and challenging... These examples will show that, even at an early age, one can reach rather deep results in a short time and starting from scratch. (Engel, 1971, p. 353)

Although it makes passing reference to pedagogical considerations, Engel's paper analyses the tasks which it presents in primarily mathematical terms, doubtless reflecting the way in which the author's craft knowledge (developed through his considerable practical experience of working with these tasks) was framed (c.f. Ruthven, 2011, p. 92).

Equally, though, a contemporary paper by Egsgard on "some ideas in geometry that can be taught from K-6" illustrates a more explicitly theorized and psychologically influenced approach to pedagogical design. This approach appeals, in particular, to Piaget's model of developmental stages and Dienes' typology of intrinsic motivation to guide the pedagogical model of *directed discovery* which frames the discussion of the student "activities" or "assignments" presented in the paper:

> In using [the directed discovery] technique opportunities and activities are provided so that a child can make his own discoveries. Careful planning of the sequence and pace of activities is essential to ensure that the child understands and learns the concepts. As children are led to discover a concept, discussion with the teacher, the class or another individual must be allowed. The primary role of the teacher is to question, to guide, to stimulate and to assess the progress made so that time is used wisely. (Egsgard, 1970, pp. 481–482)

Egsgard's paper also recognizes a need to locate tasks within a larger curricular framework: it concludes with a diagram which suggests how, over the years of elementary education, and in preparation for secondary education, such activities could underpin a systematic development of key concepts along interconnected lines of development of geometrical thinking.

This example illustrates how, from the very inception of the research field of mathematics education, task design has often formed part of a larger enterprise of *curriculum development*, a term which has figured consistently in *Educational Studies in Mathematics* since its start. Exploring potential contrasts in the connotations of task design and curriculum development calls attention to a spectrum of scale in which the micro-level of task can be differentiated from the macro-level of curriculum, with perhaps an intervening meso-level corresponding to the unit or module. Although design can refer to either process or product, development more typically refers to process: nevertheless, as an iterative conception of design research has become influential, "design" has become more strongly associated with this cyclic process of "development", as the marriage of *design as intention* and *design as implementation* discussed in Chap. 2 makes clear.

Over the last half-century, the field of mathematics education has undoubtedly become more ambitious in its aspirations to coordinate the design of teaching with the formulation of theory through a development process more closely guided by research-based techniques. The clinical interview and its evolution into the teaching experiment have been particularly influential as paradigms of method for design research in mathematics education. However, herein lies one plausible reason why attention has shifted over this period from the design of larger-scale curriculum towards smaller-scale task: within the amounts of time and levels of resourcing normally available to designers, the complexity and cost of a research-based approach to the development process renders it feasible only on a limited scale. Certainly, my own experience of conducting design research at the meso-level within such constraints is that this called for very careful focusing of the research with an eye to the core feature of the design and the associated line of theory development, leaving more peripheral aspects of the design—but ones potentially crucial to its success—to be handled more informally (Ruthven & Hofmann, 2013).

11.2 A Scheme for Design

The core of Chap. 2 surveys a range of intermediate frameworks that have been employed in task design. Each intermediate framework represents some working synthesis of grand theory and/or craft knowledge for the specific purpose of designing instructional tasks and sequences. In particular, the chapter presents each framework in a manner which articulates its central principles for design and relates these to an illustrative case of design. As the chapter notes, only some aspects of a design can be attributed to the guiding framework; art is involved as well as science. Equally, the types of learning goal on which such intermediate frameworks focus and the educational values which they express are not uniform.

Nevertheless, Chap. 2 does identify some common underlying assumptions across the intermediate frameworks presented: notably, that learning mathematics takes place through doing mathematics; that the mathematical task posed must take account of students' current understandings; and that learning depends on development of representations and models. It may be valuable, then, to push a little further by developing an organizing scheme for the analysis of intermediate frameworks and their design principles. As the chapter makes clear, these frameworks generally relate not just to the task itself, but to its staging (in the sense of *mise-en-scène*) in the classroom. Any framework for task design, then, comprises one or more of the following elements.

11.2.1 A Template for Phasing Task Activity

For example, within Japanese Lesson Study, the template for lesson activity on a problem-solving task consists of phases of teacher introduction (*donyu*), student investigation (*jiriki-kaiketsu*), class comparison (*neriage*), and teacher summing-up (*matome*). Likewise, within the Theory of Didactical Situations, the template for staging an adidactical situation calls for an opening teacher-led "didactical" phase of *devolution*, followed by "adidactical" phases of *action, formulation,* and *validation*, concluding with a further teacher-led "didactical" phase of *institutionalization*. Similarly, within Formative Assessment for Developing Problem-Solving Strategies, an initial phase of intuitive student work on a problem is followed by teacher analysis of strategies before the next lesson; that lesson then starts with a phase of student evaluation of teacher-provided samples of student work which have been chosen and annotated so as to provoke reflection on earlier strategies, leading to a phase of student modification or refinement of their own earlier strategy, and concluding with a phase of whole-class comparison of the revised strategies and reflection on them.

11.2.2 Criteria for Devising a Productive Task

As already noted, the frameworks surveyed generally stipulate that a task should pose some kind of problem to be solved; a problem which admits a range of solutions; solutions typically differing in their level of mathematical sophistication, including a level which ensures that students will be in a position to propose some kind of initial solution. Beyond this, there are some differences in the types of criteria specified within different frameworks. One type of criterion concerns the realism and potency of the task to students; for example, its origin in some out-of-school practice, which is "crucial and alive" for students, and which can be transposed into the educational system as exemplary of target mathematical concepts (Anthropological Theory of the Didactic); or its relation to a "realistic" problem situation that affords students opportunities to attach meaning to the mathematical constructs it serves to develop (Realistic Mathematics Education). Another type of criterion concerns the potential of the task to foster productive conceptual reorganization, for example, by eliciting misconceptions from students (Conceptual Change Theory); by affording students a starting approach which turns out to be unsatisfactory (Theory of Didactical Situations); or by triggering responses which highlight common mistakes (Formative Assessment for Developing Problem-Solving Strategies). By contrast, Conceptual Learning through Reflective Abstraction represents an approach to concept development which seeks to afford students the opportunity to build an abstraction from already available activity, rather than through triggering cognitive conflict and reconstruction.

11.2.3 Organization of the Task Environment

One aspect of such organization concerns the instrumentation which mediates a task. For example, in the Proof Problems with Diagrams framework, the diagram provided with a proof problem plays a crucial role in scaffolding students' search for counterexamples or non-examples, and in supporting their deductive guessing. The choice of particular representational tools plays a similar mediating function within Realistic Mathematics Education, supporting the raising of students' level of conceptualization through emergent modelling. Likewise, within Conceptual Change Theory, external representations and bridging analogies fulfil this mediating function. Another important aspect of the organization of task environment is the form of social interaction. For example, within many of the intermediate frameworks, dialogic group or class discussion is intended to support reflection on task strategies and reformulation of them. Such discussion, in turn, calls for the creation of particular interactional norms, as noted in the account of the Cognitive Apprenticeship for Productive Problem-Solving framework. Within the Theory of Didactical Situations, the organization of the task environment is conceived in terms of "the creation of a (material and social) milieu that provides students with feedback conducive to the evolution of their strategies". A more unusual variant—on the boundary between instrumentation and interaction—is the provision of annotated work samples (Formative Assessment for Developing Problem-Solving Strategies).

11.2.4 Management of Crucial Task Variables

This is a prominent aspect of the Variation Theory framework in which analysis of the variation space associated with a task or task type leads to the identification of crucial (structural rather than superficial) dimensions, and to the development of a task sequence intended to be optimally efficient in creating an enacted variation space. A similar process of analysis of variation to create a well-tempered sequence of tasks is apparent in the Conceptual Learning through Reflective Abstraction framework where:

> The task sequence starts with word problems and context-free tasks to elicit and reinforce the diagram drawing strategy. Once the student is using the intended strategy, the task sequence provokes the anticipated abstraction. For this purpose, larger numbers for the denominators and invited mental runs of diagram drawings were used.

Likewise, design within the Theory of Didactical Situations framework depends on the identification and judicious tuning of key didactical variables which influence the particular character of a task, the approaches available to it, and the pathways through which these unfold.

Thus, the "design" that emerges from a development process of this type is much more than the "task" alone—in the sense of the manifest form of the task presented to students. This design encompasses, first, any template for phasing task activity and any organization of the task environment; then the rationale for judging the task productive, for phasing activity on the task, for organizing the task environment, and for managing task variables. As the examples presented in Chap. 2 show, few if any intermediate frameworks explicitly address all these aspects of task design. However, these aspects must figure—even if only implicitly—in the design process. Thus, although intermediate frameworks and design principles serve to articulate the *science* behind a design, they are often silent on the *art* or *craft* that also contributed to it, and which may indeed have shaped crucial assumptions about the manner in which it should be staged.

11.3 Design Continues in Use

Hence, because a design is much more than the overt task, and because a design presumes more than the intermediate framework and design principles make explicit, the dissemination of task designs is far from straightforward. Recent research on patterns of use in mathematics education of textbooks (Remillard, 2005) and dynamic software (Ruthven, Hennessy, & Deaney, 2008) has shown the degree of *interpretative flexibility* that such tools afford, resulting in their being employed by teachers and students in ways very different from those envisaged by their developers. In effect, the process of design continues in use, as users appropriate the tool to address their particular purposes and adapt it to their specific context.

Accordingly, Chap. 3 focuses on the role of the teacher in (re)designing and implementing tasks. Some contributions note that—subject to teachers appreciating the rationale of its design—their adaptation of a task may enhance the result of its implementation. Exercising such a role thoughtfully calls, of course, for the teacher to (re)address—with a specific context in mind—issues that exercised the original developer. Thus, the five dilemmas and six criteria presented in Chap. 3 can be related to the scheme of design elements that has already been introduced as a means of organizing the concerns of the various intermediate frameworks and illustrative designs presented in Chap. 2. The dilemmas of context, language, structure and distribution all relate to devising a productive task, as do the criteria of epistemic, cognitive, affective and ecological suitability. Likewise, the dilemma of interaction and the criteria of interactional and mediational suitability relate primarily to organizing the task environment. The agendas provided by these dilemmas and criteria usefully enlarge on those components of the organizing scheme of design elements.

Equally, the examples presented in Chap. 3 emphasize how the interpretation and redesign of tasks by teachers are shaped by their pedagogical orientations and so by the value and instrumental rationalities informing these. At the same time, the aspiration of many researchers, developers and teacher educators involved in task design appears to be to effect particular changes in such orientations and rationalities: they intend tasks—and whatever support materials accompany them—to be "educative" for teachers (Davis & Krajcik, 2005). However, as any reader of this book will quickly appreciate, it can be hard to keep track of the differing rationales and expectations associated with a multiplicity of tasks of differing provenance, let alone to integrate them. Indeed, when every design team appears to proceed from its own distinctive position, and to operate closer to the micro-level of task rather than to the macro-level of curriculum, it is the teacher who is left to integrate the results. This challenge is all the greater if we accept the argument that teachers should display "relentless consistency" of approach in order to establish ways of working mathematically with students. This argument raises questions about achieving a sound balance and a clear relationship between—on the one hand—the often highly specified design of tasks and—on the other hand—more generic ways of working. This is one of the central challenges confronting a "re-sourcing" movement in mathematics education which advocates that schools and teachers devise their own local schemes of work through assembling, adapting and structuring materials from a variety of sources (Ruthven, 2015).

Of course, one answer to such questions is to restrict oneself—as designer and/or teacher—to a particular intermediate framework which provides—for example—an explicit and consistent phasing of task activity and organization of task environment into which students can be inducted. For example, my own experience of conducting "redesign" research to develop curricular modules—capable of implementation at scale within known systemic constraints—points to the importance of a substantial introductory module in which the task sequence specifically aims to induct teachers and students into generic ways of working (and the rationale behind them) that are then reinforced systematically in subsequent topic modules: in this case, norms and practices of "dialogic teaching" (Ruthven & Hofmann, 2013). Likewise, achieving this kind of balance appears to be a characteristic of Japanese Lesson Study, which—as Chap. 9 reports—keeps one eye on developing tasks and lessons "that bring broad educational values to life in the classroom" while the other eye attends to the mathematical topic of immediate concern.

For the teacher, then, as for the designer, one approach is to filter any task through the lens of the generic ways of working mathematically that one seeks to develop. This appears to be what is taking place in the example—described in Chap. 3—in which teachers work together to "turn a lesson upside down" in order to ensure that it will prompt certain ways of working: in this case, concerned with generalizing and justifying. There is clearly an argument, then, for placing greater emphasis on developing systems of generic heuristics to guide the staging of tasks. For example, Chap. 3 refers to a repertoire of strategies developed to guide teacher interventions while students are working on a mathematical task: these address issues such as whether or not to intervene; how to initiate an intervention; whether to withdraw or proceed with the intervention; and how to intervene to support students experiencing difficulty.

11.4 Scope for User Agency

Chapter 3 concludes that the role of the teacher in adapting tasks to context and in managing their unfolding in the classroom is unavoidable. Translating any task—however tightly specified—into classroom activity calls for a degree of interpretation and elaboration on the part of the teacher and this already grants them some scope for agency. Moreover, the teacher will seek to integrate any task into some larger system of classroom practice, and may judge it necessary to adapt the task in doing so. And finally, as Chap. 4 notes, lines of thought and action that emerge during classroom activity may encourage the teacher to further modify or extend tasks or to create new ones. There can be little doubt, then, about the scope for agency that resides with the teacher.

But the processes through which users interpret and redefine tasks are not confined to teachers. Just as between the original designer and the teacher user, many of the assumptions and expectations underpinning the design of a task—and its implementation—remain implicit, so too between the teacher proposer of a task and the student actor in response. This leads Chap. 4 to argue for the importance of accounting for student perspectives in task design, in order to understand how to reduce the gap between the intentions of teachers and the activity of students. In particular, the chapter clearly illustrates how the mathematical socialization of designers and teachers may render alternative student interpretations of a task unimaginable and incomprehensible to them: this "expert blind-spot" (Nathan, Koedinger, & Alibai, 2001) all too easily leads to a "bifurcated situation".

In searching for ways of avoiding or retrieving such situations of mismatch between teacher and student interpretations of a task, Chap. 4 examines a number of options. First, it argues that one apparent option—for the teacher to state more explicitly their expectation of the type of approach or response to the task—may be counterproductive where the aim is for students to develop creative, flexible and independent mathematical thinking. That is indeed plausible, but one might also ask whether many of the tasks presented in this book truly have—or actually realize—such an aim; often it appears that the designer already has a particular path of "creativity", "flexibility" or "independence" firmly in mind. Indeed, the chapter itself points out how intermediate frameworks tend to assume (as do classroom norms) that the knowledge to be learned through tackling a task and the anticipated learning trajectories through which that knowledge will be constructed have been determined in advance. That leaves only the option of manipulating one or more of the critical didactical variables for the task in order to manoeuvre student interpretations into closer alignment with those of the teacher and/or designer.

However, Chap. 4 explores one further option in some depth, reflecting approaches to task design which emphasize the significant role of student agency and voice in the development of mathematical thinking. One such approach seeks to bring out the utility of mathematical ideas through tasks which give students some clear and immediate purpose within the context of the lesson (in the sense of a purpose distinct from learning target mathematical ideas) but which also require students to form and/or use those target mathematical ideas in a meaningful way. The criterion for a successful design is that it achieves a strong coordination between, on the one hand, the interpretations that students form of the purposeful task and its aim, and so their approaches and responses to it, and, on the other hand, the specific teaching intentions for mathematical development that underlie the design of the task. This Design for Purpose and Utility framework appears to share the perspective of those intermediate frameworks—reviewed in Chap. 2—which emphasize the need for students to experience tasks as realistic and authentic, through resonances of the task itself with students' interests and/or a managed process of devolution in which the task is made the students' own.

11.5 An Apparatus for Design

It goes without saying that this book makes reference to many resources that—as presented more fully in their original sources—can be treated as tools for design: in particular, each of the intermediate frameworks and its design principles, but equally each task case with its potential to serve as a generic example. Nevertheless, confronted with such a diversity of intermediate frameworks, design principles, and exemplary cases, one is likely to feel a need to identify some larger order within which these can be mapped and potentially used in a more coordinated way: the organizing scheme that I set out in an earlier section represents the very simple device which I formulated in a first attempt to do just this.

I see the development of modular "design tools" as probably the most potent way to populate such an organizing scheme. This notion of a design tool was briefly referred to in Chap. 2. Its definition in the source paper (Ruthven, Laborde, Leach, & Tiberghien, 2009) is largely ostensive, through presenting and comparing examples of such tools. Nevertheless, the paper positions the design tool as one component in a system moving from the level of *grand theory* through *intermediate framework* to *design tool*, so that the latter is distinguished by its proximity to the design process and its sharpness of focus within that process. Equally, the paper characterizes a design tool in terms of its capacity to identify and address some specific aspect of task design in order to support both the initial formulation of a design and its subsequent refinement in the light of implementation. For example, the design tool of Communicative Approach identifies and addresses the specific issue of finding a suitable combination of authoritative or dialogic and interactive or noninteractive discourse in each phase of the staging of a task while the design tool of Modelling Relations identifies and addresses the specific issue of managing, over the course of a task sequence, the compound relations between everyday and disciplinary languages and between observational/ostensive and theoretical/nonostensive worlds.

Moreover, bearing in mind the way in which design continues in use, such tools might also prove more accessible to users than intermediate frameworks and more effective in guiding their implementation and adaptation of tasks, by virtue of their sharp focus on a particular aspect of task design and/or staging. Here, I find myself in sympathy with the position taken in the editorial introduction to this book to the effect that theory in task design should be clear and give meaning to phenomena in classrooms, while also having practical meaning for teachers and designers.

In my view, then, the chapters that I have discussed—and this book as a whole—provide a valuable staging post for the continuing development of task design as an area of systematic enquiry in mathematics education. By bringing together, summarizing and comparing such a rich collection of conceptual frameworks and exemplary cases, my hope is that this book will motivate ongoing analysis and further synthesis, as well as stimulating the development of more comprehensive organizing schemes to guide the process of task design.

References

Davis, E. A., & Krajcik, J. S. (2005). Designing educative curriculum materials to promote teacher learning. *Educational Researcher, 34*(3), 3–14.

Egsgard, J. C. (1970). Some ideas in geometry that can be taught from K-6. *Educational Studies in Mathematics, 2*(4), 478–495.

Engel, A. (1971). Geometrical activities for the upper elementary school. *Educational Studies in Mathematics, 3*(3–4), 353–394.

Nathan, M., Koedinger, K., & Alibai, M. (2001). *The expert blindspot: When content knowledge and pedagogical content knowledge collide*. Paper presented at the Annual Meeting of the American Educational Research Association, Seattle, WA.

Noss, R., Healy, L., & Hoyles, C. (1997). The construction of mathematical meanings: Connecting the visual with the symbolic. *Educational Studies in Mathematics, 33*(2), 203–233.

Presmeg, N. C. (1986). Visualisation and mathematical giftedness. *Educational Studies in Mathematics, 17*(3), 297–311.

Remillard, J. T. (2005). Examining key concepts in research on teachers' use of mathematics curricula. *Review of Educational Research, 75*(2), 211–246.

Ruthven, K. (2011). Conceptualising mathematical knowledge in teaching. In T. Rowland & K. Ruthven (Eds.), *Mathematical knowledge in teaching* (pp. 83–96). New York: Springer.

Ruthven, K. (2015). The re-sourcing movement in mathematics teaching: Some European initiatives. In Z. Usiskin (Ed.), *Mathematics curriculum development, delivery, and enactment in a digital world*. Charlotte, NC: Information Age Publishing (in press).

Ruthven, K., Hennessy, S., & Deaney, R. (2008). Constructions of dynamic geometry: A study of the interpretative flexibility of educational software in classroom practice. *Computers & Education, 51*(1), 297–317.

Ruthven, K., & Hofmann, R. (2013). Chance by design: Devising an introductory probability module for implementation at scale in English early-secondary education. *ZDM: The International Journal on Mathematics Education, 45*(3), 409–423.

Ruthven, K., Laborde, C., Leach, J., & Tiberghien, A. (2009). Design tools in didactical research: Instrumenting the epistemological and the cognitive aspects of the design of teaching sequences. *Educational Researcher, 38*(5), 329–342.

Scandura, J. M. (1975). How does mathematics learning take place? *Educational Studies in Mathematics, 6*(3), 375–385.

Scandura, J. M., Barksdale, J., Durnin, J. H., & McGee, R. (1969). An unexpected relationship between failure and subsequent mathematics learning. *Educational Studies in Mathematics, 1*(3), 247–251.

Tversky, A. (1964). On the optimal number of alternatives at a choice point. *Journal of Mathematical Psychology, 1*(2), 386–391.

Chapter 12
Some Reflections on ICMI Study 22

Michèle Artigue

12.1 Introduction

An ICMI Study is launched when a substantial amount of research has been carried out at an international level and there are substantial realizations in diverse social and cultural educational contexts on the study theme. It is thought timely to build a state of the art from these achievements for the benefit of the mathematics education community at large, the community that the ICMI proposes to gather and serve. Most often, the reason for the study is also that some scientific, technological, or societal moves make the theme of prominent interest at the time. It seems thus important to make clear what has been achieved or missed so far and what lessons can be drawn from the past to think about the future, to identify the challenges to be faced, and to reflect on how these might be taken up.

The ICMI Study on Task Design, which ends now with the publication of a new volume in the NISS Springer series devoted to ICMI Studies, occurred within this overall schema. The design of tasks, their transformation, and the use of tasks designed by others are the daily jobs of a teacher. Students meet, practice, and learn mathematics through tasks prescribed for them or through tasks that they contribute to defining jointly with their teachers. Tasks accompany every moment of the study process, from the first meeting with questions or mathematical objects to the summative assessments taken to establish that the expected curricular aims have been achieved. Beyond that, when stakeholders want to act on the teaching and learning of mathematics and when they want to make teaching and learning practices evolve, tasks are used systematically as a privileged lever.

Tasks are nothing new; mathematics education research has always been interested in tasks. Researchers have built theoretical frameworks to guide their design,

---

M. Artigue (✉)
Laboratoire de Didactique André Revuz, Université Paris Diderot-Paris 7, Paris, France
e-mail: michele.artigue@gmail.com

This chapter has been made open access under a CC BY-NC-ND 4.0 license. For details on rights and licenses please read the Correction https://doi.org/10.1007/978-3-319-09629-2_13

© The Author(s) 2021
A. Watson, M. Ohtani (eds.), *Task Design In Mathematics Education*,
New ICMI Study Series, https://doi.org/10.1007/978-3-319-09629-2_12

have used tasks to study students' conceptions and their evolution, have questioned the management of tasks by teachers and its effects on students' learning, have analyzed the tasks proposed in textbooks and other educational resources to identify what these resources allow students to learn or not, and have shown how technological advances open the field of possible tasks and the means for their realization, with the resulting potential for teaching and learning.

However, despite this *density* of tasks in the educational landscape, it is only more recently that communities have formed around the issue of task design, making it the focus object of their research studies. Other evolutions contribute to making this study timely:

- The technological evolution, which not only affects the range of possible tasks and the means for their realization but also substantially changes their ecology and their dissemination processes and blurs the distinction between designers and users
- The evolution toward a more collaborative vision of the relationships between teachers, teacher educators, and researchers
- The knowledge patiently accumulated about teacher professional development and teachers' practices, how these are formed and may evolve, and about the documentary work of teachers

All these evolutions contribute to partially renew the ways task design issues are perceived, expressed, and worked out. It is also clear that new needs arise related to the current epistemological vision of mathematics as a living and expanding science nourished by its multiple interactions with an increasing number of fields, a science having a significant impact on our societies, whose teaching and learning must contribute to active, reflective, and critical citizenship. This evolution is reflected, for instance, in the growing attention paid to mathematical modeling, to the development of overarching competencies and attitudes beyond mere mathematical skills, and in the mathematics curricula. These developments make it necessary to look at existing tasks differently, but also to create new ones able to support the new expectations, and to organize their sequencing, taking into account new perspectives and criteria. I had all these issues in mind when I started reading the chapters of this volume.

12.2 A Subjective Reading

Before entering into comments, focusing on Chaps. 2, 5, and 6 as suggested by the editors, I would like to make explicit some elements of my culture and experience which necessarily influence my reading of the volume.

Professionally, I grew up in a didactic culture whose main theoretical reference was the Theory of Didactical Situations (TDS) to which many references are made in the volume. From my first steps in didactics, the notion of "situation" became, thus for me, the fundamental notion, and the idea of task was inserted into it. In this emergent didactic culture, the limitations of laboratory research and the need to take into account methodologically the complexity of the functioning of didactic systems were well perceived. Moreover, in coherence with the basic principles of TDS to take into account this complexity, priority was given to the development of methodologies making it possible to control didactic systems, to construct, produce, and reproduce didactical phenomena, in line with the vision of science as *phenomenotechnique* developed by Bachelard (my work in this area is described in Chap. 8).

This explains why design perspectives have been central in the development of this didactic culture, leading to a specific concept, that of didactical engineering (cf. Chap. 8 and Artigue, 2015). For several decades, didactical engineering has systematically supported not only the production of results about the learning and teaching of specific mathematical domains but also major theoretical advances in the field. Conversely, the concept has progressively evolved in the light of these advances. It has also been enriched by the combined use in engineering work of the TDS with other theoretical frameworks. For instance, very early in engineering work, this theory was combined with the tool-object dialectics due to Douady (1986) or the theory of conceptual fields due to Vergnaud (1991). The potential offered by combinations with semiotic approaches, such as Duval's theory of representation (Duval, 1995) or the theory of semiotic mediation (Bartolini Bussi & Mariotti, 2008), was then explored. More recently, renewed visions have been proposed in terms of paths of study and research within the frame of the Anthropological Theory of Didactics (ATD) (cf. Chaps. 2 and 8) or in terms of second-generation didactical engineering to strengthen the research-development dialectic. The proceedings of the 2009 Summer School of Didactics of Mathematics provide an insightful vision of the historical development of this concept and of its current use (Margolinas et al., 2011).

Without doubt, a long-term interest in technological issues also influenced my reading of the volume and especially the sensitivity to instrumental issues that I began to develop in the mid-1990s when I was working on the integration of CAS technology into secondary mathematics education with French colleagues, which led to the emergence of the so-called Instrumental Approach (Artigue, 2002). During the last decade, in fact, my research on digital technologies mainly took place in the frame of European projects, especially the TELMA (Artigue, 2009) and ReMath projects (Artigue & Mariotti, 2014) mentioned in Chap. 6. An essential aim of these projects was the capitalization of research results, which a priori relates to the aims of an ICMI Study. As is the case in an ICMI Study, we faced the diversity of theoretical approaches, of educational contexts, and of didactic cultures, which makes capitalization so challenging. This experience increased my awareness of the incompleteness of the idea of task sensu stricto. It also showed me that, if it is generally possible and even fruitful to combine a diversity of theoretical approaches to analyze a given corpus, which was also evidenced by the networking activity we carried out in the Bremen group (Bikner-Ahsbahs & Prediger, 2014), design work obeys another logic and does not accommodate theoretical eclecticism so easily. In these projects, we created specific organizations—later on identified as networking praxeologies (Artigue, Bosch, & Gascón, 2011)—to support the building of larger coherences and to favor the capitalization of knowledge. We also exploited the potential of such organizations for studying the design of digital artifacts and of situations of use of these, trying to elucidate the exact role, operational or metaphorical, played in design by the theoretical concepts referred to by the designers themselves. For a given digital artifact, we systematically compared situations of use designed, on the one hand, by the team in charge of the artifact design and, on the other hand, by another team from another country that we called an "alien" team. This research work allowed us to understand at what point design escapes theoretical control, even when this control seems very strong. This research also showed us at what point a task designed in one culture is transformed when exploited in another culture, in which it is acceptable, and how far the comparison of its adaptations and implementations in different contexts and cultures can be enriching and insightful.

The last point is that, in recent years, I have also been involved in several of the European projects[1] which have been founded by the European Commission for supporting inquiry-based education in sciences, technology, and mathematics, after the publication of the Rocard report (Rocard et al., 2007). The design of tasks likely to support such pedagogy in classrooms or the adaptation of existing tasks has been and still is an essential dimension of these projects. It has been the source of many issues, for instance:

- Regarding the viability of tasks designed in one country but used in other contexts
- Also, more fundamentally, the ways the different partners involved in these projects conceptualize the idea of inquiry-based learning and education and how these conceptualizations relate to theoretical approaches that have developed in mathematics education over several decades without using this terminology (Artigue & Blomhøj, 2013)
- And also regarding the ways the design of tasks and sequences of tasks can jointly address both the development of specific mathematical knowledge and inquiry competences

These projects involve an important number of partners and countries. Diversity is the rule, despite the affinities leading different teams to build a joint proposal. A coherent and shared vision is not easy to build, and these shared visions differ in part from one project to another one. Participation in these projects made me aware of the predominance of isolated tasks that are not part of a substantial progression and sequence within the resources produced by these projects. There is no doubt that isolated tasks are more easily accepted and implemented by teachers and more easily transferred from one context to another one; however, their acceptance most often results just in the adding of some exotic episodes to a classroom life that remains fundamentally unchanged. Finally, these projects showed me the strength of the didactic obstacle created by any existing gap between the learning principles

---

[1] These projects are Fibonacci (www.fibonacci-project.eu), Primas (www.primas-project.eu), Mascil (www.mascil-project.eu), and Assist-Me (www.assistme.ku.dk).

expressed in the institutional discourse and any forms of evaluation which, obeying another logic, contradict these principles.

All these experiences have shaped my expectations regarding this ICMI Study and my reading of the volume that results from it, while making me especially aware of the ambition and difficulty of the task undertaken in it. In the next sections, I have organized my commentary around Chaps. 2, 5, and 6, as suggested to me by the editors. I will mention other chapters occasionally because they offer additional contributions to the questions addressed in these three chapters regarding the frameworks and principles piloting task design, the potential and limitations of text-based tasks, and the influence of tools on task design.

12.3 Frameworks and Principles for Task Design

The design of tasks, with the extended meaning given to this idea in the Study, is an essential ingredient of the educational work. It should be, thus, not surprising to observe that many approaches developed in mathematics education have taken into account task design in a more or less central way. Chapter 2, devoted to the frameworks and principles guiding task design, starts by reviewing the history of task design in mathematics education from the 1970s, taking as a filter research work carried out in the International Group on the Psychology of Mathematics Education (PME) created in 1976 or reported at its annual conferences. Even if the coauthors cannot enter into the details of this complex history, they make clear that this is really the case. The chapter opens by a quotation from Simon and reminds us that, very early, some researchers (Erich Wittmann is especially mentioned) established connections between mathematics education and the field of design sciences itself in an emerging state at the time. This historical review also shows that, within the important didactical traditions that began to develop in the 1970s during a period of intense reflection on mathematics education and curricular reforms, the vision of frameworks and principles for guiding task design went beyond the strict design of tasks. The first PME working group devoted to these issues labeled them *Principles for the design of teaching*; these addressed not only the design of tasks but also the way these should be used in teaching. These principles were built on epistemological bases regarding both mathematics as a science and mathematical learning, with an important influence of Piagetian epistemology for the latter. Regarding didactical engineering, I would add the importance of a systemic vision supported by the theoretical constructs of TDS.

There is no doubt that the relationship of the field of mathematics education with design sciences has evolved since that time, as also made clear by Chap. 2, and that the epistemologies of reference for the field of mathematics education have substantially evolved; however, we cannot forget that task design in mathematics education is the product of this history and shaped by it. This is attested by the vitality and persistent influence of design traditions, such as those of the Shell Centre at the University of Nottingham, the Freudenthal Institute and Realistic Mathematics Education (RME), or didactical engineering and TDS, all well visible in this volume.

A shared need in mathematics education, here the need of designing tasks and associated scenarios, generally results in an increasing variety of approaches. Task design does not escape this situation; this has certainly been a real challenge for the coauthors of Chap. 2 to provide the reader access to a diverse and dynamic landscape and to make sense of it and its evolution. Chapter 2 attests to the efforts made by the coauthors in taking up this challenge, through the diversity of approaches considered, the many examples used, and the many distinctions, categories, and structuring tools introduced.

The questions mainly addressed are the following: What are the frameworks and principles used, but also where do they come from, and how far do they determine task design and how? These questions are indeed crucial and, as could be anticipated, the answers provided are many, many more certainly than could be the case in the 1980s. Beyond some convergences pointed out by the coauthors, for instance, the importance attached to problem solving, divergence is more the rule.

The diversity of answers results from a diversity of factors: the nature of theoretical frameworks underpinning the design when the design is explicitly based on such frameworks, the extension given to the concept of task design, the genre of tasks that are envisaged and the scope of the design projects, the professional groups involved in the design and their respective roles, and so forth. To organize this diversity, different distinctions are introduced. Regarding theoretical frameworks, the first distinction is in terms of levels. Three levels for theories are introduced: grand frames, intermediate-level frames, and domain-specific frames. This distinction is certainly pertinent, but it does not fully solve the problem. At the same level, one can find objects of very different nature, and, as acknowledged by the coauthors themselves, task design often relies on combinations of frames situated at the same level in this classification (e.g., fruitful combinations between TDS and the theory of semiotic mediation) or at different levels. For instance, an intermediate-level frame provides a global frame to the task design, and it is complemented by frames more specific to such-and-such mathematical domain or to such-and-such dimension of the design project.

We could expect strong and operational relationships between frames and principles. The examples quoted tend to show that this is not necessarily the case. The image is perhaps biased by the limited place necessarily allocated to the description of each single case; however, in many cases, the principles listed gave me the impression that they were moderately dependent on the frames and also articulated in such general terms that they were not constraining the design very much. In contrast, what these examples make clear is that design principles go beyond the design of tasks to be proposed to students. As was already the case in the 1980s, they underpin equally, if not more, the design of scenarios for embedding these tasks. This confirms the vision of tasks as objects one cannot make sense of independently of their insertion in a scenario of implementation, a scenario which is itself dependent on the context envisaged for its implementation and not just of learning perspectives. I am fully in line with such a vision.

Another interesting distinction introduced is the distinction made between *design as intention* and *design as implementation* in reference to the work of Collins, Joseph, and Bielaczyc (2004). It is particularly used to point out the various relationships to theory that may be involved in design projects, according to the fact that the theory can primarily be seen either as a resource for design (when its function is to guide) or as a product of design (such as when the design produces learning trajectories for specific mathematical fields, which are considered as local theories). Once again, the examples used for illustration make clear that the distinction captured is relevant; however, it would be abusive to equate design as intention and theory as a resource or design as implementation and theory as a product. The relationships are, in fact, much more complex and dialectical. The examples analyzed in that section, that is, design associated to RME or to ATD, are good illustrations of these dialectical relationships. My personal experience of didactical engineering supported by TDS, in combination with other frames such as the theory of semiotic mediation or the instrumental approach, is additional evidence for me. I would like to add that, beyond the examples mentioned in this section of the chapter for which the theoretical outcomes of task design are mainly expressed in terms of learning trajectories, task design can also be engaged in the production and reproduction of didactical phenomena attached to the functioning of didactical systems, such as phenomena resulting from the paradoxes of the didactical contract, as mentioned in Chaps. 2 and 8.

As I have already pointed out, frames and principles are most often expressed in very general terms leading one to think that, whatever their influences on task design, they are far from solving all decision-making choices that any design engages. This is clearly visible in the many examples used. Frames and principles leave a large space of maneuver for designers, whether researchers, professional designers, or teachers, and many questions arise from that. How is this space apprehended or operated and with what consequences on task design? In Chap. 2, this question is related to the distinction made between two visions of design, *design as a science* and *design as art or craft*, two visions put in tension and even opposed, as is the case in Chap. 10 authored by Jan de Lange. I have to confess that I found this chapter perturbing; I am not sure, even after reading it carefully, that I have fully understood the subtlety of the message that the author wants to convey. Perhaps the reason is that I had the impression that it was addressing the concerns of a design world of professional designers to which specific demands are made, who are working under the pressure of various constraints, including hard time constraints, and also who fully embrace the idea that they are developing products for large-scale dissemination which will escape their direct influence. My design experience is that of a researcher, a university teacher, and a teacher educator whose practice has always been more modest and also a collaborative practice, particularly within structures such as the French Research Institutes in Mathematics Teaching (IREM) in which, since their creation in the late 1960s and early 1970s, teachers, teacher educators, mathematicians, and didacticians work together, coordinating the diversity of their respective expertise. This practice never negated the fact that, in task design, theoretical control has limitations, that the decisions to be taken go far beyond what is accessible to scientific rationality, and that experience, artisanal tricks, and habits shape choices as importantly as theories (from the outset, this was integrated in the idea of didactical engineering). However, I have also learned what theoretical constructions offer to design, in terms both of global vision and operational tools. I also quickly understood that task design relies on specific forms of creativity, which are not equally shared. I have always been impressed by my colleagues, teachers, and researchers, who have this form of creativity, and at times I envy them. However, I do not agree with a dichotomist vision that opposes design as science with design as art or as craft. Neither do I agree with the opposition made by de Lange between pure and applied mathematics for supporting his argumentation, using the following quotation by Hilton: "Since mathematics (analogous to educational design theory/science) incorporates a systematic body of knowledge and involves cumulative reasoning and understanding, it is to that extent a science. And since applied mathematics (analogous to the actual practice of designers) involves *choices which must be made on the basis of experience, intuition, and even inspiration, it partakes the quality of art*" (p. 95, emphasis added).

These concerns do not prevent me from finding this provocative contribution very useful. It made me realize that in this study globally, and even in Chap. 5 devoted to text-based resources which more directly concerns design processes for tasks proposed in textbooks and other educational resources, the voice of professional designers is not strongly present. The voice of teachers who daily design and adapt tasks is mostly heard indirectly through studies regarding their practices. This probably affects the vision that is proposed to the reader of the relationships between design as science and design as art or of the role of frames and principles.

Understanding the subtle alchemy at stake between invention, craftsmanship, theory, and principles in task design is not easy. As is well stressed in the study, publications leave a few traces of this alchemy and we must engage in a real archaeology of design processes for accessing it. This study has the merit of making the point clear. Beyond that, it provides the reader with various entry points to approach this difficulty and illustrates their potential with insightful examples. Moreover, in some cases, particularly in the chapters associated with plenary lectures, the study allows us to go more deeply into the design process and to access the rationale for the decisions taken along it. From this perspective, I found Chap. 9 devoted to lesson studies enlightening even if it focuses on one specific dimension of this complex process of design. Despite the fact that the reader can perceive a shared sensitivity to mathematics and its epistemology, the contrast is evident with the view of design inspired by TDS or ATD, described in Chap. 8. I hope that this chapter will help many readers to better understand these forms of design which have emerged in my own culture, what unifies them and also what differentiates them. My familiarity with these forms of design does not make me the best person for anticipating if this may really happen.

12.4 Task Design Through Text-Based Tasks

Chapter 5 is dedicated to text-based tasks, those that we have always found in textbooks but that are today available in an increasing variety of media due to technological advances. They are defined in this chapter as "the written presentation of a planned mathematical experience for a learner, which could be one action or a sequence of actions that form an overall experience."

As explained by the coeditors of the Study in Chap. 1, their initial intention was to focus on task design in textbooks and other forms of text-based communication designed to generate mathematical meaning; to compare organizations and contents according to series, countries, and cultures; and to study existing means for analyzing isolated tasks or sequences of tasks and the relationships between tasks and curricula (tasks being seen on the one hand as tools for implementing curricula and on the other hand as vectors influencing them).

Indeed, the contributions received on this theme had little connection with the questions initially set up in the Discussion Document for the Study, and the preparatory work for this chapter was reorganized around the accepted contributions. One can see here a sign of the difficulty in this ICMI study, as in others, to attract the contribution of people and institutions more at the periphery of the community or even beyond the sole community of researchers in mathematics education.

As in the other chapters produced by the working groups and fed by a variety of contributions, in this chapter, we perceive the effort made by the authors to organize the variety of objects and questions. What is proposed is to structure the study around three poles: the nature and structure of tasks, the teaching/learning goal of their design, and the intended/implemented activity embedded in them. Each pole of the triangular structure is associated with a specific substructure. For each pole, it is also proposed either to enlarge the perspective or in contrast to reduce it, "by zooming out and thinking about the overall educational context and how this affects task design, and also by zooming in to the imagined interaction between one learner and the task."

This structure provides a rather complex but a priori interesting framework. In fact, from the description of the first pole, one perceives the diversity of objects to be considered, from textual materials in which learning trajectories are incorporated (as is the case for textbooks but also for online resources offering more or less flexible learning trajectories and different forms of piloting, including by the learners themselves) to banks of tasks where teachers and even students can freely do their "shopping". It is also made clear how technological evolution contributes to this diversity, beyond the sole quantitative explosion of resources and changing conditions to access them. The same occurs in this first pole, when categories of possible authors are listed or when the nature of the mathematics that these tasks make it possible to meet is discussed.

Beyond this diversity arises the question of the educational aims of tasks and of how these aims are actually made visible both to teachers and students or could be made better visible in the specific format of text-based tasks. As the authors point out, these objectives are multiple a priori if we accept that "the aims of mathematics education are multifaceted, so that learners become knowledgeable about concepts, competent with procedures, capable and willing to select, adapt and use mathematics in a variety of familiar and unfamiliar contexts and problems." Various examples are given which show the cultural variability in this area. Categorizations are also offered, such as that of Thompson, Hunsader, and Zorin, which for assessment tasks distinguishes between the following foci: reasoning and proof, opportunities for mathematical communication, connections, representation: the role of graphics, and representation: translation of representational forms. These seem to me very general, indicating at the best genres of tasks, and, reading their description, it is not clear whether more specific criteria of analysis are associated with them.

Various important questions are raised and discussed throughout the chapter with the use of examples, such as those regarding the possibility in the context of text-based tasks:

- Of initiating a dialogical relationship with the learner via the content and form of tasks
- Of creating tasks allowing differentiated student work or supporting pattern recognition and processes of abstraction, or allowing connections between representations
- Of helping the transition from the knowledge that can potentially emerge from the resolution of open tasks toward conventional forms of knowledge
- Of supporting particular learning approaches such as those underpinning the theory of variations

The many examples used show that, despite its limitations, the context of text-based tasks is not without potential in all these areas but also that actualizing this potential usually requires substantial changes in the text of these tasks and in their traditional uses.

Due to my personal interests, I read this chapter looking specifically at its affordances regarding issues such as the design of task sequences weaving the progressive development of specific mathematical knowledge and of global competences, for instance, inquiry or modeling competences, combining mathematical learning and the development of critical citizenship, or supporting multidisciplinary or co-disciplinary work. On all these points, some insightful contributions are presented and discussed in the chapter, such as that of Movshovitz-Hadar and Edri whose objective is to propose tasks embedded in the curriculum and serving the cause of citizenship or that by Maaβ and her colleagues on the European project COMPASS whose objective was to create educational resources that support interdisciplinarity in sciences, maths, and technology. These contributions clearly show the efforts undertaken by researchers to make this genre of tasks acceptable and usable in normal school contexts. As pointed out in relation to the COMPASS project, this genre of task is a priori more appropriate to exploratory work and long-term projects; however, to extend their accessibility and use, the project partners had also to design "more structured versions to support teachers who were not confident enough to undertake long tasks." It is added that "task designers gave considerable thought to how complex materials could be made teacher friendly and easy to use." We find here the expression of ecological concerns and constraints also visible in the valuable contributions of authors whose design activity, in terms of study and research paths, is inspired by the ATD and its paradigm of questioning the world (see Chaps. 2 and 8).

However, I must admit that I remained a little hungry because these issues are only marginally worked in the chapter, probably due to lack of substantial contributions. This suggests that much more research and development work is still needed on them.

12.5 Task Design, Tools, and Technology

The use of tools to support mathematical learning is nothing new. There is no doubt that technological advances and also, from a more general perspective, the increasing attention paid by research to the semiotic dimension of mathematical activity have refocused interest on tool issues and brought new conceptual means to approach them. The existence of a specific working group in this Study on the design of tools and the content of Chap. 6 coauthored by its members well reflect this move. A tool-based task is defined in it as "a teacher/researcher design aiming to be a thing to do or act on in order for students to activate an interactive tool-based environment where teacher, students and resources mutually enrich each other in producing mathematical experiences." As the reader is reminded in the chapter introduction, the prevailing view today is that "our interaction with tools, artefacts and culture material should be considered as more than auxiliary elements" and even, more and more, that artifacts are "a constitutive part of thinking and sensing". These convergences, however, do not prevent the existence of epistemological differences between tool-based task designers influencing not only the choice of tools or their design, when such design is also part of the process, but also how a particular tool is used. Diversity is once again the rule.

The chapter is structured into three main sections—considerations in designing tasks that make use of tools, theoretical frames for designing tool-based tasks, and further design considerations and heuristics—a structure which necessarily involves some overlapping content. The important ideas of semiotic mediation and instrumentation are therefore addressed in several sections, with some variation of perspective. Reading this chapter, I was particularly interested in the implications of these ideas in terms of heuristics and principles for task design, and in the examples selected to illustrate these implications. I was also, of course, interested in the discourse developed about the changes resulting from technological evolution because, as could be expected, technology is very present in this chapter. The potential influence of a vision of tools as instruments for semiotic mediation on task design is, for instance, well illustrated. In the various references made to task design inspired by the theory of semiotic mediation, the emphasis put in such design on the transition from personal signs developed by students in their interaction with the artifact to cultural mathematical signs detached from the use of the artifact is well stressed, together with the crucial role played by the teacher in this transition. I would like to add that the cross-case studies of the use of the same artifact, here the Casyopée software (https://casyopee.math.univ-paris-diderot.fr), carried out in the ReMath project cited in this chapter, enabled us to thoroughly study the similarities and differences between designs based on different theoretical approaches. For instance, the cross-case study involving two designs using the same mathematical problem, one based on TDS and the instrumental approach and the other based on the theory of semiotic mediation, allowed us to better understand the design implications of subtle differences in the semiotic visions of the two approaches (Maracci, Cazes, Vandebrouck, & Mariotti, 2013).

The reference to Rabardel's conceptualization and to didactic research inspired by it serves as the frame for instrumental considerations. At different times in the chapter, emphasis is placed on the tensions or gaps between potential and actual uses of tools, empirical/pragmatic or epistemic. As rightly stressed, if such tension is not taken into account in the design of tasks, students' activity can remain situated at the empirical/pragmatic level. This tension is, of course, not unique to the use of technological tools, but these have especially highlighted it. As I have written in other texts, one can observe a "natural" tendency in task design involving technological tools, due to their characteristics and affordances, to favor the pragmatic potential over the epistemic potential. This trend creates, in fact, a didactic obstacle. In the chapter, several examples show that this obstacle can be overcome through careful design choices; for example, the contribution by Robotti provides an excellent case.

Two other concepts discussed in this chapter seem to me very relevant with regard to tool-based task design, those of instrumental distance and discrepancy potential of a tool, which is defined as "a pedagogical space generated by (1) feedback due to the nature of the tool or design of the task that possibly deviates from the intended mathematical concept or (2) uncertainty created due to the nature of the tool or design of the task that requires the tool users to make decisions." We perceive their potential well through the discourse developed and the examples discussed. Regarding instrumental distance, I would like to add that the research work we have carried out with Haspekian on the teaching and learning of elementary algebra with various digital artifacts (spreadsheet, CAS) shows that it is important to include also in this instrumental distance an institutional dimension, in order to better understand its possible effects (Haspekian & Artigue, 2007). Another concept, very present throughout the chapter, is that of feedback. The piloting of feedbacks, the anticipation of possible effects, the organized play on their evolution through a sequence of tasks, and the cognitive exploitation of qualitative jumps in these, which is related with the notion of informational jump in TDS, all these are essential in task design. Again, technological advances open new ways as is well shown in the examples given on the Cabri Elem textbook or the analysis of dragging feedback in dynamic geometry environments.

The question of tool-based task design, especially in technological environments, cannot be seriously approached without working on the potential offered by the multiplicity of available representations, of means of action on these representations, and of interaction between them. This is probably one affordance of digital technologies that has been most systematically studied, and it was at the heart of the ReMath project already mentioned. It is present in this chapter, but not central, particularly through the ideas of conceptual blending and multiplicity and also when alternative representations are developed for addressing specific students' needs as is the case with the original systems of representations designed by Healy and her colleagues for blind or deaf students.

As we can see, as was the case for Chaps. 2 and 5, this chapter provides a diversity of perspectives with which to approach tool-based task design, introducing and highlighting some key and overarching concepts transcending any particular approach and using them for presenting and discussing a great variety of design realizations. Task design being understood in this study in very broad terms, I was, however, expecting a bit more regarding the potential offered today by digital technologies for the design and implementation of didactic scenarios and the reality of its current use.

This question is actually thoroughly discussed in the next, very interesting chapter where Michal Yerushalmy presents and analyzes the design of an e-textbook in the field of functions. More generally, she clearly shows how, increasingly, the structured and rigid didactical organization which has traditionally been imposed on textbooks is challenged by the digital culture, leading to complex objects combining a variety of tools and media, allowing flexible use and customization by the user, either teacher or student. In such a schema, the distinction between author/designer and user fades, as also stressed by Gueudet and Trouche for instance, making unavoidable a vision of task design as a process going on in use, as conceptualized in the documentary approach (Gueudet, Pepin, & Trouche, 2013). This raises, however, the difficult question of how to ensure the epistemological and educational relevance of the increased diversity of designs based on the same matrix that results from a flexible and user-controlled situation, an issue that seems widely open at the moment.

12.6 Conclusion

Like many ICMI studies, this study is very rich. It addresses a multiplicity of issues, regarding a notion of task design conceptualized broadly, through a diversity of voices. To help make sense of this diversity, multiple frames, organizing principles, concepts, and categories are introduced throughout the different chapters and used to situate, compare, and analyze a large number of research projects and realizations corresponding to different scales, practices, and contexts of task design. This study also covers the diversity of genres of tasks that one may have to design for the teaching and learning of mathematics. The history of task design in mathematics education is traced back to the early 1970s, and most recent developments are also discussed. Students' and teachers' perspectives are similarly considered.

All this means that the study provides a good picture of the real state of knowledge in this area of task design, helps understand evolutions and their rationale, identifies research and development needs, and delineates perspectives for future work. As in other ICMI studies, for example, the ICMI Study 17 on digital technologies in which I was involved some years ago, reading the different chapters, we also perceive that while contexts and expectations are moving, while digital technologies profoundly impact the work of designers and their relationships with users, while new issues arise, still many task design issues, ordinary and basic, have not found satisfactory answers.

I started the reading of this volume with many questions in mind. Ending it, I have a more comprehensive vision of the state of the art, new conceptual tools at my disposal for organizing task design work, for comparing, analyzing, evaluating, etc. I also understand that issues worrying me are shared and worked on by many researchers all over the world. For instance, I worry about appropriate levels for the description of tasks and task sequences that allow teachers to express their creativity and to adapt tasks to their particular context while preserving their essence and learning potential, in other words the complex relationships between task authors and users. I am concerned about the profound changes resulting from technological evolution in that respect and how task design can support the coherent development of knowledge, skills, inquiry competences and attitudes, and how to align the design of assessment tasks with educational values and so on. Promising constructions and realizations are presented and discussed, but it would be an exaggeration to say that the study fully answers these issues. It shows that these issues, as is generally the case in mathematics education, are addressed in a diversity of ways according to contexts and educational cultures; it offers valuable knowledge and develops tools to connect them. This certainly reflects what was accessible in the Study within the current state of international knowledge on task design and paves the way toward future research and development.

References

Artigue, M. (2002). Learning mathematics in a CAS environment: The genesis of a reflection about instrumentation and the dialectics between technical and conceptual work. *International Journal of Computers for Mathematical Learning, 7*, 245–274.

Artigue, M. (Ed.). (2009). Connecting approaches to technology enhanced learning in mathematics: The TELMA experience. *International Journal of Computers for Mathematical Learning, 14*(3), 217–240.

Artigue, M. (2015). Perspectives on design research: The case of didactical engineering. In A. Bikner-Ahsbahs, C. Knipping, & N. Presmeg (Eds.), *Approaches to qualitative research in mathematics education* (pp. 467–496). Dordrecht: Springer.

Artigue, M., & Blomhøj, M. (2013). Conceptualizing inquiry-based education in mathematics. *ZDM: The International Journal on Mathematics Education, 45*(6), 797–810.

Artigue, M., Bosch, M., & Gascón, J. (2011). Research praxeologies and networking theories. In M. Pytlak, T. Rowland, & E. Swoboda (Eds.), *Proceedings of the Seventh Congress of the European Society for Research in Mathematics Education* (pp. 281–290). Rzeszów: University of Rzeszów.

Artigue, M., & Mariotti, M. A. (2014). Networking theoretical frames: The ReMath enterprise. *Educational Studies in Mathematics, 85*(3), 329–356.

Bartolini Bussi, M. G., & Mariotti, M. A. (2008). Semiotic mediation in the mathematics classroom: Artifacts and signs after a Vygotskian perspective. In L. English, M. Bartolini Bussi,

G. Jones, R. Lesh, & D. Tirosh (Eds.), *Handbook of international research in mathematics education* (2nd ed., pp. 746–805). Mahwah: Lawrence Erlbaum (revised edition).

Bikner-Ahsbahs, A., & Prediger, S. (Eds.). (2014). *Networking of theories as a research practice in mathematics education*. New York: Springer.

Collins, A., Joseph, D., & Bielaczyc, K. (2004). Design research: Theoretical and methodological issues. *Journal of the Learning Sciences, 13*, 15–42.

Douady, R. (1986). Jeux de cadres et dialectique outil-objet [Games between settings and tool-object dialectics]. *Recherches en Didactique des Mathématiques, 7*(2), 5–32.

Duval, R. (1995). *Sémiosis et pensée humaine [Semiosis and human thinking]*. Bern: Peter Lang.

Gueudet, G., Pepin, B., & Trouche, L. (2013). Textbooks' design and digital resources. In C. Margolinas (Ed.), *Task design in mathematics education: Proceedings of ICMI Study 22* (pp. 325–336). Oxford. Available from http://hal.archives-ouvertes.fr/hal-00834054

Haspekian, M., & Artigue, M. (2007). L'intégration de technologies professionnelles à l'enseignement dans une perspective instrumentale: le cas des tableurs [The integration of professional technologies in teaching from an instrumental perspective: the case of spreadsheets]. In G. Baron, D. Guin, & L. Trouche (Eds.), *Environnements informatisés et ressources numériques pour l'apprentissage* (pp. 37–63). Paris: Hermès.

Maracci, M., Cazes, C., Vandebrouck, F., & Mariotti, M. A. (2013). Synergies between theoretical approaches to mathematics education with technology: A case study through a cross-analysis methodology. *Educational Studies in Mathematics, 84*(3), 461–485.

Margolinas, C., Abboud-Blanchard, M., Bueno-Ravel, L., Douek, N., Fluckiger, A., Gibel, P., et al. (Eds.). (2011). *En amont et en aval des ingénieries didactiques. XVe école d'été de didactique des mathématiques*. Grenoble: La Pensée Sauvage Editions.

Rocard, M., Csermely, P., Jorde, D., Lenzen, D., Walberg-Henriksson, H., & Hemmo, V. (2007). *Rocard report: Science education now: A new pedagogy for the future of Europe* (Technical Report). European Commission.

Vergnaud, G. (1991). La théorie des champs conceptuels [The theory of conceptual fields]. *Recherches en Didactique des Mathématiques, 10*(2–3), 133–170.

Correction to: Task Design In Mathematics Education

Correction to:
A. Watson, M. Ohtani (eds.), *Task Design In Mathematics Education*, New ICMI Study Series, https://doi.org/10.1007/978-3-319-09629-2

The original version of the book was published in 2015 with exclusive rights reserved by the Publisher. As of March 2021 it has been changed to an open access publication: © The Editor(s) (if applicable) and The Author(s) 2021.

All chapters in the book are licensed under the terms of the Creative Commons Attribution-NonCommercial-NoDerivatives 4.0 International License.

Any third party material is under the same Creative Commons license as the book unless specified otherwise below.

The updated version of this book can be found at https://doi.org/10.1007/978-3-319-09629-2

Index

A
Abstraction, 49–51, 60, 157, 163–164, 173, 216, 220, 314, 330
Acquisition, 21, 65, 135, 159, 192, 194, 221
Activity Theory, 20, 39, 54, 58, 101–105, 202–203
Adaptation, 123, 134, 157–159, 166, 181–183, 204
Adidactical situation, 32–33, 135, 198, 250, 260
Agency, 123, 134–136
Anthropological Theory of Didactics (ATD), 43–45, 60–68, 261–262, 269
Anticipation, 34, 40, 86, 94–95, 105–108, 110, 121, 161–162, 274–275, 282–283
Application, 98–100, 129, 151, 157, 174, 263–264, 290, 295
Artifact, 191–193, 199–204, 215, 233–235
Assessment, 54–56, 60–62, 116, 160, 165–167
Authority, 84, 102, 148–152, 164, 182, 204, 229–232

B
Bricolage, 30
Bridging, 26, 30, 49, 65, 73, 145, 195, 197, 208, 212, 216, 253

C
CAS. *See* Computer algebra system (CAS)
Central concept design, 296
Chronogenesis, 265, 267
Classroom culture, 37, 105–108, 172, 229

Cognitive apprenticeship, 26, 37
Cognitive conflict, 22, 24, 37
Cognitive load theory, 26, 135, 179, 211
Cognitively Guided Instruction, 88
Coherence, 28, 134, 148–149, 154, 163–167, 232, 251, 290
Commognitive Theory, 32, 198
Common Problem Solving Strategies as links between Mathematics and Science (COMPASS), 147, 174
Computer algebra system (CAS), 197, 199, 202–203
Concept study, 183
Conceptual blending, 206, 214–218
Conceptual Change Theory, 47–49
Conceptual field, 24, 71
Connected Maths, 4
COREM, 258–259, 262
Craft, 32–36, 56, 287, 294
Critical aspect, 47, 170, 174, 179, 216
Cultural-Semiotics theory, 32, 200
Culture, 71, 84, 96, 103–105, 157–159, 183, 191, 231, 266
Curriculum developers, 144

D
Design
 as art/science, 22, 62–64, 70, 72, 249, 295
 community, 22–25, 32, 62–64, 68–70, 203, 269, 287, 294
 experiment, 20, 25–26, 28–29
 as implementation/intention, 28–29, 34–38, 41, 72
Devolution, 33, 250, 257

Didactical contract, 118–119, 122–124, 127, 137
Didactical engineering, 25, 43, 173, 249–272
Didactical obstacle, 193, 195, 220
Didactical variable, 33, 198, 209–210, 219
Digital Mathematics Environment, 204–208
Digital textbook, 231–232, 245
Discrepancy potential, 212–215, 221
Documentational genesis, 86, 200
Dragging, 206, 209, 210, 214, 217
Dynamic digital tools, 170, 195, 206, 209, 212

E
Emergent perspective, 40, 122, 130, 133–134, 137, 204
Enactivist approach, 130, 132
Epistemological distance, 196, 203
Epistemological obstacle, 193, 195, 219–220
e-textbook, 229–247
Evaluation, 34–36, 59, 183, 284–285
Example space, 205–206, 218, 234–235

F
Fast design, 294
Feedback, 24, 33, 54, 119, 144, 151, 172, 198, 203, 209–212, 219–220, 237, 292
Formative assessment, 54–56, 60, 146, 219
Frames
 domain-specific, 36–41
 grand, 30–31
 intermediate level, 31–36
Free design, 301
French Research Institutes in Mathematics Teaching (IREM), 327
Freudenthal Institute, 4, 22, 204, 298

G
Gakushushido-an, 275
GEMAD. *See* Masters in Didactical Analysis (GEMAD)
Grain size, 73, 171–175, 185
Graphics, 105, 160, 177–178, 301
Guided inquiry, 229–247

H
Habits of mind, 172
Helice, 154, 165
Herbart, 163, 261
Hypothetical learning trajectory, 40, 69, 85, 130, 132

I
Institutionalization, 33, 167–168, 250
Instrumental genesis, 193, 197, 200, 208–209
Instrumentation, 199–200
Interactive diagram, 233–237
Interactive tool, 192, 218, 241
Interdisciplinary, 148, 172–175
International Society for Design and Development in Education (ISDDE), 4, 11
Intuition, 53, 54, 63, 290, 295–296
IREM. *See* French Research Institutes in Mathematics Teaching (IREM)
ISDDE. *See* International Society for Design and Development in Education (ISDDE)

J
JumpMaths, 148

K
KOSIMA project, 167–168
Kyozaikenkyu, 35–36, 59, 67, 183, 277–279, 281, 285

L
Learning management systems, 146, 160, 185, 231
Learning trajectories, 274, 276, 292. *See also* hypothetical learning trajectory
Lesson Study, 32–36, 56–72, 87, 183, 273–286

M
Masters in Didactical Analysis (GEMAD), 107
Mathewerkstatt, 151
MATH taxonomy, 160
Matome, 35, 276
MATRIZMAT, 201
Mesogenesis, 265, 267
Milieu, 33, 118–122, 135, 198, 250, 254–255, 262, 266
Modelling, 30, 40, 52–53, 60, 67, 97, 102, 116, 132, 136, 173, 175, 194, 204, 252, 261–269

N
Neriage, 35, 59, 167, 276, 283

O

Open tasks
 open-ended, 67, 92, 93, 107, 135, 219
 open-middled, 93
Ostensive, 198–9, 208–9

P

Phenomenology, 23, 53, 124, 170, 204
Phenomenotechnique, 323
Praxeology, 199, 208, 251, 261, 263, 269

Q

QUASAR, 4
Quasi-variable, 283

R

Realistic Mathematics Education (RME), 40, 51–53, 60–61, 67–69, 163, 170, 204, 294
ReMath project, 71, 193, 203–204, 221
Research lesson, 34–36, 72, 274–277, 281–282
Resources for Learning and Development Unit (RLDU), 149–153, 180

S

Semiotic mediation, 104, 193, 200–201, 212
Semiotic potential, 193, 200
Sequencing, 26–29, 48–53, 67–72, 85, 94, 104, 123, 146–147, 156–159, 164, 183, 199, 211, 219, 242–245, 259–261, 265, 275, 279–280, 296, 301
Sesamath, 149, 154
Shell Centre, 4, 21, 24, 54
Slow design, 63–64, 289–294
Socioconstructivist, 30–31
Sociocultural, 25, 30, 34, 37
Space of learning, 170, 199
Systematization, 168, 173, 184, 233

T

Task banks, 147, 149, 182
Teaching experiments, 25, 49
Technology Enhanced Learning in Mathematics (TELMA), 193, 203–204, 221
Text-based task, 143–188
Textbook, 4, 35, 64, 101, 143–145, 147–148, 150–156, 158, 166–167, 177, 182, 185, 229–234, 279–281
Theory of Didactial Situations (TDS), 32–33, 249–270
TIMSS Video Study, 12, 273
Tool-based tasks, 192–225
Tool ergonomic theory, 197, 203
Topaze Effect, 127
Topogenesis, 265, 267

U

University of Chicago School Mathematics Project, 151, 163
Utility, 73, 132
Utilization, 86–90, 95, 193, 197–198, 200, 208

V

Variation Theory, 45–47, 61, 66, 89, 103, 170–171, 206
Vertical mathematization, 64, 167
Visual features, 154–155, 176–181, 195, 208, 211, 232, 245, 288. *See also* Graphics
Visualization, 195, 208, 211, 288, 291
Voice, 123, 134–136, 151–152
Vygotskian, 25, 31, 104, 193, 200

W

W
Wikitext, 149, 182
Word problems, 116–122, 137
Worked-out examples, 145
Worksheets, 143, 151–152, 166–167, 182–183

Z

Zone of proximal development (ZPD), 8, 110, 193

MIX
Papier aus verantwortungsvollen Quellen
Paper from responsible sources
FSC® C105338

If you have any concerns about our products,
you can contact us on
ProductSafety@springernature.com

In case Publisher is established outside the EU,
the EU authorized representative is:
**Springer Nature Customer Service Center GmbH
Europaplatz 3, 69115 Heidelberg, Germany**

Printed by Libri Plureos GmbH
in Hamburg, Germany